T0135303

Studies in Computational Intelligence

Volume 751

Series editor

Janusz Kacprzyk, Polish Academy of Sciences, Warsaw, Poland
e-mail: kacprzyk@ibspan.waw.pl

The series "Studies in Computational Intelligence" (SCI) publishes new developments and advances in the various areas of computational intelligence—quickly and with a high quality. The intent is to cover the theory, applications, and design methods of computational intelligence, as embedded in the fields of engineering, computer science, physics and life sciences, as well as the methodologies behind them. The series contains monographs, lecture notes and edited volumes in computational intelligence spanning the areas of neural networks, connectionist systems, genetic algorithms, evolutionary computation, artificial intelligence, cellular automata, self-organizing systems, soft computing, fuzzy systems, and hybrid intelligent systems. Of particular value to both the contributors and the readership are the short publication timeframe and the world-wide distribution, which enable both wide and rapid dissemination of research output.

More information about this series at http://www.springer.com/series/7092

Yaxin Bi · Supriya Kapoor
Rahul Bhatia
Editors

Intelligent Systems and Applications

Extended and Selected Results from
the SAI Intelligent Systems Conference
(IntelliSys) 2016

 Springer

Editors
Yaxin Bi
School of Computing
Ulster University at Jordanstown
Newtownabbey, County Antrim
UK

Rahul Bhatia
The Science and Information
  (SAI) Organization
Bradford
UK

Supriya Kapoor
The Science and Information
  (SAI) Organization
Bradford
UK

ISSN 1860-949X   ISSN 1860-9503   (electronic)
Studies in Computational Intelligence
ISBN 978-3-319-88745-6   ISBN 978-3-319-69266-1   (eBook)
https://doi.org/10.1007/978-3-319-69266-1

Printed on acid-free paper

This Springer imprint is published by Springer Nature
The registered company is Springer International Publishing AG
The registered company address is: Gewerbestrasse 11, 6330 Cham, Switzerland

Editor's Preface

The SAI Intelligent Systems Conference (IntelliSys) 2016 was held on 21–22 September 2016 in London, UK. This conference is a prestigious annual conference on areas of intelligent systems and artificial intelligence and their applications to the real world, which builds on the success of previous IntelliSys conferences also held at London.

This conference not only presented state of the art methods and valuable experience from researchers in the related research areas, but also provided the audience with a vision of further development in the field. The event was a two day program comprised of twenty six presentation sessions (paper and poster presentations). The themes of the contributions and scientific sessions ranged from theories to applications, reflecting a wide spectrum of artificial intelligence.

Out of 168 papers published in the proceedings, 22 papers, which received highly recommended feedback, were selected, and the extended versions are published as chapters in this book. We believe this edition will increase the visibility of rescarch results presented in the conference, and certainly help further disseminate new ideas and inspire more international collaborations.

It has been a great honor to serve as the Program Chair for the SAI Intelligent Systems Conference (IntelliSys) 2016 and to work with the conference team. The conference would truly not function without the contributions and support received from authors, participants, keynote speakers, program committee members, session chairs, organizing committee members, steering committee members, and others in their various roles. Their valuable support, suggestions, dedicated commitment and hard work have made the IntelliSys 2016 successful. Finally, we would like to thank the conference's sponsors and partners: HPCC Systems, IEEE and IBM Watson AI XPrize.

Newtownabbey, UK   Yaxin Bi

The original version of the book was revised:
For detailed information please see Erratum.
The erratum to the book is available at
https://doi.org/10.1007/978-3-319-69266-1_23

Contents

Pattern Sets for Financial Prediction: A Follow-Up

Mattias Wahde[✉]

Chalmers University of Technology, SE-412 96, Göteborg, Sweden
mattias.wahde@chalmers.se

Abstract. As a follow-up to an earlier investigation, a true forward test has been carried out by applying a previously developed financial predictor (in the form of a so called *pattern set*, optimized using an evolutionary algorithm) to a new data set, involving data for 200 stocks and covering a time period from February 2016 to the end of that year. Despite being applied to previously unseen data, the pattern set generated a set of trades with an average one-day return of 0.394%. Moreover, the pattern set's total trading return (excluding transaction costs) over the entire period covered by the new data, when applied as a trading strategy with a simple m–day holding period for each trade, was 15.9% for $m = 1$, 24.9% for $m = 3$, and 61.6% for $m = 6$, compared to 16.2% for the benchmark index (S&P 500) over the same period.

1 Introduction

In the academic literature on finance, there is a strong, and generally well-motivated, scepticism against claims of predictability of financial time series. The so-called *efficient market hypothesis* (EMH) (see e.g. [1,3]) summarizes this scepticism by suggesting that all relevant information is already contained in the price of a financial instrument (for example a stock), such that consistent, profitable prediction of future prices is impossible. Even though many studies exist that support such a conclusion, the EMH has also been challenged by considering the irrational and emotional behavior often seen among market participants, manifested in the form of herd-like behavior resulting in overreactions, both positive and negative, to events that impact the price of a financial instrument [6]. Thus, some studies have reported short-term predictability under certain circumstances; see e.g. [2,4–6]. In [9], an evolutionary algorithm (EA) was applied to a set of daily stock market data in order to find predictors, so called *pattern sets*, with the best possible performance, measured as the Sharpe ratio of the one-day returns, i.e. the average one-day return (minus the risk-free rate of return) divided by its standard deviation. The results showed that pattern sets could be found that gave consistent, strongly positive results.

© Springer International Publishing AG 2018
Y. Bi et al. (eds.), *Intelligent Systems and Applications*,
Studies in Computational Intelligence 751,
https://doi.org/10.1007/978-3-319-69266-1_1

However, there is a general and fundamental problem with any approach involving optimization of financial predictors, namely the fact that, even for completely random data, structures that appear to be non-random can almost always be found, if one looks hard enough. The very act of optimizing pattern sets, i.e. setting the parameters so as to get the best possible performance, involves searching through thousands of possible pattern sets, discarding all those that do *not* generate a high Sharpe ratio or any other measure used to gauge the performance, only keeping the ones that happen to give a good result for the data set used during optimization. Thus, per definition, the retained pattern sets will handle *that* data set well, but not necessarily other data sets. In other words, it is common that predictors exhibit *overfitting* to the inevitable noise present in the data set.

An obvious remedy is to test an optimized predictor on previously unseen data. This procedure was indeed carried out in [9], and the pattern sets presented there were those that did well both over the training set (used during optimization) and the validation set (applied once optimization had been completed). Still, even in that case, there remains a possibility that the results were more due to chance than anything else. For example, even though the training and validation sets were completely disjoint, they did cover the same 11-year period of time (from early January 2005 to early February 2016). Thus, one cannot discard the possibility that the overall correlation between stock prices was strong enough to give positive results also for the validation data.

In order to investigate the true performance of a pattern set (or any other form of financial predictor) one should therefore carry out a pure forward test, applying the pattern set, without any tweaking, to new data, gathered after the pattern set was generated. The purpose of this paper is simply to make just such an analysis, by applying the best pattern set found while writing [9] to new data, collected after the completion of that paper.

This chapter is organized as follows: Sect. 2 gives a brief general introduction to pattern sets and their use in financial trading, as well as a specific description of the particular pattern set considered here. The data are described in Sect. 3. The results are presented in Sect. 4 and discussed in Sect. 5. The conclusions are presented in Sect. 6.

2   Pattern Sets

In [9], the author presented an optimizable structure, referred to as a pattern set, for (one-day) prediction of (daily) stock data, along with a process for optimization of pattern sets based on non-length preserving evolutionary algorithms. The pattern sets, in turn, consist of sequences of so called patterns, which will now be described briefly; a complete description is given in [9].

It is assumed that, for each day, the price information for an instrument contains four elements: open, high, low, and close. With these four elements, one can form a so-called candlestick representation, as illustrated in Fig. 5. A *pattern*, then, concerns the ratio between two aspects of the price, for example the closing prices on two different days, or the high price and the low price on a given day. Each pattern, in turn, can be encoded using seven parameters that determine, for example, the exact days (relative to the current day) from which data should be taken when forming any of the ratios just mentioned, as well as the range (upper and lower limit) in which a ratio must fall in order for the conditions defining the pattern to be fulfilled.

By combining a sequence of patterns, a pattern set can be made to represent a large variety of events in a financial time series. For any day in which the price variation over the preceding days (up to and including the current day) is such that the conditions defined by the patterns are *all* fulfilled, the pattern set triggers a long trade, i.e. a trade in which one makes a profit if the price goes up over the coming day(s). If at least one condition is not fulfilled, no trade is triggered. The trades, if any, are executed just before the market closes for the day. As an illustration, consider the following simple pattern set, consisting of three patterns:

Pattern 1: (Close[-3]/Close[-1]) in [0.995, 1.000]
Pattern 2: (Close[-4]/Close[-2]) in [0.995, 1.000]
Pattern 3: (Close[-5]/Close[-3]) in [0.995, 1.000]

In this case, Pattern 1 would be fulfilled if the ratio between the close price on day -3 (relative to the current day) and the close price on day -1 (also relative to the current day) were to fall in the range $[0.995, 1.000]$ etc. As mentioned above, the pattern set would trigger a trade only if all three patterns were to be fulfilled, something that, in this particular example, would require a (rather vague) rising trend in prices.

The aim in [9] was to capture the overreactions of traders [7,8], which are often manifested in the stock price time series, but also difficult to quantify. In this approach, the user normally defines a vaguely defined starting point such as, for example, a downtrend interrupted by a sudden upward price jump. Next, an optimization run is carried out, using a fairly standard evolutionary algorithm in which each individual encodes the $7 \times N$ parameters for pattern sets consisting of N patterns, and where two-point crossover is used in order to allow for structural modification (i.e. length variation) of the pattern sets. Thus, the evolutionary algorithm is allowed to modify both the parameters and the structure of the pattern sets, with the objective of maximizing the Sharpe ratio (see Sect. 1) while also carrying out sufficiently many predictions (at least 500, in [9]) in order for the results to be reliable.

2.1   Selected Pattern Set

The best-performing pattern set evolved for applications involving large-cap stocks on NYSE and NASDAQ was denoted PS1. Its performance (one-day Sharpe ratio) over the training and validation sets was 0.243 and 0.258, respectively, with average one-day returns of 0.758% and 0.775% over the two sets, respectively. From the same optimization run, a slightly better (but very similar) pattern set was found already in connection with the writing of [9], but it was not included in the paper since the number of trading instances (over the validation set) was 493, slightly below the chosen cutoff of 500. However, since the cutoff is somewhat arbitrary, that pattern set (henceforth denoted PS1+) was, in fact, selected for the analysis presented here, in view of its slightly better performance over the validation set: A Sharpe ratio of 0.288 and an average one-day return of 0.823%.

The pattern set PS1+, the detailed structure of which will not be given here, consisted of five patterns, and was generated by starting from a loosely defined downtrend, which was then (rather strongly) modified by the evolutionary algorithm, partly by changing the detailed aspects of the downtrend, and partly by adding a few more patterns. Some examples of trades generated with this pattern set are shown in Fig. 5.

2.2 Trading Strategy

Here, the main aim is to investigate the trading performance of the selected pattern set. By itself, the pattern set (PS1+) provides only *entry* points for the trades. In order to assess trading performance, one needs a complete trading strategy, supplying not only the entry points but also the *exit* points for each trade. There are various ways of managing a trade and selecting the exit point: For example, one can use a fixed lower stop (stop loss) and a fixed upper stop, each at a given percentage distance from the entry price. As an alternative to a fixed lower stop, one can use a trailing stop, such that the price at which the trade would be exited is raised whenever a new maximum profit (for the trade in question) is reached. Of course, one can also devise much more complex exit rules.

However, even with the simple fixed stops, there will be additional parameters to set, thus both increasing the risk of overfitting and potentially muddling the

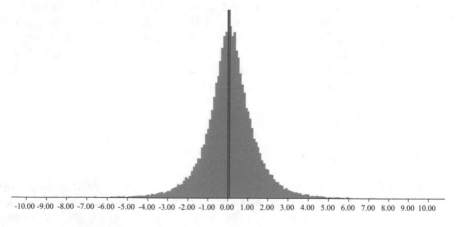

-10.00 -9.00 -8.00 -7.00 -6.00 -5.00 -4.00 -3.00 -2.00 -1.00 0.00 1.00 2.00 3.00 4.00 5.00 6.00 7.00 8.00 9.00 10.00

Fig. 1. The distribution of the 44118 one-day returns over the entire $\mathcal{D}_3$ data set. A vertical line indicates the average one-day return (0.0829%). The vertical scale is somewhat arbitrary, but if the values are normalized by the total number of samples (44118), the distribution can be taken as a probability distribution

assessment of the pattern set's performance, since the outcome of the trade will then also depend on the specific exit parameters used. Thus, here, an even simpler exit rule has been applied: Each trade lasts for precisely m days, where $m \geq 1$ is an integer. This is the simplest possible exit rule and it also has the advantage of including the one-day returns as a special case ($m = 1$).

3   Data

The main data set ($\mathcal{D}_1$) used in [9] covered the period from January 2, 2005 until Feb. 1, 2016, for the 50 largest stocks (by market capitalization) on NYSE and NASDAQ, a set that was then further divided into a training set with 25 stocks and a validation set with the remaining 25 stocks. In addition, a data set ($\mathcal{D}_2$) involving price data for the stocks of 50 companies with small market capitalization was also considered in [9].

By contrast, the data set used here, henceforth referred to as $\mathcal{D}_3$, consisted of the 200 largest stocks (by market capitalization) on NYSE and NASDAQ, with data collected from the end of January, 2016 until the end of December, 2016. Since the selected pattern set (PS1+) requires price data for the preceding three weeks in order to decide whether or not to enter a trade, the first possible trading day in $\mathcal{D}_3$ turned out to be Feb. 17, 2016. The last trading day was Dec. 30, 2016.

Thus, not only does this data set consist entirely of new data, non-existent at the time of the writing of [9], and without any (trading) overlap with the data sets considered in that paper, but it also contains a majority (150 of the 200) of stocks that were not included in either the training set or the validation set used when generating and selecting the pattern sets in the earlier paper.

The total number of trading days from Feb. 17 to Dec. 30 (inclusive) was 222. With 200 stocks in the $\mathcal{D}_3$ data set, the number of possible one-day returns would be $221 \times 200 = 44200$. The actual number of one-day returns turned out to be slightly smaller, namely 44118, since one company was acquired by another company (and thus delisted) during the time period in question. The average one-day return over those 44118 instances turned out to be 0.0829 %, i.e. close to 0, albeit somewhat larger than the 0.0526% found for the data set $\mathcal{D}_1$ in [9]. The distribution is shown in Fig. 1.

4   Performance

The performance of the trading strategy described in Sect. 2.2 above was evaluated by (i) studying the distribution of results and (ii) carrying out sequential trading over the period from Feb. 17 to Dec. 30 (inclusive).

Case (i) includes all trades generated by the strategy, regardless of whether or not different trades (partially or fully) overlap in time. By contrast, case (ii) takes into account the fact that not all of the possible trades will be carried out in a given portfolio. For example, if all available cash has been committed on day d in one or several m-day trades, any trade triggered before day $d + m$ will be excluded, since there is no cash available for the trade in question. Moreover, on any given day when $k > 0$ entry points are triggered by the PS1+ pattern set, the available cash (if any) is distributed *equally* among the k stocks. If no cash is available, no action is taken. All trading is assumed to take place at or near the close of the trading day. Note that cash *is* considered to be available on a given day if there is a trade exit on that same day. In such cases, if the pattern set also triggers one or several entry points, both the exits (to make the cash available) and the entry into new trades are considered to take place almost simultaneously at the end of the trading day. If the liquidity of the stocks considered is low, such trading may be difficult to carry out in practice, an issue that is briefly considered in the discussion (Sect. 5).

It should also be noted that, here, the performance is computed using *linear* compounding rather than exponential compounding, so as to avoid inflating the results obtained. In order to illustrate the procedure, let the available cash at the first trading day be equal to 1. If the first trade (or set of trades, if there are several simultaneous entry points) results in, say, a net 2% profit, the available cash at the time of exit would be 1.02. The next trade(s), however, would be carried out again starting from 1 unit of cash, not 1.02. Thus, if the second trade (or set of trades) also gives a net 2% profit, the total profit would be $1.04 - 1 = 0.04$, *not* $1.02 \times 1.02 - 1 = 0.0404$ etc.

4.1   Trade Distributions

The distributions of trades, i.e. case (i) above, were computed for $m = 1$ (one-day returns), $m = 3$, and $m = 6$. The average returns were: 0.394% ($m = 1$, 174 trades), 1.18% ($m = 3$, 173 trades), and 2.00% ($m = 6$, 172 trades). The corresponding distributions are shown in Figs. 2, 3, and 4. As in Fig. 1, the vertical scale is a bit arbitrary, and has therefore been omitted in the plots:

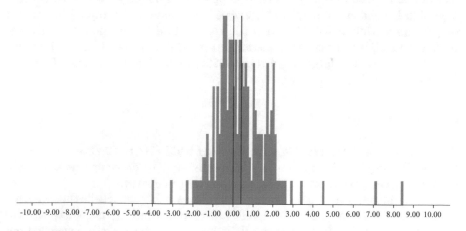

Fig. 2. The distribution of the 174 one-day trades generated by pattern set PS1+ over the $\mathcal{D}_3$ data set. The average trade result (0.394%) is indicated by a vertical line

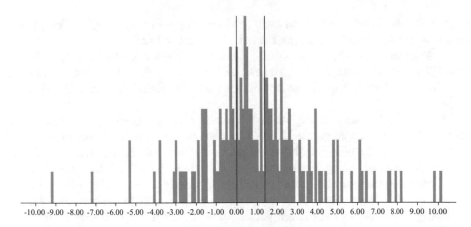

-10.00 -9.00 -8.00 -7.00 -6.00 -5.00 -4.00 -3.00 -2.00 -1.00 0.00 1.00 2.00 3.00 4.00 5.00 6.00 7.00 8.00 9.00 10.00

Fig. 3. The distribution of the 173 three-day trades generated by pattern set PS1+ over the $\mathcal{D}_3$ data set. The average trade result (1.18%) is indicated by a vertical line. The four best trades, with results exceeding +10%, are not shown

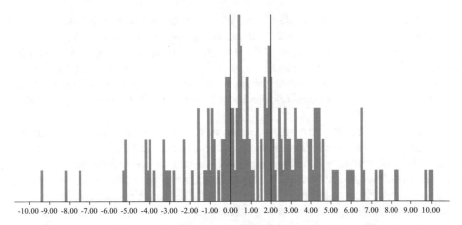

-10.00 -9.00 -8.00 -7.00 -6.00 -5.00 -4.00 -3.00 -2.00 -1.00 0.00 1.00 2.00 3.00 4.00 5.00 6.00 7.00 8.00 9.00 10.00

Fig. 4. The distribution of the 172 six-day trades generated by pattern set PS1+ over the $\mathcal{D}_3$ data set. The average trade result (2.00%) is indicated by a vertical line. The nine best trades, with results exceeding +10%, are not shown

The histogram bars show the *number* of trades in bins of width 0.1%, but those values can of course be normalized by the total number of trades so that one can consider the resulting distribution as a probability distribution.

As is evident from those figures, the distributions are quite noisy, which is to be expected given that the number of trades is much smaller than the 44118 one-day returns shown in Fig. 1. In order to facilitate comparison between the three cases, the horizontal range (−10 to 10%) is the same in all figures. Thus, while all negative instances *are* shown in all three figures, some positive instances (for $m = 3$ and $m = 6$) did not fit in this range: For $m = 3$ there were four trades with a result above +10% (the best trade giving +13.9%), whereas for $m = 6$ there were nine such trades (the best trade giving +18.8%).

It is clear from the three figures that the one-day performance of $PS1+$ (0.394%) greatly exceeds the market performance of 0.0829%, even though it is quite a bit lower than the performance (0.823%) found for the validation set in [9].

The three-day and six-day returns are also strongly positive. For the three-day trades the average result (1.18%) implies an average *daily* performance of $1.18/3 = 0.393\%$, whereas the average result for the six-day trades (2.00%) implies an average daily performance of $2.00/6 = 0.333\%$. Thus, the average daily return appears to be quite insensitive to the duration of the trade, at least up to $m = 6$. Larger values of m were tried, but the average *daily* return started dropping above $m = 6$. For example, for $m = 10$, the average 10-day return was 2.35%, giving an average of 0.235% daily.

4.2 Sequential Trading

Sequential trading, i.e. case (ii) above, gives a more realistic view of the results that would be obtained in actual trading. Here, too, the cases $m = 1$, $m = 3$, and $m = 6$ were considered. Over the time period considered here, the comparison index (S&P 500) rose from 1926.82 to 2238.83, i.e. an increase of 16.2%. Trading as described in the presentation of case (ii) above, the results obtained with pattern set $PS1+$, by contrast, were 15.9% ($m = 1$), 24.9% ($m = 3$), and 61.6% ($m = 6$).

Seeing these results, an obvious question to ask is: How can the trading with $m = 1$ give a *worse* result than the overall market, given that the average trade returns 0.394% whereas the average *market* return was only 0.0829%? Of course, since the trading period was quite short (less than a year) the poor performance for $m = 1$ could be attributed to bad luck (after all, the *distribution* of trade returns is quite wide, even for $m = 1$). However, if this were the full explanation, one would expect a rather poor performance also for $m = 3$ and $m = 6$ where, even though the trades last longer, the entry point sets are of course subsets of the entry points used for $m = 1$. This is not seen: On the contrary, the result for $m = 3$ is quite good, and the result for $m = 6$ is exceptional.

An alternative explanation can be found by considering how the trades are distributed in time over the data set. Let k denote the number of trade entry points (or, equivalent, one-day trades) on a given day, and let $n(k)$ denote the number of days with k trade entry points. Furthermore, let $p(k) = n(k)/N$, where N is the total number of trading days in the data set. These distributions are shown in Table 1 As is clear from the table, on most days (132 out of 221)[1] there were, in fact, *no* trades. In the light of this information, the performance of the $m = 1$ trading strategy is not so bad: The PS1+ pattern set achieved

---

[1] Note that entry points are ignored for the last (222nd) day of the data set, since the outcome of a trade entered on the very last day would not be computable. Thus, the effective number of days is 221.

a performance roughly equivalent to the performance of the comparison index, despite trading only on around 40% of the available days.

Table 1. The distributions $n(k)$ and $p(k)$ of the number and fraction of trade entry points (k), respectively, generated by the selected pattern set (PS1+) over the data set $\mathcal{D}_3$. For example, there were 132 days with no trade entry points, 58 days with one trade entry point etc.

k	$n(k)$	$p(k)$
0	132	0.597
1	58	0.262
2	19	0.086
3	3	0.014
4	4	0.018
5+	5	0.023

Still, the performance for $m = 1$ is, in fact, a bit below what one would expect: Given the distribution in Table 1, one can compute the expected result at the end of the time period defined by the $\mathcal{D}_3$ data set, as follows: For every day, determine the number of trades (k) by randomly sampling the distribution in Table 1. Then, randomly sample the distribution in Fig. 2 k times, and compute the average for that day. Repeat the procedure for a total of 221 times, and then sum the results thus obtained. This analysis was carried out, repeating the random sampling 100,000 times, giving an average expected outcome of 35.0%, with 95% confidence interval of $[11.6, 60.8]\%$. Simplifying even more, one can obtain an analytical estimate of the average performance: For each of the trading days, either set the return to 0 (i.e. no trade) with probability 0.597 or to 0.394% (the average one-day return) with probability $1 - 0.597 = 0.403$. In this case, the expected total profit would be 35.1%, very close to the estimate obtained numerically. Thus, in conclusion, the results obtained for $m = 1$ are within the 95% confidence interval of expected results, but close to its lower limit.

Turning now to the trades with $m = 3$ and $m = 6$, one can observe that, for both cases, the performance exceeded the return of the comparison index over the same time period. The better performance for $m > 1$ is easy to explain: As mentioned above, for $m = 1$, trades were carried out only on around 40% of the available days. Now, the average *daily* performance for $m = 3$ and $m = 6$ is roughly the same as for $m = 1$ (see above). Moreover, for $m > 1$, for any given day, it is more likely that a trade would be in progress than for $m = 1$: Simplifying a bit, one can note that when a six-day trade takes place, it will generate an average profit of around 2.00%, whereas, over the same six-day period, trading with $m = 1$, one can only expect an average profit of around $6 \times 0.394 \times (1 - 0.597) = 0.953\%$. As shown in Table 1, for $m = 1$, trading took place only on 40.3% of all available days. By contrast, for $m = 3$ trades were in progress for 61.1% of all available days, whereas for $m = 6$ the corresponding figure was 74.7%.

One can make a simplified analytical estimate of the expected average performance in this case as well: For $m = 3$, with an average daily return (see above) of 0.393% and trading take place on 61.1% of all days, the average total result over the entire 221 days would be 53.1%. For $m = 6$, with an average daily return of 0.333% and with trading taking place on 74.7% of all days, the average total result would be 55.0%. Of course, also in these cases, the distribution of the total results will be rather wide, as was found numerically for the case $m = 1$. Returning to the actual total results obtained, namely 24.9% for $m = 3$ and 61.6% for $m = 6$, one can observe that the result for $m = 3$ is quite a bit below the average expected result, whereas for $m = 6$ the result is slightly *above* the expected average.

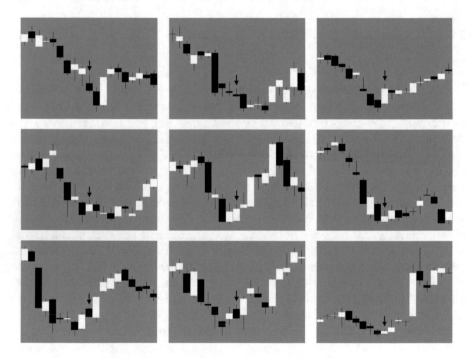

Fig. 5. Schematic view of nine representative entry points obtained by the pattern set PS1+. The entry points are indicated by an arrow in each panel. Here, the time series are shown in candlestick format, with one candlestick (rectangle) per trading day, such that a white rectangle indicates that the closing price (for a given day) was above the opening price, and a black rectangle indicates the opposite situation. Thus, for a white rectangle, the bottom represents the opening price and the top represents the closing price, and vice versa for a black rectangle. The narrow lines emanating from the top and the bottom of (some of) the rectangles show the low and high prices for each day

As an example, Fig. 5 shows nine of the 174 entry points generated for $m = 1$. For this case, around 57% of the trades gave a positive result whereas 43% gave a negative or zero result. The figure shows five entry points with a positive one-day result and four with a negative one-day result, i.e. roughly the same ratio between positive and negative one-day trades as that found among all 174 trades. It should be noted that the fraction of positive trades improves as m is increased: For $m = 3$, around 69% of the trades gave a positive result whereas the corresponding number for $m = 6$ was around 71%. This effect can indeed be seen in Fig. 5. For example, the entry point in the upper left panel would give a negative one-day result, but positive three-day and six-day results.

5   Discussion

The most important, and somewhat surprising, result presented above is that the PS1+ pattern set does indeed achieve a rather good trading result over previously unseen data. This is interesting, as one could have expected that the effects of overfitting (when generating the pattern set) would quickly become visible, as they often do in trading algorithms as soon as new data are considered.

As mentioned above, the PS1+ pattern set was generated starting from a loose downtrend, which was then modified and tuned by the evolutionary algorithm, but still retained its basic characteristics. Thus, one can view PS1+ as a short-term contrarian strategy, in which a long trade is triggered at the end of a downtrend lasting a few days. It is interesting to note that other recent studies [5, 10] have also found similar results, namely that positive short-term results can be obtained by following a contrarian strategy after a short-term downtrend.

However, as shown Sect. 4, the actual results obtained do show some variation relative to the expected average result (estimated either numerically or in a simplified analytical way). This is not surprising, since the *distribution* of expected results is rather wide. For the particular data set considered here, the one-day and three-day trades gave results a bit below the expected average, whereas the six-day trades gave a result above the expected average.

In the case considered here, the benchmark index rose by 16.2% which, in fact, is quite a lot for a period of less than a year. Even though the three-day and six-day trading strategies gave results well above the benchmark, an interesting investigation would be to check the performance of the strategies when the benchmark index instead drops. Consider, for example, a period in which the S&P 500 were to *drop* by 16.2% instead. Would the six-day trading strategy then maintain its edge over the index, or would it perhaps give a *negative* return of the same order of magnitude as its positive return over the period considered here? This remains to be seen, and the testing of the strategies will of course continue during 2017. Of obvious interest would be to investigate, in detail, *why* the pattern set manages to do so well. However, the author chooses to maintain a degree of scepticism, and to defer a detailed analysis of the pattern set until (i) more trades are available and (ii) the pattern set has been thoroughly exposed to a period of overall *falling* prices. Once those two conditions are fulfilled, if the pattern set still does well, a detailed analysis of its properties will follow.

Determining whether or not a trading strategy obtains results that outperform some benchmark index is a standard procedure. However, of even greater importance (at least in the author's view) is the ability of the strategy to generate positive returns *regardless* of the return of the benchmark index over the same period. Using the numerical approach described in Sect. 4, one can at least estimate the probability that the one-day strategy would give a negative absolute return. To that end, another 100,000 samplings were made over a 252-day period (i.e. one trading year). For a period of this length, the expected average was 40.0%, with a 95% confidence interval of $[15.1, 67.6]$%. The probability of getting a negative absolute return was as low as 0.068%, and the probability of getting a return below 10% (i.e. roughly the historical average annual return of the S&P 500) was found to be 0.86%. Thus, provided that the distributions of trades (Fig. 2) and entry points ($p(k)$ in Table 1) remain largely constant, the probability of obtaining a negative result is very small, as is the probability of doing worse than the expected 10% annual return of the benchmark index.

The choice of holding periods considered here, i.e. one, three, or six days, is somewhat arbitrary. However, since the pattern set PS1+ was *trained* using one-day returns, in makes perfect sense to include that case also for the performance evaluation considered here. Moreover, it appears that the entry points found by the pattern set PS1+ remain useful for around one week. After that interval, the average daily return drops, as discussed in Sect. 4.1.

In the analysis presented in Sect. 4, transaction costs were not included, partly because the trading costs may depend somewhat on the chosen service provider. Thus, rather than arbitrarily selecting a specific level of transaction costs, those costs were ignored in the analysis above. Moreover, over the last decades, there has been a strong negative pressure on transaction costs. In fact, some service providers even offer cost-free trading for retail investors (at least up to a certain maximum investment) and larger investors pay very low fees in general. On the other hand, there are other factors that may affect the price at which a trade is entered or exited. Here, it has been assumed that all trading takes place at the closing price. However, as there is always a certain spread between bid and ask and also, of course, price movements until the very last trade of the day, the actual price at which a trade takes place may differ a bit from the closing price. Now, one can argue that the effect of last-minute price movements may be positive (i.e. give a slightly lower price when entering a trade, or a slightly higher price when exiting) just as often as they are negative. On the other hand, the spread between bid and ask will always work against the investor but, for the stocks considered here, all related to companies with large or very large market capitalization, the spread is typically very small. A common price level is around 50–250 USD, with a spread of 0.01 or 0.02 USD. Overall, the effects of trading costs, bid-ask spread, and the difficulty in achieving the exact closing price, should be rather small. Still, an analysis was carried out in which the aggregate effect of those factors was set to (a loss of) 0.05% per trade action (entry or exit). The total results dropped from 15.9 to 6.86% for the one-day strategy, from 24.9 to 20.4% for the three-day strategy, and from 61.6 to 58.8% for the six-day strategy. Thus, the deterioration in the results for three-day and, especially, six-day trading is quite small, while the drop in performance would be quite painful for one-day trading, illustrating the fact that frequent transactions tend to reduce performance.

Another difficulty would be to carry out near-simultaneous trades (i.e. first exiting ongoing trades, then entering new trades) in those cases where one or several new trades are triggered on the same day when exits occur. However, given the large liquidity of the stocks considered here, namely the largest 200 stocks in the US markets, it is generally possible (even manually) to carry out first exits and then entries in the last 10–20 s of the trading day, at least if the positions are not too big. Moreover, at least for the six-day strategy, such events are not very common: The most common situation is a gap of one or two days between a trade exit and the subsequent trade entry.

6   Conclusion

The main conclusion is that the pattern set PS1+, developed in connection with the writing of [9], does indeed generate positive one-day returns, with an average of around 0.394% for the data set used here (compared to 0.823% for the validation set used in [9]). Of particular interest is the fact that the pattern set does so well for holding periods exceeding the single day used during optimization. For example, for trades lasting six days, the PS1+ pattern set achieved a total result (excluding transaction costs) of 61.6% over the period from Feb. 17, 2016 to the end of that year, and around 58.8% including transaction costs and other effects that might influence the price at which transactions take place.

Even though the results are promising, the pattern set and the associated trading strategy should also be tested in a period of generally *falling* prices. Moreover, the various factors that can impact negatively on the trades (for example transaction costs, the bid-ask spread, and the difficulty in carrying out trades at or near the market closing time), whose magnitudes were only estimated here, must be investigated in actual trading.

References

1. Ang, A., Goetzmann, W.N., Schaefer, S.M.: Review of the efficient market theory and evidence—implications for active investment management. Found. Trends Financ. **5**, 157–242 (2011)
2. Caginalp, G., Laurent, H.: The predictive power of price patterns. Appl. Math. Financ. **5**, 181–205 (1998)
3. Fama, E.F.: The efficient market hypothesis: a review of theory and empirical work. J. Financ. **25**, 383–417 (1970)
4. Lu, T.-H., Shiu, Y.-M., Liu, T.-C.: Profitable candlestick trading strategies—the evidence from a new perspective. Rev. Financ. Econ. **21**, 63–68 (2012)

5. Lu, T.-H., Shiu, Y.-M.: Can 1-day candlestick patterns be profitable on the 30 component stocks of the DJIA? Appl. Econ. **48**, 3345–3354 (2016)
6. Park, C.-H., Irwin, S.H.: The profitability of technical analysis: a review. Technical Report AgMAS Project Research Report 2004-04 (2004)
7. Subrahmanyam, A.: Behavioural finance: a review and synthesis. Eur. Financ. Manag. **14**, 12–29 (2008)
8. Thaler, R.H.: The end of behavioral finance. Financ. Anal. J. **55**, 12–17 (1999)
9. Wahde, M.: A framework for optimization of pattern sets for financial time series prediction. In: Proceeding of the SAI Intelligent Systems Conference (Intellisys), pp. 31–38 (2016)
10. Wu, M., Huang, P., Ni, Y.: Investing strategies as continuous rising (falling) share prices released. J. Econ. Finan. https://doi.org/10.1007/s12197-016-9377-3 (2016)

Optimum Wells Placement in Oil Fields Using Cellular Genetic Algorithms and Space Efficient Chromosomes

Alexandre Ashade L. Cunha(✉), Giulia Duncan, Alan Bontempo, and Marco Aurélio C. Pacheco

ICA, PUC-RJ, Rua Marquês de São Vicente, 225, Gávea - Rio de Janeiro, Gávea, RJ, Brazil
{giuliaduncan,alanbontempo}@gmail.com, alexandre@ashade.com.br

Abstract. The present work introduces a new approach to the optimum wells placement problem in oil fields using evolutionary computation. In particular, our contribution is twofold: we propose an efficient algorithm for initialisation of highly constrained optimisation problems based on Monte-Carlo sampling and we propose a new optimisation technique that uses this population sampling scheme, a space-efficient chromosome and the application of cellular genetic algorithms to promote a large population diversity. Usually, authors define a domain representation having oil wells placed at any arbitrary position of the chromosome. On the other hand, the proposed representation enforces a unique relative wells position for each combination of wells. Therefore, the suggested scheme diminishes the problem size, thus making the optimisation more efficient. Moreover, by also employing a cellular genetic algorithm, we guarantee an improved population diversity along the algorithm execution. The experiments with the UNISIM-I reservoir indicate an enhancement of 6 to 10 times of the final NPV when comparing the proposed representation and the traditional one. Besides, the cellular genetic algorithm with the suggested chromosome performs better than the classical genetic algorithm by a factor of 1.5. The proposed models are valuable not only for the oil and gas industry but also to every integer optimisation problem that employs evolutionary algorithms.

1 Introduction

The problem of optimising wells placement in oil fields is essential for oil companies. Many engineering and geological variables affect the reservoir and produce complicated constraints. Accordingly, decision making is not a simple task and finding solutions to minimise cost and maximise profits is an essential and challenging problem. In this context, optimisations are an automated process to seek solutions for oil well locations, trajectories and types, providing safe reservoir exploration plans.

© Springer International Publishing AG 2018
Y. Bi et al. (eds.), *Intelligent Systems and Applications*,
Studies in Computational Intelligence 751,
https://doi.org/10.1007/978-3-319-69266-1_2

The present text handles nonconventional wells, which are arbitrary wells regarding slope, shape and type [1]. There are numerous works on this matter [1–9], most of them using commercial reservoir simulators in conjunction with proved optimisation heuristics to determine a suitable wells placement alternative.

Articles [1–4] use classical genetic algorithms with chromosomes that include the location of each well and their type. Both locations and types are integer genes constrained to simple domain boundaries, without any direct relationship. Furthermore, these cited works also use activation bits to denote whether a well is present in the decoded solution, thus allowing a variable number of wells.

There are some possible disadvantages to these models. For instance, the search-space size increases exponentially whenever the maximum number of wells in the chromosome increases. As a result, the number of simulated scenarios required to achieve a proper solution grows too large. Since simulation of oil fields is a computationally intensive task, any optimisation algorithm that would require too many simulations to reach the optimum is impractical.

Another disadvantage of the above models is the redundancy of their chromosome representation: many distinct chromosome instances decode to the same concrete implementation. In particular, any permutation of the wells within a given chromosome produces the same physical solution. Consequently, the genetic algorithm tends to fragment its population, since many different solutions have the same fitness, which, in turn, slows down the convergence rate.

An additional limitation of standard genetic algorithms is the lack of control of their population diversity. As the algorithm iterates, highly adapted individuals rapidly replace less adapted ones, resulting in undesirable convergence to local minima. The literature has proposed many variations of the classical genetic algorithm. In particular, the cellular genetic algorithm is an adaptation of standard genetic algorithm that enforces diversity by defining geographic locations of the individuals and applying recombination exclusively between neighbours [10].

Lastly, the standard genetic algorithm models in the literature do a naive population initialisation: they randomly sample individuals until the required number of feasible individuals is found. While this scheme works for simple optimisation problems, it is not efficient for the kind of black-box nonlinear constraints present in the optimisation of oil fields. Therefore, we also propose a new way to find a random feasible initial population using Monte-Carlo sampling.

The following sections describe a new model to solve the optimal wells placement problem using evolutionary heuristics and geophysics simulations. Our model derives from the work of [3] and enforces a unique domain representation for each physical implementation. Therefore, the proposed model has less redundancy when compared to those in the literature, and it should result in less fragmented populations with better evolution curves. Moreover, we apply a proved cellular genetic algorithm (CGA) with online diversity control, and compare it to the standard genetic algorithms. Additionally, the proposed model adapts the Genocop III technique to make it meaningful with CGA and integer genes. Finally, we also describe how to use Monte-Carlo sampling to efficiently find a feasible initial population to start the evolution process.

2   Problem Description

This text addresses the problem of optimising reservoir exploration alternatives by deciding its wells locations, trajectories, and types to maximise the NPV (net present value) of the oil field.

The NPV is calculated using the per well oil production and costs, acquired from a commercial reservoir simulator [11]. The commercial simulator needs a 3-dimensional discrete geological representation of the petroleum reservoir, provided beforehand. An example of freely available reservoir model is the UNISIM-I [12] provided by UNICAMP, a state university from Brazil.

Considering that the petroleum reservoir model is a discrete grid of cells, then the wells trajectories assume a finite set of values, represented by the coordinates of the grid blocks. Hence, we represent them using a single line segment, and only the endpoints of the well are required.

Regarding well types, the adopted simulation model sets the available choices. The black oil model [13–15] allows two types of wells: water injector and oil producer. The first type represents a well that injects water into the reservoir, whereas the latter type accounts for a well that extracts oil from the reservoir. On the other hand, the compositional model [16–18] is a more complex scheme that allows additional well types, including the water alternating gas (WAG) type and the cyclic well type. Since the compositional model uses more well types, the search-space is usually bigger and, consequently, the optimisation of wells alternatives become more demanding.

Additionally, the optimiser should also be able to define the optimal number of wells in a field, considering the possibility of pre-existing wells. Usually, this is accomplished by setting a maximum number of wells in the oil field and by including auxiliary binary variables that define which wells are active. Then, only active wells are considered a part of the concrete implementation. This approach increases the problem domain by a factor of 2^N, where N is the maximum number of wells in the oil field. This work depicts an alternate chromosome representation that reduces this factor from 2^N to N.

Finally, it is useful to be able to optimise the wells placement under uncertainty. In particular, the industry is interested in finding a wells placement alternative that is robust to poorly-known parameters. These parameters are typically related to geological or geophysical properties of the oil field, which are modelled by statistical distributions. Since some of the problem parameters are random variables, the resulting NPV is a random variable. Consequently, we try to maximise the *expected value* of the NPV. This work also shows how to use the proposed optimisation algorithm to approach this kind of problem.

3   Methodology

The purpose of this section is to present the mathematical formulation of the optimisation problem and the proposed solution model.

At first, we express the optimisation as a black-box nonlinear integer programming problem. Thus all decision variables are integer numbers, and all mathematical functions are analytically unknown, but numerically computable. In particular, the numerical computation of the fitness function is time-consuming, deterministic and robust. Conversely, the numerical calculation of all constraints is relatively easy.

Next, this section presents two algorithms to solve it: a conventional genetic algorithm (GA) and a cellular genetic algorithm (CGA). The CGA as applied to the optimal wells placement problem is an innovation, and we also propose novelties to the classical CGA present in the current literature. These novelties include a black-box constraint handling scheme for CGA based on the Genocop III [19] heuristic, an efficient initial population sampler based on the Metropolis-Hastings algorithm [20,21] that guarantees that all sampled individuals are feasible, and a space-efficient chromosome representation, which reduces the search space size by removing redundancy. We also propose a recombination operator and a mutation operator for this new representation. Finally, we proposed an adaptation of the model to handle uncertainties in the definition of the reservoir geophysical parameters.

3.1   Mathematical Formulation

The standard representation uses six integer variables to represent the discrete coordinates of two blocks in the well ends locations, one binary integer to indicate that a specific well exists (the activation bit) and another integer to define the well type.

Let $\bar{i}, \bar{j}, \bar{k}, \underline{i}, \underline{j}, \underline{k}$ denote, respectively, the (i,j,k) coordinates of the initial block and the (i,j,k) coordinates of the final block of the well. Let a denote the activation bit and τ denote the well type. If N is the maximum allowed wells count, then the decision vector $\mathbf{x}$ is:

$$
\begin{aligned}
\mathbf{x} = (& a_1, a_2, \ldots, a_N, \\
& \bar{i}_1, \bar{j}_1, \bar{k}_1, \underline{i}_1, \underline{j}_1, \underline{k}_1, \tau_1, \\
& \bar{i}_2, \bar{j}_2, \bar{k}_2, \underline{i}_2, \underline{j}_2, \underline{k}_2, \tau_2, \ldots, \\
& \bar{i}_N, \bar{j}_N, \bar{k}_N, \underline{i}_N, \underline{j}_N, \underline{k}_N, \tau_N)
\end{aligned}
\tag{1}
$$

where the numeric subscripts indicate the index of the well. For simplicity, we write $w_k = (\bar{i}_k, \bar{j}_k, \bar{k}_k, \underline{i}_k, \underline{j}_k, \underline{k}_k, \tau_k)$ and then the Eq. (1) reads:

$$
\mathbf{x} = (a_1, a_2, \ldots, a_N, w_1, w_2, w_N).
\tag{2}
$$

The objective function is defined over all values of $\mathbf{x}$. The relative order of the wells in Eq. (1) is not relevant, that is, the NPV depends only on the values of the wells position, its types, and activation bits. Therefore, if, for example, $\mathbf{x}' = (a_2, a_1, \ldots, a_N, w_2, w_1, \ldots, w_N)$ and f is the objective function, then $f(\mathbf{x}) = f(\mathbf{x}')$. More generally, the NPV is invariant to any permutation of the wells. Hence, there is a large number of points in the problem domain with the same evaluation, which might lead to many different optimal solutions that actually represent the same physical solution. Moreover, since it is interesting to apply

genetic algorithms, the redundancy in the problem domain makes the search
space unnecessarily large, so the algorithm tends to converge much slower to a
relevant solution.

To solve this issue, the present work proposes a new strategy for representing
the decision vector $\mathbf{x}$, where the order of the vectors $w_1, w_2, ..., w_N$ is unique.
Therefore, for a given set of distinct wells, there is only one representation of $\mathbf{x}$
having

$$w_1 \leq w_2 \leq \ldots \leq w_N. \tag{3}$$

We specify the relation "$\leq$" for any pair of wells in the algorithm 1.

The algorithm 1 works by sequentially comparing the coordinates of the
vectors w_1 and w_2 until they differ or all the coordinates are compared. The
proposed model starts by comparing the initial i coordinate, namely $\bar{i}$, and if $\bar{i}_1 <
\bar{i}_2$ then $w_1 < w_2$. Clearly, if $\bar{i}_1 > \bar{i}_2$, then $w_2 < w_1$, Finally, if the coordinates are
equal, then it repeats the comparison on the next coordinate. Hence, for instance,
the wells $w_1 = (1, 0, 2, 2, 2, 2, injector)$ and $w_2 = (1, 0, 3, 2, 2, 2, injector)$ satisfy
$w_1 \leq w_2$. The presented model defines $injector < producer$.

Another proposal to reduce even more the search space is to replace the per-
well activation bits by a single integer variable that represents the number of
active wells. Therefore, by rearranging the chromosome such that the active wells
appear before the inactive ones, the new chromosome has only N distinct possibilities, as opposed to the former 2^N possibilities of the chromosome presented
in Eq. (1), where is the maximum wells count. Then, our final space-efficient
chromosome reads:

$$\mathbf{x} = \big(\eta, w_1, w_2, \ldots, w_N\big) \tag{4}$$

where w_1 to w_η are active wells.

The objective function is the NPV of the platform. The presented model
assumes there is only one platform, and it has all the active wells. There are some
models in the literature for calculating the NPV. The model from [1] uses discrete
time-steps from the simulator outputs, which reports the total production of oil
or gas and the total injection of water for each well and each period. After that,
the authors of [1] calculate the well profits per produced or injected volumes
and multiply them by the outputs of the simulator to determine the total profit
of each time-step. Ultimately, the NPV is the sum of all discounted time-step
profits. This model, however, considers only vertical or horizontal wells, which is
an oversimplification of the problem in question. Furthermore, we need to take
into account other costs associated with the wells, as the abandonment costs,
the costs depending on the wells length, the flowline costs, drilling complexity
costs, and others.

The proposed model is based on the work of [3], which models the NPV as
the sum of the NPV of all wells minus the platform cost. The platform cost is
the total expense of building a platform on the reservoir and it is a constant
specified beforehand. Conversely, the NPV of the well is dependent upon the

decision variable $\mathbf{x}$ and considers many aspects. The Eq. (5) depicts this model.

$$NPV = \sum_{k=1}^{N} NPV_w(k) - C_P \tag{5}$$

In the Eq. (5), NPV and $NPV_w(k)$ are, respectively, the platform NPV and the NPV of the kth well. Additionally, C_P is the total platform cost. The NPV of the well is the difference between the total present value of the income and the well costs:

Algorithm 1 Boolean function $\leq (w_1, w_2)$.

1: **procedure** $\leq (w_1, w_2)$
2:     $\triangleright w_k = (\bar{i}_k, \bar{j}_k, \bar{k}_k, \underline{i}_k, \underline{j}_k, \underline{k}_k, \tau_k)$
3:     **if** $\bar{i}_1 < \bar{i}_2$ **then**
4:         **return true**
5:     **else if** $\bar{i}_1 > \bar{i}_2$ **then**
6:         **return false**
7:     **if** $\bar{j}_1 < \bar{j}_2$ **then**
8:         **return true**
9:     **else if** $\bar{j}_1 > \bar{j}_2$ **then**
10:         **return false**

11:         $\vdots$
12:     **if** $\underline{k}_1 < \underline{k}_2$ **then**
13:         **return true**
14:     **else if** $\underline{k}_1 > \underline{k}_2$ **then**
15:         **return false**
16:     **if** $\tau_1 \leq \tau_2$ **then**
17:         **return true**
18:     **else**
19:         **return false**

$$NPV_w(k) = (1 - I) \cdot \sum_{t=1}^{T} \frac{R(k,t) - C_o(k,t)}{(1+D)^{y_t}} - C_w(k), \tag{6}$$

where t is the discrete time, T is the number of time-steps, $R(k,t)$ is the revenue between times $t - 1$ and t, $C_o(k,t)$ is the operational cost of the well between times $t - 1$ and t, D is the annual discount rate and y_t is the number of years measured from the start of the reservoir operation to time t. Furthermore, I is the tax rate, and $C_w(k)$ is the cost of the kth well.

The revenue between times $t - 1$ and t is:

$$R(k,t) = O_p(k,t) \cdot P_o(t) + G_p(k,t) \cdot P_g(t) \tag{7}$$

where $O_p(t)$, $P_o(t)$, $G_p(t)$, and $P_g(t)$ are, respectively, the oil production, the oil price, the gas production and the gas price between times $t - 1$ and t. The productions are a simulation output, whereas the prices are pre-specified quantities.

The operational costs include the fixed costs of the well, the maintenance expenses in the time-step, the variable expenses in the time-step, the royalties, and the costs associated with the amount of fluid production or injection. The Eq. (8) shows the general equation.

$$
\begin{aligned}
C_o(k,t) = &\left[C_M \cdot (y_t - y_{t-1}) \right] \\
&+ C_{vf} + R_y \cdot R(k,t) \\
&+ \Big(O_p(k,t) \cdot O_{pc} + G_p(k,t) \cdot G_{pc} + \\
&\quad W_p(k,t) \cdot W_{pc} + G_i(k,t) \cdot G_{ic} + \\
&\quad W_i(k,t) \cdot W_{ic} \Big)
\end{aligned}
\tag{8}
$$

In the Eq. (8), C_M is the maintenance cost per year, C_{vf} is a constant cost, and R_y is the royalties percentage. Moreover, O_{pc} is the oil production cost per volume of oil, G_{pc} is the gas production cost per volume of gas, $W_p(k,t)$ is the water production of the kth well between times $t-1$ and t, and W_{pc} is the water production cost per volume of water. Finally, $G_i(k,t)$ is the gas injected between times $t-1$ and t, G_{ic} is the gas injection cost per unit of gas volume, $W_i(k,t)$ is the amount of water injected into the kth well between times $t-1$ and t, and W_{ic} is the water injection cost per unit of water volume.

The well development cost, $C_w(k)$, is a complicated non-linear function of the well length, the well position, the well inclination, and the well type. This function includes the drilling costs, the distance between the kth well and the platform, and the cost of shutting down the kth well. For simplicity, we choose to omit this function herein.

To guarantee the physical meaning of the solution $\mathbf{x}$, we should define suitable restrictions involving the wells length, the wells pairwise distances and the total well count. The positive integer constant N specifies the maximum number of wells the platform could handle. Since this solution uses activation bits, a_k, to indicate whether the kth well exists or not, the decision variable $\mathbf{x}$ always have N wells and the solution may have any number of wells from 0 to N.

For operational reasons, the length of each well should not exceed a maximum constant length, L. This problem models the wells as line segments whose ends are the centre points of the wells start and end blocks. Therefore, the well length $l(k)$ is simply the Euclidean distance between the line segments ends and, for each well k, $1 \leq k \leq N$, we have:

$$
l(k) \leq L
\tag{9}
$$

Similarly, there is a minimum wells distance, that is, it is not possible to place two wells closer than a minimum distance d_{min}. Hence, for each pair of wells k_1 and k_2, $k_1 \neq k_2$, we enforce the restriction:

$$
d(w_{k_1}, w_{k_2}) \geq d_{min}.
\tag{10}
$$

Since the problem models the wells as line segments (Fig. 1), the distance between two wells, $d(w_{k_1}, w_{k_2})$, is the minimum distance between the two line segments that geometrically represent the wells w_{k_1} and w_{k_2}. In [22], the author explains this problem in detail.

In conclusion, we seek the solution $\mathbf{x}$, as defined in the Eq. (4), that maximises the NPV in Eq. (5), subject to the nonlinear restrictions (9) and (10). The following section describes the algorithm this paper employs for solving this problem efficiently on a digital computer.

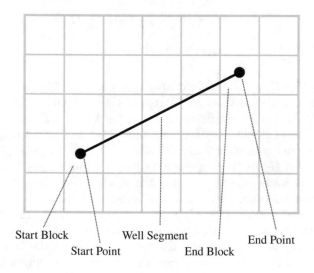

Start Block   Well Segment   End Point
      Start Point   End Block

Fig. 1. Representation of well in a grid

3.2   Solution to the Optimisation Problem

This work uses evolutionary algorithms to solve the optimisation problem of the Sect. 3.1. In particular, we use a classical genetic algorithm and a more modern approach, the cellular genetic algorithm [23, 24]. The classical GA utilises a chromosome representation that does not enforce the wells order and has one activation bit for each well. On the other hand, the proposed CGA model uses the space-efficient representation as mentioned earlier, which reduces the search space by ensuring a certain wells order and by using a single gene to represent the number of active wells, as opposed to activation bits.

The current work employs the chromosome of the Eq. (2). The fitness function is the NPV of the platform, as the Eq. (5) exhibits.

Both the cellular and classical genetic algorithms require mutation and recombination operators. Since there are two possible chromosome representations, the optimisation model needs to develop these operators according to each representation.

Mutation Operators

Mutation operates on a single individual, possibly generating a new (mutated) individual. Our model employs two types of mutation: the activation mutation and the uniform mutation. The former operates on the activation bits, thereby not changing the relative order of the wells in the chromosome. The latter influences the position and type genes. Hence, it is possible that a solution satisfying the order criterion Eq. (3) do not keep satisfying it after mutation.

The activation mutation is a simple random bit mutation. For each activation bit, we sample a random number between 0 and 1 using a uniform distribution, and if this random number is less than or equal to a mutation probability, the activation bit is flipped. This mutation type is only meaningful if the representation uses activation bits.

Algorithm 2 Activation mutation.

1: **procedure** ACT_MUTATE($\mathbf{x}$, p)
2:     ▷ $\mathbf{x} = (a_1, \ldots, a_N, w_1, \ldots, w_N)$.
3:     ▷ p is the gene mutation probability.
4:     $\mathbf{x}_{new} \leftarrow \mathbf{x}$
5:     **for** $k \leftarrow 1 \ldots N$ **do**
6:         $r \leftarrow$ RANDOM(0,1)
7:         **if** $r \leq p$ **then**
8:             Flip a_k of $\mathbf{x}_{new}$
9:     **return** $\mathbf{x}_{new}$

Contrary to the activation mutation, the uniform mutation needs to distinct between the two chromosome representations. For the orderless chromosome representation, the algorithm 3 depicts the process.

The algorithm 3 first selects the gene to mutate. This gene can be a position or a type gene. After that, it samples a random integer in the range of possible values of the selected gene. Finally, it returns a copy of the original individual with the new mutated gene. This algorithm, however, does not enforce the relative order of the wells of the individual, as in the equation Eq. (3). Therefore, it takes a small modification to render this algorithm useful for the order-sensitive representation.

The algorithm 4 shows how to guarantee that the mutated individual satisfies the criterion (3) provided the original individual satisfy it. First, it tries to mutate using the algorithm 3. If the mutation result does not meet the order, then it attempts to mutate again. By doing so, it guarantees that all ordered individuals have approximately equal probability of generation.

Recombination

Recombination operates on two inputs and generates two more individuals. There are two types of recombination employed in the present work: the single point crossover and the arithmetical crossover. The algorithms 5 and 6 describe these two methods.

The function ROUND($\mathbf{v}$) rounds each element of the vector $\mathbf{v}$ to its nearest integer. The authors of [3] explain the single point crossover and the arithmetic crossover in detail. These operators are valid only for the representation that does not emphasise the relative order of the wells in the chromosome.

Algorithm 3 Uniform mutation.

1: **procedure** UNIF_MUTATE($\mathbf{x}$)
2:     $\triangleright \mathbf{x} = (a_1, \ldots, a_N, w_1, \ldots, w_N)$.
3:     $k \leftarrow$ RANDOMINT$(1, N)$
4:     $i \leftarrow$ RANDOMINT$(1, 7)$
5:     $\mathbf{x}_{new} \leftarrow \mathbf{x}$
6:     $min \leftarrow$ minimum of the ith gene of the kth well.
7:     $max \leftarrow$ maximum of the ith gene of the kth well.
8:     $r \leftarrow$ RANDOM$(min, max)\triangleright$ new gene value.
9:     Replace the ith gene of the kth well of $\mathbf{x}_{new}$ by r.
10:     **return** $\mathbf{x}_{new}$

Algorithm 4 Uniform mutation (order-aware representation).

1: **procedure** UNIF_MUTATE_ORDER($\mathbf{x}$)
2:     **repeat**
3:         $\mathbf{x}_{new} \leftarrow$ UNIF_MUTATE(x)
4:     **until** $\mathbf{x}_{new}$ satisfies the Eq. (3)
5:     **return** $\mathbf{x}_{new}$

Algorithm 5 Single Point Crossover

1: **procedure** SINGLE_CROSS($\mathbf{x}_1, \mathbf{x}_2$)
2:     Randomly choose an index i, $1 \leq i \leq 8N$
3:     left($\mathbf{x}_1$) $\leftarrow$ genes of $\mathbf{x}_1$ having index $\leq i$.
4:     left($\mathbf{x}_2$) $\leftarrow$ genes of $\mathbf{x}_2$ having index $\leq i$.
5:     right($\mathbf{x}_1$) $\leftarrow$ genes of $\mathbf{x}_1$ having index $> i$.
6:     right($\mathbf{x}_2$) $\leftarrow$ genes of $\mathbf{x}_2$ having index $> i$.
7:     $\mathbf{x}'_{new} \leftarrow$ left($\mathbf{x}_1$) concatenated with right($\mathbf{x}_2$).
8:     $\mathbf{x}''_{new} \leftarrow$ right($\mathbf{x}_1$) concatenated with left($\mathbf{x}_2$).
9:     **return** $(\mathbf{x}'_{new}, \mathbf{x}''_{new})$

Algorithm 6 Arithmetic Crossover

1: **procedure** ARITH_CROSS(x_1, x_2)
2:    $\alpha \leftarrow$ RANDOM$(0, 1)$
3:    $x'_{new} \leftarrow$ ROUND$(\alpha x_1 + (1 - \alpha)x_2)$
4:    $x''_{new} \leftarrow$ ROUND$((1 - \alpha)x_1 + \alpha x_2)$
5:    **return** (x'_{new}, x''_{new})

Algorithm 7 Arithmetical Crossover for order-aware representation.

1: **procedure** ARITH_CROSS_ORDER($\mathbf{x}_1, \mathbf{x}_2$)
2:    **repeat**
3:        $(\mathbf{x}'_{new}, \mathbf{x}''_{new}) \leftarrow$ ARITH_CROSS$(\mathbf{x}_1, \mathbf{x}_2)$
4:    **until** $\mathbf{x}'_{new}$ and $\mathbf{x}''_{new}$ satisfy the Eq. (3)
5:    **return** $(\mathbf{x}'_{new}, \mathbf{x}''_{new})$

For the case of the order-aware representation, we modify the arithmetical crossover similarly to the adaptation of the uniform mutation. We apply the crossover until a pair of individuals that satisfy Eq. (3) is found. The algorithm 7 displays this procedure.

Population Initialisation

Initialising the population is the process of creating random individuals for the first generation of the evolutionary algorithm. Since the Genocop III technique requires a fully feasible initial population, we need to find a way to sample individuals from the feasible set of solutions. In other words, the sampling method should be able to choose individuals from the set of all feasible individuals uniformly.

A naive solution would be to uniformly sample individuals until the required number of feasible individuals is found. Therefore, if a sampled individual does not satisfy one of the constraints, then it is discarded, and the process continues to sample new individuals. Algorithm 8 depicts this method.

The main drawback of algorithm 8 becomes clear whenever the feasible search space is small when compared to the whole search space. In this case, the probability of uniformly choosing a chromosome that satisfies all the constraints is small; thus it might take too long to find a valid initial population. Therefore, this article proposes to sample individuals based on a distribution function that has small probability in infeasible subsets of the search space and has a higher (and almost uniform) probability within feasible subsets of the search space.

The chosen algorithm to draw samples from a pre-specified custom distribution is the Metropolis-Hastings method [20,21]. This algorithm requires the definition of the target density function $\Pi(\mathbf{x})$ and a conditional density function $Q(\mathbf{x}|\mathbf{x}')$ (so-called the candidate's proposal or kernel). Given a current sampled individual $\mathbf{x}_t$, a new sample is drawn from the conditional Q density, and it is kept if it has likelihood greater than the current sample likelihood. Algorithm 9 explains in detail the Metropolis-Hastings scheme.

Algorithm 8 Naive Sampling

1: **procedure** NAIVE_SAMPLING(pop_size)
2:     $pop\_size\_counter \leftarrow 0$
3:     **repeat**
4:         sample new individual $\mathbf{x}$
5:         **if** $\mathbf{x}$ is feasible **then**
6:             $pop\_size\_counter \leftarrow pop\_size\_counter + 1$
7:     **until** $pop\_size\_counter = pop\_size$

Algorithm 9 Metropolis Hastings Sampling

1: **procedure** METROPOLIS_HASTINGS_SAMPLING(pop_size)
2:     $pop\_size\_counter \leftarrow 0$
3:     $t \leftarrow 0$
4:     $\mathbf{x}_0 \leftarrow$ sample from $\Pi(\mathbf{x})$
5:     **repeat**
6:         draw sample $\mathbf{Y}$ from $Q(\mathbf{x}|\mathbf{x}_t)$
7:         $a_1 \leftarrow {\Pi(\mathbf{x})}/{\Pi(\mathbf{x}_t)}$
8:         $a_2 \leftarrow {Q(\mathbf{x}_t|\mathbf{x})}/{Q(\mathbf{x}|\mathbf{x}_t)}$
9:         $a \leftarrow a_1 \cdot a_2$
10:         **if** $a \geq 1$ **then**
11:             $\mathbf{x}_{t+1} \leftarrow Y$
12:         **else**
13:             $\mathbf{x}_{t+1} \leftarrow \begin{cases} \mathbf{Y} & \text{with probability } a \\ \mathbf{x}_t & \text{with probability } 1 - a \end{cases}$
14:         $t \leftarrow t + 1$
15:         **if** $\mathbf{x}_{t+1}$ is feasible **then**
16:             $pop\_size\_counter \leftarrow pop\_size\_counter + 1$
17:     **until** $pop\_size\_counter = pop\_size$

We use the algorithm 9 to sequentially draw samples from the distribution $\Pi(\mathbf{x})$ until the required number of feasible individuals is found, similarly to the case of the random sampling mentioned earlier. However, since our method designs the distribution $\Pi(\mathbf{x})$ so that it has higher probabilities within feasible subsets of the search space, this proposed scheme tends to be faster and more reliable than the naive solution. The next few paragraphs explain in detail how to design the function $\Pi(\mathbf{x})$ based on the maximum well length constraint of Eq. (9) and the minimum pairwise wells distance constraint of Eq. (10).

Hence, we write the density function $\Pi(\mathbf{x})$ as a product of two density functions, namely $\Pi_d(\mathbf{x})$ and $\Pi_l(\mathbf{x})$:

$$\Pi(\mathbf{x}) = \Pi_d(\mathbf{x}) \cdot \Pi_l(\mathbf{x}) \tag{11}$$

The density Π_d does not value individuals that have pairs of wells too close to each other. More clearly, if the minimum allowed pairwise wells distance is d_{min}, then:

$$\Pi_d(\mathbf{x}) = \prod_{(r,s)} \Pi_{dw}(w_r, w_s)$$

$$\Pi_{dw}(w_r, w_s) = \begin{cases} A \tanh\left(\zeta(\lambda - 1)\right) + 1 & \lambda \leq 5 \\ \nu e^{5-\lambda} & \lambda > 5 \end{cases} \tag{12}$$

where

$$\nu = 1.9$$
$$\zeta = 10.0$$
$$A = \frac{\nu - 1}{\tanh(4\zeta)}$$
$$\lambda = \frac{d(w_r, w_s)}{d_{min}} \tag{13}$$

Thus, as the wells distance of a particular pair of wells of the individual $\mathbf{x}$ approaches 0, the value of $\Pi_d(\mathbf{x})$ decreases, as expected. The same happens if the distance between a pair of wells approaches infinity because the function $\Pi_d(\mathbf{x})$ should be integrable. If all pairs of wells of the individual $\mathbf{x}$ satisfy $1 \leq \lambda \leq 5$, the value of $\Pi_d(\mathbf{x})$ is high and almost constant, as expected (see Fig. 2).

The next function is $\Pi_l(\mathbf{x})$, which should be high for chromosomes whose wells are large, but smaller than the maximum well size. The proposed function is:

$$\Pi_l(\mathbf{x}) = \prod_{r=1}^{N} \Pi_{lw}(w_r)$$

$$\Pi_{lw}(w_r) = \begin{cases} \left[1 + \left(\frac{l(r)}{L} - 1\right)^2 \right]^{-1} & l(r) \leq L \\ \exp -\zeta\left(\frac{l(r)}{L} - 1\right) & l(r) > L \end{cases} \tag{14}$$

where $l(r)$ is the length of the well w_r and L is the maximum allowed length of a well.

As the length of the well increases towards the maximum allowable size L, the likelihood approaches the maximum value 1. Furthermore, as the well length increases above its maximum allowed value, the value of $\Pi_l(\mathbf{x})$ vanishes, which guarantees both the low probability of having large sized wells and the integrability of $\Pi_l(\mathbf{x})$. The Fig. 3 shows how Π_{lw} behaves.

In this work, we compared the naive approach to the proposed Metropolis-Hastings generator. The Sect. 4 specifies the results in detail.

The Evolutionary Algorithms

The article [3] explains the classical genetic algorithm (GA). It uses the Genocop III (Genetic Algorithm for Numerical Optimization of Constrained Problems III) technique [19] to handle black-box constraints. Additionally, for generating an

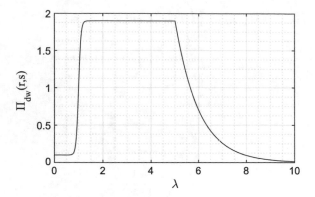

Fig. 2. Likelihood function of the wells distance

Algorithm 10 Description of the CGA algorithm.

1: **procedure** EVOLVE_CGA(*gen*)
2:     **repeat**
3:         *pop* ← random feasible population
4:         **for each** individual *ind* in *pop* **do**
5:             *parent*$_1$ ← random neighbor of *ind*
6:             *parent*$_2$ ← random neighbor of *ind*
7:             *child* ← RECOMBINATE(*parent*$_1$, *parent*$_2$)
8:             *ind*$_{new}$ ← MUTATE(*child*)
9:             EVALUATE(*ind*$_{new}$)
10:            **if** *ind*$_{new}$ evaluation ¿ *ind* evaluation **then**
11:                Replace *ind* by *ind*$_{new}$
12:        **until** *gen* ≥ *gen*$_{max}$
13:        **return** Best Solution

---

initial population, the algorithm randomly creates chromosomes until it finds a sufficient number of feasible individuals.

On the other hand, the cellular genetic algorithm (CGA) is a variation of the standard GA that enforces a unique geographic location for every individual. As a result, the selection operation takes the individuals locations into account, so the algorithm restricts the recombination to neighbour individuals. Moreover, the substitution operates only on individuals with the same geographic location, so a solution is replaced by a new one only if the new individual is better and has the same geographic position. The authors of [23] make a comparison between classical genetic algorithms and cellular genetic algorithms. Additionally, [10] explains in detail how CGA works. The Fig. 4 and the algorithm 10 illustrates the CGA workflow.

For handling black-box constraints, we propose in the next few paragraphs a variation of Genocop III suitable for the concept of geographic locations the CGA uses.

The Genocop III works by maintaining two separate populations: a search population and a reference population. The reference population is generated at the start of the optimisation during the initial population sampling, and it should not contain infeasible individuals. Conversely, the search population consists of individuals used by the optimisation. Whenever an invalid individual appears, it goes through a repair process that converts it into a feasible one.

Repairing infeasible individuals consists of selecting one of the reference individuals and applying arithmetic crossover between the infeasible and the reference individuals until a new feasible individual is known. Equivalently, the individuals can be interpreted as points so that $\mathbf{r}$ is the reference point and $\mathbf{s}$ is the search point. Thereby, the process creates a segment between $\mathbf{s}$ and $\mathbf{r}$ and chooses a random point $\mathbf{z}$ where $\mathbf{z} = a\mathbf{s} + (1 - a)\mathbf{r}$ and a is a random number between 0 and 1. If $\mathbf{z}$ is infeasible, then the process is repeated until a valid $\mathbf{z}$ is known. After that, $\mathbf{z}$ replaces the infeasible point $\mathbf{s}$ (see Fig. 5).

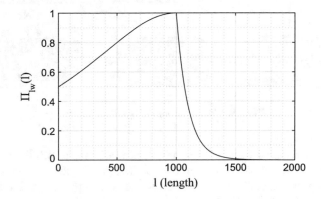

Fig. 3. Likelihood function of the well length

The CGA, unlike the traditional GA, arranges each individual to a determined geographic location. The purpose of the geographic locations is to maintain a healthy population diversity during the evolution process. Therefore, using a random reference individual to repair the infeasible individuals would completely disregard the locations of the individuals, since the reference individual of the Genocop III technique does not emphasise position. Hence, we propose to discard the reference population and to employ as the reference the last known feasible individual in the same location. This scheme maintains the property that each individual is influenced only by its nearby neighbours.

3.3 Algorithm Adaptation to Problems with Uncertainties

To model well placement problems having uncertainties in the geophysical parameters, we use the concept of *geological scenarios*. Geological scenarios are versions of the geological model constructed from samples of the geophysical parameters distribution.

Typical geophysical parameters are modelled as random variables, usually uniformly distributed from a possible range of values. These variables include the reservoir porosity and the reservoir permeability. The geological scenarios are constructed from samples of the porosity and permeability samples, and each scenario is a concrete reservoir model ready to simulate.

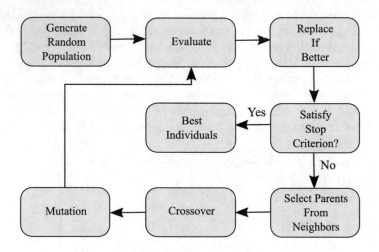

Fig. 4. Sketch of the CGA algorithm

Therefore, the uncertainty in a specific parameter is modelled as a set of geological scenarios. Since each scenario has its specific net present value and its specific probability of occurrence, the all-scenario net present value is defined as the expected value of the *NPV* calculated as:

$$E[NPV] = \sum_{i=1}^{N} p_i \cdot NPV_i^s, \tag{15}$$

where NPV_i^s is the *NPV* of the geological scenario i and p_i is its probability of occurrence. The value p_i is provided beforehand.

Consequently, the fitness evaluation now consists of not only one, but many simulations, because we need many geological scenarios to represent the reservoir. Additionally, the number of *NPV* computations is also multiplied by the number of geological scenarios, since we need to calculate a *NPV* for each scenario and the compute Eq. (15). After these modifications, the algorithms proposed in Sect. 3.2 are still valid and their result is now an optimal wells placement alternative that is robust to geological uncertainties.

4   Experiments and Results

This section splits the experiments into two parts: the initialisation part and the optimisation part. The initialisation experiments concern on comparing the performance of the naive and the enhanced initialisation algorithms described in Sect. 3.2, whereas the optimisations compare the effect of the optimisation models on the final solution.

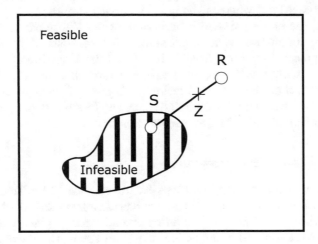

Fig. 5. Sketch of the Genocop III heuristic

4.1   Population Initialisation Experiments

To compare the performance of the Metropolis-Hastings method to the performance of the naive method, we generated 1000 individuals using both algorithms and counted how many of them were feasible. The model used was the UNISIM-I [25], from UNICAMP. We varied the minimum wells distance from 100 to 500 meters and the maximum wells length from 500 to 2000 meters, and we set the number of wells of each individual to 20.

Table 1. Count of feasible individuals found after 1000 samples using the naive and the proposed algorithms for a given maximum allowed well length and minimum allowed pairwise wells distance

Max. length (m)	Min. distance (m)	Naive method	Proposed method
500	100	1	77
2000	100	108	649
500	500	13	381
2000	500	46	302

Table 1 shows the results of the experiment. As the minimum wells distance increases (second column), is becomes harder to find an alternative that respects the minimum wells distance for every pair of wells (there are 190 pairs of wells). However, the Metropolis-Hastings approach still can find a significant number of feasible alternatives, whereas the naive approach cannot.

Moreover, the same pattern is observable when the maximum well length (first column) is constrained to small values. For example, after 1000 samples, the naive approach found one feasible individual for a maximum well length of 500 m (row 1), while the Metropolis-Hastings approach found 77.

We conclude that the proposed technique is more efficient to find the initial population when it requires that all individuals be feasible. From now on, all the following experiments use the proposed technique.

4.2   Optimisation Experiments

The experiments were divided into three classes: classical GA with orderless representation, classical GA with order-aware representation and CGA with order-aware representation. Since the orderless chromosome implies a bigger search space due to increased redundancy, it is expected better results with the order-aware chromosome.

The experiments aimed to maximise the NPV of exploring the UNISIM synthetic reservoir model [25]. This model has two flavours: the black-oil version, which simulates in the IMEX simulator [11], and the compositional model version, which simulates in the GEM simulator. The typical simulation time ranges from 2 to 10 min. Therefore, possible optimisations cannot have much more than a thousand fitness function evaluations, or it would take too long to complete.

This work placed the experiments in the OCTOPUS 2 reservoir management platform [2], by developing a new optimisation plug-in. This way, we were able to focus solely on the scientific aspects of the experiments, namely the evolutionary algorithms and the optimisation results.

Both the classical GA and the CGA used binary tournament selection, where the winning probability is proportional to the fitness value. Further, the algorithms utilised the appropriate arithmetic crossover for the chromosome and, more specifically, the classical GA with orderless representation also adopted the single point crossover. Finally, both algorithms employed a "replace if better" substitution principle, where the new individuals replace the older ones only if they have a better fitness.

Additionally, the CGA algorithm employed an adaptive grid scheme to maintain a healthy population diversity. This paper uses the technique of [10] to control the entropy of the population. Whenever the entropy is decaying too fast, it changes the grid to a more narrow shape, thus making it harder to propagate the best individuals. Conversely, whenever the entropy is decaying too slowly (or decaying at all), it reshaped the grid to make it more square, thus allowing a faster convergence rate. Hence, it avoided convergence to local maxima and maximised the chances of finding the global maximum of the problem.

Table 2 summarises the parameters used in each experiment. In particular, the CGA needs the threshold ϵ, which controls the grid shape switching procedure. As we show, we tried to make the experiments as even as possible, so we believe the comparisons among their results are fair.

Table 2. Summary of the optimization parameters

	Classical GA	Cellular GA
Population size	24	48
Generations	50	25
Mutation rate	10%–70%	10%–70%
Mutation	Uniform and activation	Uniform and activation
Recombination	Single Pt. and arithmetic	Arithmetic
Threshold ϵ	–	0.05
Max. wells (N)	20	20
Wells radii	0.0762 m	0.0762 m

In addition to the optimisation parameters, a typical economic scenario is shared for all experiments. Specifying herein all the constants of the Eqs. (5)–(7) would be impractical, so Table Table 3 exhibits only a few of them.

The Figs. 6, 7 and 8 illustrate the results, respectively, of the classical GA with orderless and ordered chromosomes and the CGA with ordered chromosome. All experiments executed 10 times, and the plots display the averaged *NPV*. Moreover, all three experiments had about 800 evaluations of the fitness function, and thus they took about the same total time to complete. Since the CGA

Table 3. Economic scenario parameters

Quantity	Unit	Value
Platform cost (C_p)	billion US$	1.48
Oil price (P_o)	US$/m^3	250.00
Gas price (P_g)	US$/m^3	0.05
Oil prod. cost (O_{pc})	US$/m^3	40.00
Water prod. cost (W_{pc})	US$/m^3	2.00
Water inj. cost (W_{ic})	US$/m^3	2.00
Gas inj. cost (G_{ic})	US$/m^3	0.002
Gas prod. cost (G_{pc})	US$/m^3	0.002
Tax rate (I)	%	34.00
Discount rate (D)	%	9.00

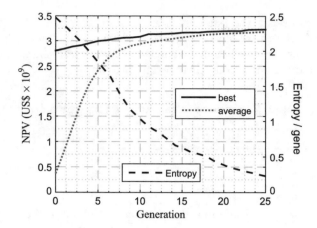

Fig. 6. Average of 10 runs of the CGA algorithm using the order-aware representation

method used an entropy based control of diversity, the entropy is a secondary axis in the plot in Fig. 6.

On average, the proposed representation using order-aware chromosomes reaches an optimum with roughly 6 to 10 times bigger *NPV*. In particular, the CGA version performs better than the GA with order-aware chromosome by a factor of 2, which was concluded by comparing the curves "best" of Figs. 6 and 8. Also, its is possible to note that the average individual of the CGA with the ordered representation reaches an 1 billion *NPV* in generation 3, whereas the best solution of the classical GA with orderless representation does not find this value at all. Hence, we tend to think that the chromosome representation that

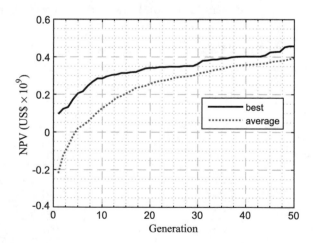

Fig. 7. Average of 10 runs of the GA algorithm using the orderless representation

enforces the relative wells order is more efficient than the traditional orderless representation.

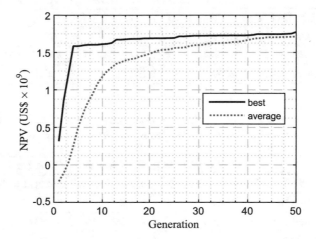

Fig. 8. Average of 10 runs of the GA algorithm using the order-aware representation

Table 4 compares the final results of the three optimisation models. The "relative" results use the formula $[NPV(end)-NPV(1)]/|NPV(1)|$, where $NPV(end)$ is the final NPV and $NPV(1)$ is the initial NPV. As it can be checked from the row "Best NPV", the cellular GA gives the higher final result, with over 3 billion US$ for the best individual and the average population fitness. On the other hand, traditional GA with orderless representation yields the worst result of the three models. Additionally, it can be observed that the relative improvements are also bigger whenever we employed the proposed order-aware chromosome. Hence, these findings reflect the consequences of integer optimisation with smaller search spaces.

When comparing conventional to cellular GA with the proposed representation, we concluded that the CGA is better than the standard GA, for the CGA finds more adapted individuals on an average of 10 experiments. This fact is observable by comparing the "average" curves of Figs. 6 and 8. In theory, this behaviour relates to the controlled diversity nature of the CGA. In our experiment, the population entropy is controlled during the evolution, thereby avoiding local maxima, which is a weakness of the classical GA.

Finally, we compared the performance of the proposed method and the conventional GA for a typical optimisation under uncertainty scenario. Our experiment used the UNISIM-I oil field with 5 different geological scenarios, with probabilities of occurrence equal to 10%, 10%, 20%, 20%, and 50%. Therefore, there were 2 low probability scenarios, 2 medium probability scenarios, and a high probability one. The expected NPV was calculated using Eq. (15).

Table 4. Comparison among the three models after 10 runs. The rows "relative" show the relative improvement of the algorithm

	GA (orderless)	GA (ordered)	CGA (ordered)
Best NPV(US$)	0.45786×10^9	1.7747×10^9	3.2341×10^9
Relative best	377%	462%	15%
Average NPV	0.39411×10^9	1.7137×10^9	3.1861×10^9
Relative average	280%	879%	760%

Table 5. Comparison among the three models after 10 runs for the multi-scenario optimisation. The rows "relative" show the relative improvement of the algorithm

	GA (orderless)	GA (ordered)	CGA (ordered)
Best NPV(US$)	0.33572×10^9	1.28813×10^9	1.48871×10^9
Relative best	402%	382%	69%
Average NPV	0.39411×10^9	1.7137×10^9	3.1861×10^9
Relative average	280%	488%	255%

The Table 5 shows that the CGA with order-aware chromosome finds the solution with highest expected NPV. Since the multi-scenario experiment has some poor geological scenarios, the maximum expected NPV is smaller than the NPV on Table 4. We conclude that the proposed CGA + order-aware model is the best option for optimising the wells placement problem, even in the presence of uncertainties.

5   Conclusions

This work presented two new approaches for the wells placement and type optimisation problem using evolutionary algorithms. These are the CGA algorithm with an adapted version of the Genocop III algorithm and the space-efficient order-aware chromosome model. We also presented a new efficient way to find a random feasible initial population based on Monte-Carlo sampling.

The Sect. 2 depicted the fundamental problem approached, explaining its discrete nature, the need for a reservoir simulator and the adopted idea of maximising the net present value of the reservoir under analysis. Then, the Sect. 3 proposed the order-aware representation, in contrast to the classical chromosome utilised for representing the wells alternatives. Next, the text describes the conventional and cellular genetic algorithms, emphasising the CGA is a new approach to this kind of optimisation problem. Finally, the Sect. 4 presented the experiments in two parts.

The first part compared the new population initialisation model to the traditional naive initialisation. The proposed Monte-Carlo based initialisation per-

formed better than the traditional naive approach, especially for highly constrained optimisations. Therefore, we conclude that it should be used to find a
feasible initial population on the experiments to follow.

Then, we presented three optimisation experiments: classical GA with the
traditional representation and with the proposed order-aware representation and
the CGA with the proposed order-aware representation. The findings showed
that the order-aware chromosome is better, for the experiments that used it converged to higher NPV. We believe that this better behaviour is due to reduced
search space since the proposed order-aware chromosome reduces the redundancy
of individuals because there is only one possible representation of each decoded
physical implementation. Moreover, we also observed that the CGA performed
better than the traditional GA, for the average population and the best individual of the CGA evolved to a higher NPV. We credit it to the population diversity
control that is a natural part of the CGA algorithm, and that is absent from the
classical GA. Hence, the CGA features a smaller probability of hanging in local
maxima of the fitness function than the standard GA.

The presented model is widely applicable beyond the area of oil field optimisation. In particular, the concept of an order-aware chromosome that is space
efficient is relevant for any integer optimisation problem using evolutionary algorithms. Additionally, being able to efficiently find random feasible initial populations on highly constrained optimisation problems is always a challenge in the
field of evolutionary algorithms, and our model based on Metropolis-Hastings
sampling worked fairly well. Finally, it is important to notice that any improvement in the area of oil field optimisation increases the economic viability of
reservoirs and is of particular concern to top oil companies in the world. Therefore, we consider this work highly relevant for the oil & gas industry.

References

1. Yeten, B., Durlofsky, L.J., Aziz, K., et al.: Optimization of nonconventional well
   type, location and trajectory. In: SPE Annual Technical Conference and Exhibition. Society of Petroleum Engineers (2002)
2. Lima, R., Abreu, A.C., Pacheco, M.A., et al.: Optimization of reservoir development plan using the system octopus. In: OTC Brasil. Offshore Technology Conference (2015)
3. Emerick, A.A., Silva, E., Messer, B., Almeida, L.F., Szwarcman, D., Pacheco,
   M.A.C., Vellasco, M.M.B.R., et al.: Well placement optimization using a genetic
   algorithm with nonlinear constraints. In: SPE Reservoir Simulation Symposium.
   Society of Petroleum Engineers (2009)
4. Morales, A.N., Gibbs, T.H., Nasrabadi, H., Zhu, D., et al.: Using genetic
   algorithm to optimize well placement in gas condensate reservoirs. In: SPE
   EUROPEC/EAGE Annual Conference and Exhibition. Society of Petroleum Engineers (2010)
5. Bittencourt, A.C., Horne, R.N., et al.: Reservoir development and design optimization. In: SPE Annual Technical Conference and Exhibition. Society of Petroleum
   Engineers (1997)

6. Nasrabadi, H., Morales, A., Zhu, D., Well placement optimization: a survey with special focus on application for gas/gas-condensate reservoirs. J. Nat. Gas Sci. Eng. **5**, 6–16 (2012)

7. Jesmani, M., Bellout, M.C., Hanea, R., Foss, B.: Well placement optimization subject to realistic field development constraints. Comput. Geosci. **20**(6), 1185–1209 (2016)

8. Siavashi, M., Tehrani, M.R., Nakhaee, A.: Efficient particle swarm optimization of well placement to enhance oil recovery using a novel streamline-based objective function. J. Energy Resour. Technol. **138**(5), 052903 (2016)

9. Al Dossary, M.A., Nasrabadi, H.: Well placement optimization using imperialist competitive algorithm. J. Pet. Sci. Eng. **147**, 237–248 (2016)

10. Bernabe Dorronsoro, E.A.: Cellular Genetic Algorithms. Springer (2008)

11. Three-Phase, black-oil reservoir simulator, CMG (Computer Modeling Group Ltd.) (2015). https://www.cmgl.ca/uploads/files/pdf/SOFTWARE/2015ProductSheets/IMEX_Technical_Specs_15-IM-04.pdf

12. Gaspar, A.T., Avansi, G.D., dos Santos, A.A., von Hohendorff Filho, J.C., Schiozer, D.J.: Unisim-id: Benchmark studies for oil field development and production strategy selection. Int. J. Model. Simul. Pet. Ind. **9**(1) (2015)

13. Trangenstein, J.A., Bell, J.B.: Mathematical structure of the black-oil model for petroleum reservoir simulation. SIAM J. Appl. Math. **49**(3), 749–783 (1989)

14. Rankin, R., Riviere, B.: A high order method for solving the black-oil problem in porous media. Adv. Water Res. **78**, 126–144 (2015)

15. Kozlova, A., Li, Z., Natvig, J.R., Watanabe, S., Zhou, Y., Bratvedt, K., Lee, S.H., et al.: A real-field multiscale black-oil reservoir simulator. SPE J. (2016)

16. Thiele, M.R., Batycky, R.P., Blunt, M.J., et al.: A streamline-based 3d field-scale compositional reservoir simulator. In: SPE Annual Technical Conference and Exhibition. Society of Petroleum Engineers (1997)

17. Coats, K.H., et al.: An equation of state compositional model. Soc. Pet. Eng. J. **20**(05), 363–376 (1980)

18. Qiao, C., Khorsandi, S., Johns, R.T., et al.: A general purpose reservoir simulation framework for multiphase multicomponent reactive fluids. In: SPE Reservoir Simulation Conference. Society of Petroleum Engineers (2017)

19. Michalewicz, Z., Nazhiyath, G.: Genocop iii: a co-evolutionary algorithm for numerical optimization problems with nonlinear constraints. In: 1995, IEEE International Conference on Evolutionary Computation, vol. 2, pp. 647–651. IEEE (1995)

20. Griffin, J.E., Walker, S.G.: On adaptive metropolis–hastings methods. Stat. Comput. **23**(1), 123–134 (2013). http://dx.doi.org/10.1007/s11222-011-9296-2

21. Yildirim, I.: Bayesian inference: metropolis-hastings sampling. Department of Brain and Cognitive Sciences, Univ. of Rochester, Rochester, NY (2012)

22. Eberly, D.: Robust computation of distance between line segments. Geometric Tools, LLC, Technical report (2015)

23. Gong, Y.-J., Chen, W.-N., Zhan, Z.-H., Zhang, J., Li, Y., Zhang, Q., Li, J.-J.: Distributed evolutionary algorithms and their models: a survey of the state-of-the-art. Appl. Soft Comput. **34**, 286–300 (2015)

24. Zhao, Y., Chen, L., Xie, G., Zhao, J., Ding, J.: Gpu implementation of a cellular genetic algorithm for scheduling dependent tasks of physical system simulation programs. J. Comb. Optim. 1–25 (2016)

25. Avansi, G.D., Schiozer, D.J.: Unisim-i: Synthetic model for reservoir development and management applications. Int. J. Model. Simul. Pet. Ind. **9**(1) (2015)

Transient Stability Enhancement Using Sliding Mode Based NeuroFuzzy Control for SSSC

Rabiah Badar$^{(\boxtimes)}$ and Jan Shair

Department of Electrical Engineering, COMSATS Institute of Information
Technology, Islamabad, Pakistan
rabiah.badar@comsats.edu.pk, janshair@outlook.com

Abstract. Voltage Source Converters (VSCs) based Flexible AC Transmission
Systems (FACTS) are popular for speedy regulation of different network
parameters, thus being a strong candidate for transient stability enhancement by
damping Low Frequency Oscillations (LFOs). Static Synchronous Series
Compensator (SSSC) is a series FACTS controller with built in capability to
absorb or deliver reactive power. SSSC may damp LFOs by installation of
efficient supplementary damping control (SDC). Due to recent advancements in
the field of Soft Computing (SC), there is a growing realization of their contribution to damping control design for FACTS. The direct focus of this chapter
is to exploit the potential of a hybrid control, obtained from assorted domains
such as NeuroFuzzy and Sliding Mode Control (SMC). SMC technique is the
most lucrative choice to design SDC due to its optimal performance, delivery in
critical applications with low complexity and high precision. The contributions
of this framework are the damping performance improvement for single and
multimachine power system with fast convergence speed.

Keywords: Sliding mode control · NeuroFuzzy · Adaptive control
Power system · SSSC

1 Introduction

Technological advancement and geographical expansion of human civilization has
resulted in increasing power demand which in turn stressed the installed transmission
capacity of the transmission network. However, the expansion of existing transmission
network is constrained by geographical and economical factors. In pursue of remedial
measures, designers found more practical methods rather exploitation of existing
network.

It has been found that traditionally phase shifting transformers and series or shunt
compensation may increase the power transfer capability and thus improve the power
system stability. On the other hand, modern methods include Flexible AC Transmission System (FACTS) controllers like Static Synchronous Series Compensator (SSSC),
Static Synchronous Compensator (STATCOM), Thyristor based FACTS and Series
Shunt FACTS etc. [1–3]. Among these FACTS, SSSC is a dynamic and fast switching
FACTS controller which may not only increase the power transfer capability but also
improve the stability of power system. However, its capability to improve the dynamic

© Springer International Publishing AG 2018
Y. Bi et al. (eds.), *Intelligent Systems and Applications*,
Studies in Computational Intelligence 751,
https://doi.org/10.1007/978-3-319-69266-1_3

performance of the system by damping Low Frequency Oscillations (LFOs) greatly depends upon a suitably designed Supplementary Damping Control (SDC).

Many conventional and advanced control techniques have been proposed as SDC for SSSC. One of the simplest and practical technique is Proportional Integral (PI) controller. However, due to its linear and local design orientation its performance is not robust for various operating conditions. Other control techniques include non-linear controller design which usually require mathematical description of a system which in case of power system is very difficult and sometimes impossible task, depending upon the network structure [4].

In last few decades, Sliding Mode Control (SMC) has emerged as powerful control technique due to its robustness for parametric uncertainties and disturbances occurring in the system [5, 6]. SMC has found applications in many areas of control including robotics, motor control, magnetic levitation, flight control etc. [7–11]. In power system, SMC and its variants with fuzzy and Neural Networks (NN) have been used for control of Static VAR Compensator (SVC), Thyristor Controlled Series Compensator (TCSC) and SSSC [12–14]. However, these applications of SMC suffered problems like chattering phenomena, getting stuck in local minima and global convergence.

In this work, the aforementioned drawbacks of SMC have been overcome by integration of NeuroFuzzy structure. The proposed control technique is basically a design paradigm shift from conventional Takagi Sugeno Kang (TSK) NeuroFuzzy structure to SMC based NeuroFuzzy.

The rest of the chapter has been organized as follows: Sect. 2 summarizes SSSC basic operation, modeling and control, Sect. 3 gives the details of the proposed control scheme, Sect. 4 presents the simulation results and their analysis. Finally, Sect. 5 concludes the findings of this research and defines some future dimensions of this work.

2 SSSC Modeling, Operation and Control

SSSC comprises a boosting transformer with a leakage reactance x_{ss}, a three phase IGBT based Voltage Source Converter (VSC) backed by a DC capacitor. SSSC can be modeled as synchronous AC source $V_{ss} \angle \theta_{ss}$ [15]. $\angle \theta_{ss}$ is the phase of the injected voltage and is kept in quadrature with the line current $I_l \angle \theta$ ignoring the inverter losses. Therefore, SSSC can be controlled dynamically to change the compensation level by changing the injected voltage and thus improve power system dynamic stability. Figure 1 shows SSSC installed between nodes m and n of an n-machine power system.

The phasor model of SSSC neglecting the switching dynamics, with modulation ratio m_{ss} and firing angle θ_{ss} is given as [16]:

$$\bar{V}_{ss} = m_{ss}kV_{DC}(\cos \theta_{ss} + j \sin \theta_{ss}) \tag{1}$$

$$\bar{I}_l = I_{ld} + jI_{lq} \tag{2}$$

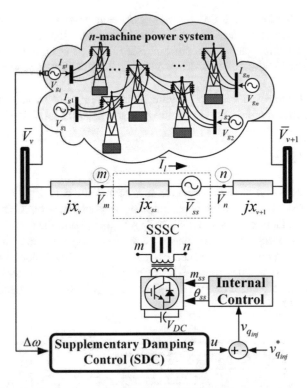

Fig. 1. Power system installed with SSSC and its internal control

$$\dot{V}_{DC} = \frac{m_{ss}k}{C_{DC}} \left(I_{ld} \cos\theta_{ss} + jI_{lq} \sin\theta_{ss}\right) \tag{3}$$

where, k is the fix ratio between AC and DC voltages. The energy exchange between DC link and AC system, during transient periods, is governed by (3).

Since, SSSC is installed between buses with voltages $\bar{V}_v$ and $\bar{V}_{v+1}$ the line current and both end voltages are linked as follows:

$$\bar{V}_v = \bar{V}_{ss} - j(x_{le} + x_{ss})\bar{I}_l + \bar{V}_{v+1} \tag{4}$$

where, x_{le} is the total series external reactance. Then the active power flow can be calculated as:

$$P = \frac{V_v V_{v+1}}{j(x_{le} + x_{ss})} \sin\delta \mp \frac{V_v V_{ss}}{j(x_{le} + x_{ss})} \sin(\delta - \theta_{ss}) \tag{5}$$

where, δ is the angle between sending and receiving end voltages and the second term ΔP represent the change in power transfer due to installation of SSSC.

Mainly SSSC has two controls the main/internal control and SDC. The internal control not only ensures the quadrature relationship between the injected voltage and line current but also keeps the DC voltage at constant level in steady-state, using the changes in reference variables. Internal control inputs are the three phase voltage of both ends and the three phase line current. SSSC can be operated in capacitive and inductive modes using the Pulse Width Modulation (PWM) parameters provided by the internal control.

The external control is used to provide reference injected voltage monitoring the changes in system variables. In what follows is the detail of SDC design.

3   Design of SDC

The overall closed-loop system structure has been shown in Fig. 2. Power system installed with SSSC has been considered as plant, whereas, control block represents the proposed online Adaptive NeuroFuzzy SMC (ANFSMC) and the conventional control schemes used for comparative evaluation. Switching logic system can be used for selection among different control schemes. The output of SDC block is the reference signal to modulate injected voltage. The SDC output $u \in \{u_S, u_T, u_L\}$ for proposed ANFSMC, Adaptive NeuroFuzzy TSK Control (ANFTSC) and Lead-Lag Control (LLC), respectively.

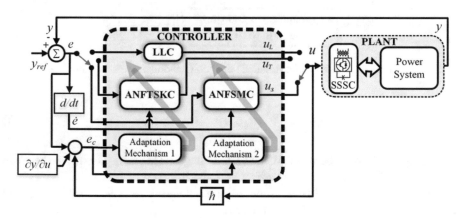

Fig. 2. Closed-loop control structure

3.1   Fundamentals of SMC

SMC is, a nonlinear control method, well known for its robustness against varying conditions of a highly nonlinear complex and dynamic systems. SMC enables to design a robust control scheme for higher order dynamical systems. It eliminates the need of exact modeling of dynamic system as it is insensitive to plant parameters variations. However, SMC designed for such a system might face frequency switching, called chattering, in the control loop. Chattering effect is unacceptable in any control strategy and leads to low accuracy of control or might even damage the controlled structure. The reason behind chattering phenomenon is that switching frequency of the different control structures is so fast that it cannot be realized in practical systems. Different methods have been developed to reduce chattering phenomenon. But these methods reduce chattering on the cost of increased complexity and decreased robustness. SMC is a variable structure control method in which many control structures are designed. The overall control trajectory is not in one control structure but it switches itself to a different adjacent control structures. In this way, SMC forces the states of systems to stay on the normal behavior of the system. The geometrical locus of the system states, where it exhibits normal behavior, is called sliding surface. SMC constrains the system states to reach the surface where $s(x_i) = 0$, irrespective of initial conditions. After reaching to the sliding surface, SMC ensures that control action is able to maintain the states of a system at $s(x_i) = 0$.

Generalized SMC law is given as:

$$u = k\text{sgn}(s) \tag{6}$$

where, $k > 0$ is gain and 'sgn' is the signum function given below:

$$\text{sgn}(s) = \begin{cases} -1 & if \quad s \leq 0 \\ 1 & if \quad s > 0 \end{cases} \tag{7}$$

Here, s is the sliding surface defined as, $s = \dot{e} + e\lambda_j + c_j$. The use of conventional *sign* function in SMC law often results in chattering phenomena. One remedy to avoid this phenomena is to introduce a boundary layer around the switching surface defined as following function:

$$sat\left(\frac{s}{\phi}\right) = \begin{cases} \frac{s}{\phi} & if \left|\frac{s}{\phi}\right| \leq 1 \\ sign\left(\frac{s}{\phi}\right) & if \left|\frac{s}{\phi}\right| > 1 \end{cases} \tag{8}$$

A variation to the above saturation function is to use a tangent hyperbolic function. Then SMC switching function becomes:

$$S = k\tanh\left(\frac{\dot{e} + e\lambda + c}{\phi}\right) \tag{9}$$

where, λ, c and ϕ are the slope, offset and the width of the boundary layer of the sliding surface, respectively. k is the constant gain factor and its value is normally kept high for asymptotic stability.

3.2 Proposed SMC Based NeuroFuzzy Control

The proposed Multiple Input Single Output (MISO) architecture of SMC based NeuroFuzzy control is governed by the following fuzzy rules;

IF x_1 is η_{1j} and x_2 is η_{2j} and $\ldots x_i$ is η_{ij} and $\ldots$ and x_n is η_{nm} THEN y_j is f_j

Here, $f_j \in \{S_j, T_j\}$ such that S_j and T_j denote the output of the consequent part for ANFSMC and ANFTSC, respectively. T_j is a linear polynomial function given as:

$$T_j = \sum_{\substack{i=0 \\ j=1}}^{n,m} b_{ij} x_i \tag{10}$$

where, b_{ij} denotes the adaptive coefficients of the polynomial. The consequent part S_j of SMC is given as:

$$S_j == w_j \sum_{i,j=1}^{n,m} k \tanh\left(\frac{\dot{x}_i + x_i \lambda_j + c_j}{\phi_j}\right) \tag{11}$$

where, λ_j, c_j and ϕ_j are the slope, offset and the width of the boundary layer of the sliding surface, respectively. w_j are the adaptive weight for jth rule. The generalized NeuroFuzzy layered architecture has been depicted in Fig. 3. The input layer 1 directs the inputs to next layer for fuzzification and to layer 4, simultaneously. Layer 2 fuzzifies the inputs using globally-tuned Gaussian membership functions, given as:

$$\eta_{ij} = e^{-\left(\frac{x_i - g_{ij}}{\sigma_{ij}}\right)^2} \tag{12}$$

Here, $i = 1, 2, \ldots, n$ and $j = 1, 2, \ldots, m$. g_{ij} and σ_{ij} are mean and variance of the Gaussian membership function, respectively.

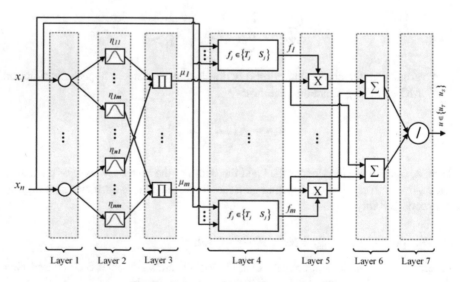

Fig. 3. SMC based NeuroFuzzy network

Layer 3 uses T-norm product operator to calculate Degree of Fulfillment (DOF) for each fuzzy rule.

$$\mu_j = \prod_{i=1}^{n} \eta_{ij} \tag{13}$$

Layer 4 of NeuroFuzzy network, being the first layer of the consequent part, contains SMC switching function. The multiplication of outputs of layers 3 and 4 is carried out in layer 5. Finally, Center of Gravity (COG) defuzzification method is used in layers 6 and 7 to compute the output of the proposed control scheme. The output of layer 7 is given as:

$$u = \frac{\sum_{j=1}^{m} \mu_j f_j}{\sum_{j=1}^{m} \mu_j} \tag{14}$$

3.3 Adaptation Mechanism

The controller parameters have been adapted using Gradient Decent (GD) optimization technique with back propagation algorithm. The parameters update involves the minimization of the following cost function:

$$J = \frac{1}{2}\left[\left(y_{ref} - y \right)^2 + \hbar u^2 \right] \tag{15}$$

where, y_{ref} and y are the reference and desired rotor speed deviations. $\hbar$ is a constant gain factor. The inclusion of second term involving the controller output has been made to ensure that the control effort should also be minimized and must go to zero as the error becomes negligible in steady-state.

The GD update law has the following form:

$$\Omega_{ij}(n+1) = \Omega_{ij}(n) - \zeta \frac{\partial J}{\partial \Omega_{ij}} \tag{16}$$

where, $\Omega_{ij} = \begin{bmatrix} g_{ij} \, \sigma_{ij} \, w_j \, \lambda_i \, c_j \, \phi_j \, b_{ij} \end{bmatrix}$ is the adaptation parameter vector and ζ is the learning rate.

The gradient of cost function, for both ANFTSC and ANFSMC, can be found using the following chain rules;

$$\frac{\partial J}{\partial g_{ij}^u} = \frac{\partial J}{\partial y} \frac{\partial y}{\partial u} \frac{\partial u}{\partial \mu_j} \frac{\partial \mu_j}{\partial \eta_{ij}} \frac{\partial \eta_{ij}}{\partial g_{ij}^u} \tag{17}$$

$$\frac{\partial J}{\partial \sigma_{ij}^u} = \frac{\partial J}{\partial y} \frac{\partial y}{\partial u} \frac{\partial u}{\partial \mu_j} \frac{\partial \mu_j}{\partial \eta_{ij}} \frac{\partial \eta_{ij}}{\partial \sigma_{ij}^u} \tag{18}$$

Since the antecedent part for both ANFSMC and ANFTSC is the same. Therefore, the same chain rules will be used for antecedent part parameters and the superscript u has been introduced to distinguish between the two.

$$\frac{\partial J}{\partial w_j} = \frac{\partial J}{\partial y} \frac{\partial y}{\partial u_S} \frac{\partial u_S}{\partial S_j} \frac{\partial S_j}{\partial w_j} \tag{19}$$

$$\frac{\partial J}{\partial \lambda_j} = \frac{\partial J}{\partial y} \frac{\partial y}{\partial u_s} \frac{\partial u_s}{\partial S_j} \frac{\partial S_j}{\partial \lambda_j} \tag{20}$$

$$\frac{\partial J}{\partial c_j} = \frac{\partial J}{\partial y} \frac{\partial y}{\partial u_s} \frac{\partial u_s}{\partial S_j} \frac{\partial S_j}{\partial c_j} \tag{21}$$

$$\frac{\partial J}{\partial \phi_j} = \frac{\partial J}{\partial y} \frac{\partial y}{\partial u_s} \frac{\partial u_s}{\partial S_j} \frac{\partial S_j}{\partial \phi_j} \tag{22}$$

$$\frac{\partial J}{\partial b_{ij}} = \frac{\partial J}{\partial y} \frac{\partial y}{\partial u_T} \frac{\partial u_T}{\partial T_j} \frac{\partial T_j}{\partial b_{ij}} \tag{23}$$

Due to direct adaptive and model free nature of closed-loop strategy, it can be assumed that, $\partial y/\partial u = 1$ [17]. After simplifying (17)–(23) and using the results in (16) the final update laws, for antecedent and consequent parameters of ANFTSC and ANFSMC, are given as:

$$g_{ij}^u(t+1) = g_{ij}^u(t) - \gamma \wp \frac{\mu_j}{\sum\limits_{j=1}^{m} \mu_j} \frac{2\left(x_i - g_{ij}^u\right)\left(f_j - u\right)}{\left(\sigma_{ij}^u\right)^2} \tag{24}$$

$$\sigma_{ij}^u(t+1) = \sigma_{ij}^u(t) - \gamma \wp \frac{\mu_j}{\sum\limits_{j=1}^{m} \mu_j} \frac{2\left(x_i - g_{ij}^u\right)^2\left(f_j - u\right)}{\left(\sigma_{ij}^u\right)^3} \tag{25}$$

$$w_j(t+1) = w_j(t) - \gamma \wp \frac{\mu_j}{\sum\limits_{j=1}^{m} \mu_j} \tanh\left(\frac{\dot{e} + \lambda_j e + c_j}{\phi_j}\right) \tag{26}$$

$$\lambda_j(t+1) = \lambda_j(t) - \gamma \wp \frac{\mu_j}{\sum\limits_{j=1}^{m} \mu_j} ew_j \text{sech}^2\left(\frac{\dot{e} + \lambda_j e + c_j}{\phi_j}\right) \tag{27}$$

$$c_j(t+1) = c_j(t) - \gamma\wp\,\frac{\mu_j}{\sum\limits_{j=1}^{m}\mu_j}\,w_j\mathrm{sech}^2\left(\frac{\dot{e}+\lambda_j e+c_j}{\phi_j}\right) \tag{28}$$

$$\phi_j(t+1) = \phi_j(t) + \gamma\wp\,\frac{\mu_j}{\sum\limits_{j=1}^{m}\mu_j}\,w_j\left(\frac{\dot{e}+\lambda_j e+c_j}{\phi_j^2}\right)\mathrm{sech}^2\left(\frac{\dot{e}+\lambda_j e+c_j}{\phi_j}\right) \tag{29}$$

where, $\wp = \frac{\partial J}{\partial y}\frac{\partial y}{\partial u} = (-e+\hbar u)$ and $e = y_{ref} - y$.

4 Simulation Results and Performance Evaluation

SSSC auxiliary controller design and the nonlinear time domain simulations are carried out in MATLAB/SIMULINK environment using SimPowerSystem toolbox. SimPowerSystem is a modern toolbox in MATLAB, used by the engineers and scientists to simulate test power systems and design models. All the simulations are carried out on an Intel® Core™ i3-4010U CPU @ 1.7 GHz having 4.0 GB RAM and a Microsoft Windows 10 operating system. The *powergui* block from the SIMULINK library is used for the machine initialization and load flow analysis of the test power systems. The machine initialization tool is used to modify the machine loading conditions and to select swing bus. The load flow analysis tool assures that the power system simulation start in steady-state.

Nonlinear time domain simulations are carried out with continuous variable solver (ode15 s) with a step size of 0.001 s. Phasor solution method is used for studies involving power system stability. The test system, under consideration, is a power system comprising of large synchronous generators with slow oscillation modes. The purpose of nonlinear time domain simulations is to study these slow modes of electromechanical oscillations. In this research, fast oscillation modes are of no interest that are produced due to resonance of resistance, capacitance and inductance of the transmission lines. The fast oscillation modes are generally above the fundamental frequency of the system and they have no interference with those of electromechanical oscillations of slow modes. The differential equations of the network are replaced by the algebraic equations to ignore fast modes. Phasor models of the components are sufficient to study LFOs.

SMIB and multi-machine test power systems are developed to investigate the performance of the proposed control strategy. The performance of different control schemes has been validated for each test system under different contingencies and operating conditions.

Although, nonlinear time domain simulations is an effective tool to evaluate the performance of a control scheme in transient and steady-state region. However, it becomes a tedious and error prone task to compare the performance for different control schemes when the improvement margin is small. Therefore, different performance indices (PIs) have been used to further quantify the comparative evaluation for the proposed control technique. PIs are defined as the integral of the product of time and error and are given as:

$$PI = \int_{t=0}^{t_s} t^p |e|^q (t) dt \tag{30}$$

where, t_s is the total simulation time, p and q are constants, such that $(p, q) \in \{(0, 1), (1, 1), (0, 2), (1, 2)\}$ which correspond to Integral Absolute Error (IAE), Integral Time Absolute Error (ITAE), Integral Square Error (ISE), and Integral Time Squared Error (ITSE), respectively. The generalized error term of PIs for SMIB and multi-machine test power systems is given as, $e = \sum_i (|\Delta\omega_{L_i}| + |\Delta\omega_{I_i}|)$ with L and I representing the local and interarea modes of oscillations, respectively, and i represents the mode number.

4.1 SMIB Test System

SMIB test system has been considered to analyze the performance of the proposed controller for local mode of oscillation. The single line diagram of SMIB test system is shown in Fig. 4. The system consists of a generating unit of 2100 MVA and a power transformer to step up the generated voltage of 13.8–500 kV. A 300 km double circuit transmission line connects the unit to an infinite bus. The generating unit has a Hydraulic Turbine and Governor (HTG) system and DC Type-1 excitation system installed with a generic PSS. A SSSC is connected in series between buses B1 and B2. A 250 MW load center has been modeled at bus B1. All the relevant parameters for generator, transformer, transmission line and SSSC can be found in [18].

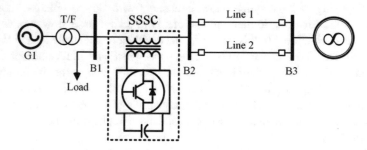

Fig. 4. SMIB test system

Different fault scenarios are contemplated to check the competence of the proposed SSSC damping controller. The performance results for the proposed control scheme has been compared with a conventional LLC and ANFTSC, as SSSC auxiliary damping controllers. The results are also compared when no damping control signal is provided to SSSC. LLC control is conventional two stage lead-lag control structure with gain and washout blocks. he inputs to the controller are the relative rotor speed deviation and its derivative.

4.1.1   Scenario-I: Nominal Loading

The efficacy of the proposed control scheme has been evaluated at nominal loading by applying a severe 3-phase, 5 cycles, self-cleared fault at the middle of transmission line 2, at $t = 1$ s. Figure 5, shows the nonlinear time domain simulation results for rotor

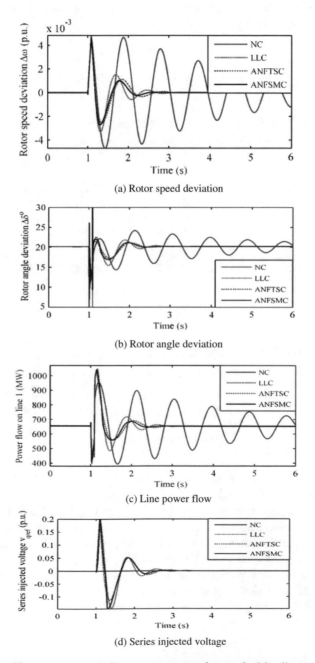

(a) Rotor speed deviation

(b) Rotor angle deviation

(c) Line power flow

(d) Series injected voltage

Fig. 5. Scenario-I. System parameters for nominal loading

speed and angle deviation, power flow on line 1 and control effort. It can be observed that ANFTSC gives better performance results in damping oscillations as compared to no control and LLC, however, ANFSMC further improves the performance results. The settling time performance improvement for ANFSMC on the basis of damping rotor speed deviation, is 24 and 22% as compare to LLC and ANFTSC, respectively. The update parameters variation plots for both antecedent and consequent parts of ANFSMC are shown in Fig. 6, respectively.

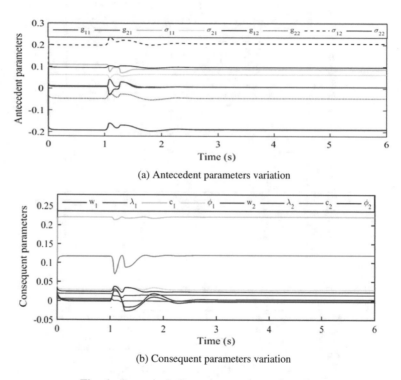

(a) Antecedent parameters variation

(b) Consequent parameters variation

Fig. 6. Scenario-I. Control parameters adaptation

Different PIs have been calculated using (30). Figure 7 shows PI plots for IAE, ISE, ITAE and ITSE. The ITAE and ITSE PIs involve multiplication with time, thus giving more weight to steady-state behavior. On the other hand, IAE and ISE focuses the performance in transient region. Figure 7, clearly shows the performance improvement for ANFSMC for both regions. It can be observed that slope of the curve for ANFSMC is almost zero as compared to ANFTSC and LLC in steady-state for IAE and ITAE which shows the superior performance of ANFSMC in steady-state. Furthermore, PIs curves show that ANFSMC has sharp transition from transient region to steady-state which confirms its faster performance as compared to other controls.

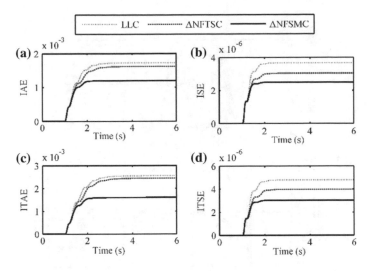

Fig. 7. PIs for scenario-I

4.1.2 Scenario-III: Heavy Loading

In this case, the system loading condition is varied from nominal to heavy loading by and the same fault has been applied at the middle of transmission line 2. The system was restored to its original operating condition when after fault removal. This more stressed scenario leads the system to instability if no supplementary damping control is applied to SSSC. If SSSC is provided with LLC based auxiliary control, it damps out the growing oscillations at almost 4.5 s. ANFSMC significantly performs better as compared to LLC and ANFTSC. Figure 8 shows the nonlinear time domain simulation results for rotor speed and angle deviation, power flow through line 1 and control effort, respectively. Settling time performance improvement for ANFSMC is 35 and 27% w.r.t. LLC and ANFTSC respectively, in terms of damping rotor speed deviation. The parameter variations of the proposed control for both antecedent and consequent parts are given in Fig. 9. Figure 10 shows the IAE, ITAE, ISE and ITSE curves for heavy loading. The results reveal that the transition from cliff region to steady-state region is prolonged for LLC and ANFTSC, showing their slow convergence as compared to ANFSMC.

4.2 Mutlimachine Test System

To further assess the performance of the proposed control structure, in a more practical and stressed scenario, the test system is extended to multimachine power system. The multimachine test power system is a three machines, six bus system comprising three generating units with SSSC installed in series with the transmission line connecting buses B5 and B6, as shown in Fig. 11. The generator G1 has 4200 MVA nominal power, while the other two generators, G2 and G3, have the same 2100 MVA nominal power. G2 and G3 are connected to each other through transformers TF2 and TF3, respectively. G1 is connected to bus B1 through a 4200 MVA step up transformer. This

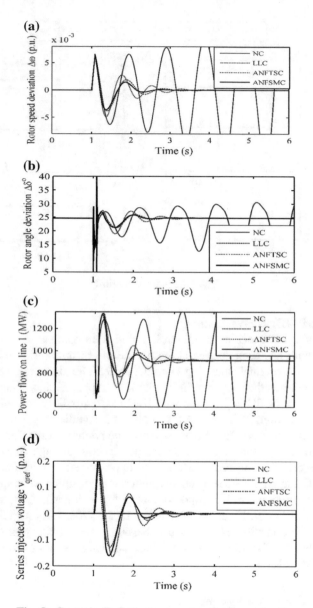

Fig. 8. Scenario-II. System parameters for heavy loading

arrangement creates a two area power system making it suitable for study of both local and interarea modes of oscillations. Area 1 contains G1 and area 2 consists of G2 and G3. Each generating unit is equipped with HTG, IEEE DC Type-I excitation system and a generic PSS. Load 1 is connected at bus B1 to model the load center for area 1, whereas, loads 2, 3 and 4 lie in area 2. The detail of system parameters can be found in [18].

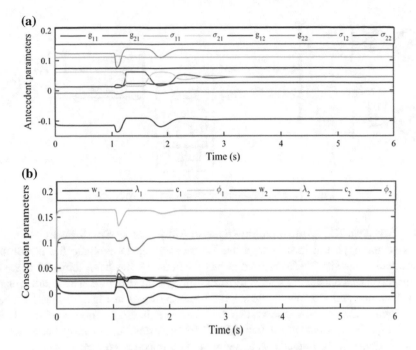

Fig. 9. Scenario-II. Control parameters adaptation

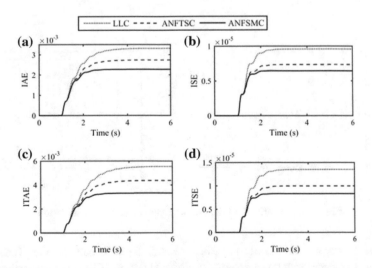

Fig. 10. PIs for scenario-II

Different fault scenarios are developed for the purpose of performance validation of the proposed SSSC damping controller for multimachine system.

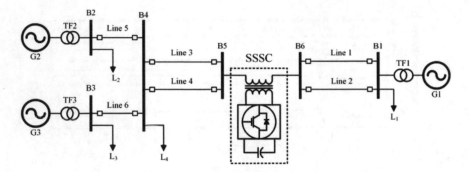

Fig. 11. Multimachine system installed with SSSC

4.2.1   Scenario-III: Multi-machine System Subjected to Small Fault

In this scenario, partial load outage has been examined to evaluate the performance under small fault. The 250 MW load at bus B4 in area 2 has been removed at $t = 1$ s. and the reconnected after six cycles restoring the system structure. The results for rotor speed deviation and line power flow have been presented in Figs. 12 and 13, respectively. Figure 12a, b show the results for local and interarea modes of oscillations, respectively. The results reveal that ANFSMC performs significantly better for both modes in transient and steady-state regions. The injected voltage results have been shown in Fig. 13b. It has been observed that parameters variation for ANFSMC is very small in this case.

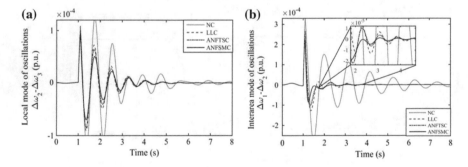

Fig. 12. Scenario-III. Local and interarea modes of oscillations

Figure 14 shows the results for PIs for small fault. It can be observed that the performance improvement gap between ANFSMC and ANFTSC is greater than that of ANFTSC and LLC showing the superior performance improvement in case of ANFSMC. The improvement margin for ITSE and ISE is larger as compared to that of IAE and ITAE showing more pronounced effect of ANFSMC in transient region.

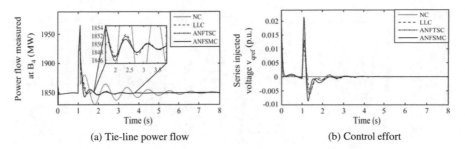

(a) Tie-line power flow   (b) Control effort

Fig. 13. Scenario-III. Line power flow and control effort

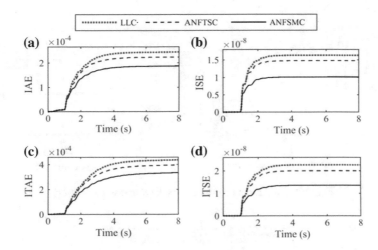

Fig. 14. PIs for Scenario-III

4.2.2 Scenario-IV: Multi-machine System Subjected to Large Fault

To simulate system under large disturbance, a 3-phase self clearing fault is applied between bus 1 and 6, on line 1, at $t = 1$ s. Figure 15a, b show the results for local and interarea modes of oscillations, respectively. The results for tie-line power flow and control effort are shown in Fig. 16. The results reveal that the system shows poorly damped oscillatory behaviour after the fault application. Figure 15 shows that the proposed ANFSM has significant performance difference in transient region and performance improvement is more pronounced for both transient and steady-state regions in interarea mode.

The comparative results for PIs are shown in Fig. 17. The large improvement difference between the proposed and conventional control schemes for ITSE and ISE curves show that the performance improvement of ANFSMC is large in transient region as compared to steady-state region. However, ANFTSKC being adaptive and intelligent shows competitive results for steady-state region as depicted in ITAE and IAE cureves.

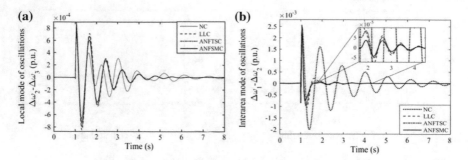

Fig. 15. Scenario-IV. Local and interarea modes of oscillations

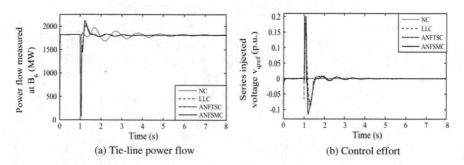

(a) Tie-line power flow   (b) Control effort

Fig. 16. Scenario-IV. Line power flow and control effort

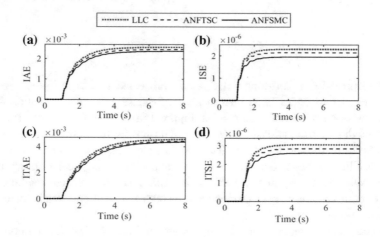

Fig. 17. PIs for Scenario-IV

5 Conclusion and Future Work

This work presents the application of synergistic NeuroFuzzy SMC paradigm to power system. The performance of the proposed controller has been evaluated for both SMIB and multimachine test cases. It can be concluded that the proposed ANFSMC performs better in transient and steady-state region not only for nominal operating condition but also maintains its superior performance for change in operating conditions and fault location. It can also be concluded that the proposed ANFSMC performs better in more stressed scenarios. Although, ANFTSC also shows competitive results for some cases because it is also optimized and adaptive not the conventional TSK with fixed parameters architecture. The work presented in this chapter can further be extended to large power systems installed with multi-type FACTS.

References

1. Hingorani, N.G., Gyugyi, L.: Understanding FACTS concepts and technology of flexible AC transmission systems. IEEE Press, New YorK (2000)
2. Bainsla, U., Mohini, R.K., Gupta, A.: Transient stability enhancement of multi machine system using static VAR compensator. Int. J. Eng. Res. Gen. Sci. (3), 1068–1076 (2015)
3. Yazdani, A., Iravani, R.: Voltage Sourced Converters in Powers Systems. IEEE Press, New Jersey (2010)
4. Nayeripour, M., Narimani, M.R., Niknam, T., Jam, S.: Design of sliding mode controller for UPFC to improve power oscillations damping. Appl. Soft Comput. 11, 4766–4772 (2011)
5. Slotine, J.J.E., Li, W.: Applied Nonlinear Control. Prentice-Hall, Englewood Cliffs (1991)
6. Ellis, D.T.: Sliding Mode Control (SMC): Theory, Perspectives and Industrial Applications. Nova Science Publishers (2015)
7. Utkin, V., Guldner, J., Shi, J.: Sliding mode Control in Electro-Mechanical Systems. CRC Press, (2009)
8. Chen, F., Dunnigan, M.W.: Sliding-mode torque and flux control of an induction machine. IEE Proc. Electric Power Appl. 150, 227–236 (2003)
9. Jafarov, E.M., Tasaltin, R.: Robust sliding mode flight control systems for the uncertain MIMO aircraft model F-18. IEEE Trans. Aerosp. Electron. Syst. 36, 1–15 (2000)
10. Al-Muthairi, N.F., Zribi, M.: Sliding mode control of a magnetic levitation system. Math. Probl. Eng. 2004, 93–107 (2004)
11. Chen, C.T., Peng, S.T.: Design of a sliding mode control system for chemical processes. J. Process Control 15, 515–530 (2005)
12. El-Sadek, M.Z., El-Saady, G., Abo-El-Saud, M.: A variable structure adaptive neural network static VAR controller. Electr. Power Syst. Res. 45, 109–117 (1998)
13. Ghazi, R., Azemi, A., Badakhshan, K.P.: Adaptive fuzzy sliding mode control of SVC and TCSC for improving the dynamics of power systems. In: IEE International Conference AC-DC Power Transmission, pp. 331–337 (2001)
14. Qi, W., Tao, W., Q, Chao, J.: Study on fuzzy sliding mode controller for PWM based static var compensators. J. Power Syst. Technol. (China) (28), 46–50 (2004)
15. Jamali, S., Shateri, H.: Locus of apparent impedance of distance protection in the presence of SSSC. Eur. Trans. Electr. Power 21, 398–412 (2010)

16. Ndongmo, J.D.N., et al.: A simplified nonlinear controller for transient stability enhancement of multimachine power systems using SSSC device. Electr. Power Energy Syst. **24**, 650–657 (2014)
17. Lui, Z.: Self tuning control of electrical machines using gradient descent optimization. Optim. Control **28**, 77–93 (2006)
18. Panda, S.: Robust coordinated design of multiple and multi-type damping controller using differential evolution algorithm. Electr. Power Energy Syst. (33), 1018–1030 (2011)

A Multi-objective Genetic Algorithm for Path Planning with Micro Aerial Vehicle Test Bed

H. David Mathias[✉] and Vincent R. Ragusa

Florida Southern College, Lakeland, FL, USA
hmathias@flsouthern.edu, vincent.ragusa.94@gmail.com

Abstract. The problem of robotic path planning is relevant to many applications that have led to extensive study. This has only increased as autonomous robotic vehicles have become more affordable and varied. Optimal solutions to this problem can be computationally expensive, leading to the need for efficiently achievable approximate solutions. In this work, we present a genetic algorithm to solve the path planning problem. The algorithm operates offline but runs onboard the micro aerial vehicle (MAV). This is accomplished by mounting a single-board computer on the vehicle and integrating it with the flight control board. In addition, we evaluate the effectiveness of two genetic operators: crossover and mass extinction. Results demonstrate that a standard, single-point crossover operator is largely ineffective. Mass extinction, an operator that has been used rarely in previous work, is explored within the framework of a genetic algorithm utilizing only mutation and selection. Based on initial results, mass extinction may have some utility for path planning, however, due to large number of parameters and potential implementations, additional experimentation is needed.

1 Introduction

Robotic vehicles have been the subject of, or a platform for, a great deal of research within computer science. Though work with these vehicles spans decades, interest has intensified in recent years due to the increased availability and decreased cost of micro aerial vehicles (MAVs) and ground-based rovers. Problems of interest cross areas within computing. Problems and applications of interest include cooperative behaviors and swarming, localization and mapping, learning, autonomous operation, path planning, and various problems that involve computer vision, such as tracking, inspection in inhospitable or remote environments, and search and rescue.

The study of path planning predates by centuries the effort to develop autonomous vehicles. Perhaps the best known early example is the Bridges of Königsberg problem, which was solved by Euler in the first half of the 18th century. First mention of the Traveling Salesman problem appear around the same

© Springer International Publishing AG 2018
Y. Bi et al. (eds.), *Intelligent Systems and Applications*,
Studies in Computational Intelligence 751,
https://doi.org/10.1007/978-3-319-69266-1_4

time and it has occupied mathematicians, and later, computer scientists, since. Over the centuries, many models and techniques have been brought to bear in an effort to find efficient solutions.

Robotic path planning, a modern variant of earlier problems, is similar in many ways. An agent, which we will call a robot, must devise an efficient path from a known source to a known destination while avoiding obstacles in the environment. *Offline* path planning takes place when the agent has full knowledge of the environment in advance. In some instances, the environment is bounded. This is often a concession to reality since physical autonomous agents have limited range. While reaching the destination is a necessary condition for success, solutions to path planning problems must also optimize some measure known as an *objective*. Historically, path length is the most common of these. As is the case in this work, it may be necessary to try to optimize multiple objectives simultaneously. These may include the number of waypoints along the path from the source to the destination, the magnitude of the changes in direction necessary to navigate the path, and the number of obstacles with which the path collides. For each of these objectives, an algorithm would attempt minimization.

Multi-objective optimization, attempting to optimize multiple objectives, is a substantially more difficult problem. This arises in part from the fact that objectives often conflict with each other. Path planning provides an obvious example. The shortest distance between two points is a straight line. Thus, path length is optimized by a straight line path from the source to the destination. Of course, such a path would certainly collide with obstacles in any interesting instance of a path planning problem. Thus, the straight line path is not optimal with respect to obstacle collisions. Any attempt to improves the value of the obstacle collision objective will, naturally, result in a increased value for path length.

Researchers have applied numerous techniques to path planning problems, an indication of their importance. Successful approaches to the problem include heuristic search (Ferguson et al. 2005), neural networks (Glasius et al. 1995), bug algorithms (Buniyamin et al. 2011), cell decomposition (Choset and Pignon 1997), potential fields (Al-Sultan and Aliyu 2010), and, most relevant here, evolutionary algorithms (Goldberg 1989). The work solution presented here employs evolutionary computation, specifically, a genetic algorithm. Genetic algorithms provide a great deal of flexibility through a variety of operators and numerous parameters.

The path planning variant considered here is more general than most found in the genetic algorithms literature. Whereas most results consider discrete environments, the domain for this work is a very general, continuous environment containing circular and polygonal obstacles. The polygons may be non-axis-aligned and concave, capable of modeling real-world scenarios.

Robotic path planning algorithms that are deployed on physical vehicles are rare in the genetic algorithms literature. Some are run within a simulator, though many are exclusively theoretical. The algorithm presented in this work is an exception. It runs immediately prior to flight on the micro aerial vehicle that will fly the evolved path. Once a path has been successfully calculated, the algorithm communicates with the flight control board to facilitate completion of a mission. Because the capabilities of the computer incorporated into the vehicle are modest, computational efficiency is of more than theoretical interest.

Genetic algorithms are, in effect, encodings of Darwinian evolution. Over a number of generations, a population evolves through a process of "natural selection." In this case, the population is comprised of potential solutions to some problem. Members of the population are represented by chromosomes, typically a binary encoding or sequence of real numbers. Biological evolution is driven by three processes: mating, random mutation, and survival of the fittest. The algorithmic analogs are *crossover, mutation*, and *selection*. Given generation i of the population, generation $i + 1$ is created through some sequence of these three operations. In most cases, a fixed population size is maintained.

As a fundamental operator in evolutionary computation, crossover is typically employed in genetic algorithms for path planning. There has not, however, been significant research into the efficacy of that operator for this problem. In one study, Tonupunuri (2008) examined the effect of crossover for pathfinding in a mobile sensor network. In that work, however, an algorithm using crossover was compared to an algorithm using only selection. Mutation was then added to the algorithm with crossover. There was no comparison of an algorithm using crossover against an algorithm using mutation.

In biological evolution, environmental factors may significantly affect a species' evolutionary thread. Examples include pandemic, severe climate change and cataclysmic events such as meteor impacts and volcanic eruptions. Each of these events can result in the extinction, or near extinction, of a species. Biological research suggests that when a species survives such an event, albeit in greatly reduced numbers, there may arise a period of accelerated evolution (Lehman and Miikkulainen 2015). If we seek to model evolution in our algorithms, we must then consider incorporating such events (Jaworski et al. 2012). Mass extinction can be modeled as randomly occurring events that may be triggered by some measure (such as a prolonged period with little evolutionary progress) or, instead, with some probability at a fixed interval. Each extinction event kills a large fraction of the population. Repopulation can take place through a variety of mechanisms including crossover, mutation and creation of random members and may occur quickly or over some prolonged period. Through the introduction of a significant amount of new genetic material, it may be possible to invigorate the evolutionary process. Of course, some strong members of the population will certainly be lost.

In this work, we describe a multi-objective genetic algorithm for offline path planning in an autonomous vehicle, in particular a micro aerial vehicle. Previous work in this area is largely limited to discrete environments and implementations that are tested in simulators rather than the physical world. We deploy our system on an actual vehicle that operates in a highly unconstrained physical and, therefore, continuous, environment that may also include intermediate destinations. The algorithm runs on a single-board computer onboard the vehicle allowing completely autonomous flight. We undertake a preliminary study of the impact on evolution offered by crossover and mass extinction. Neither operator is well understood with respect to path planning. The remainder of the paper outlines the problem addressed and describes the algorithm and the vehicle. Subsequent sections analyze results, describe conclusions, and present opportunities for further investigation.

2 Related Work

The genesis of the application of evolutionary methods to robotic path planning dates to at least 1992 (Page et al. 1992; Lin et al. 1994). Work since has addressed many aspects of the problem (Hermanu et al. 2004; Jun and Qingbao 2010; Siddiqi et al. 2013; Zheng et al. 2004; Sedighi et al. 2004; Ahmed and Deb 2011).

Burchardt and Salomon (2006) implement their algorithm on physical, ground-based robots. The purpose of their system is to enable robots to play soccer; their ultimate goal is to enable a robotic team to compete against human teams by 2050. Zheng et al. (2004) created a system for planning paths for multiple, cooperating aerial vehicles. However, their impressive algorithm runs only in a simulator. Siddiqi et al. (2013) have addressed the need for efficiency in algorithms, genetic or otherwise, for path planning in embedded systems, an important consideration given that embeddable computers cannot compete with larger computers with respect to memory and computing power.

Hasircioglu et al. (2008) consider a problem similar to that examined in this work: offline path planning in a continuous environment. While there are similarities in the genetic encoding used, there are also important differences between this work and theirs. Their system operates in a largely natural (as opposed to man-made) 3D environment for which topographical data may provide accurate 3D descriptions of obstacles. For member evaluation, they use a scalar objective function. Their algorithm is not deployed to a physical system.

In 1997, Xiao et al. (1997) outlined an adaptive approach to path finding using a genetic algorithm. Their two-stage approach, consisting of finding an initial, possibly imperfect, path which is refined during the mission, is likely more relevant today than when originally published due to the small size and light weight of hardware now available.

Konak et al. (2006) catalog a number of genetic strategies for multiobjective optimization (Fonseca and Fleming 1993; Knowles and Corne 1999; Srinivas and Deb 1994; Zitzler and Thiele 1999; Zitzler et al. 2001; Yun and Lu 2003). The work in this area by Deb et al. (2002) is among the most seminal in the field. Their algorithm, known as NSGA-II, and in particular, their method of non-dominated sorting to create a partial order on the population and help ensure genetic diversity, informed design decisions made in the work presented here. A later result, NSGA-III, more effectively handles larger numbers of objectives (Deb and Jain 2014; Jain and Deb 2014).

Applying the NSGA-II framework to the problem of path planning, Ahmed and Deb (Ahmed and Deb 2011) examine the impact of chromosome representation. While they experiment with a binary encoded chromosome, their best results are achieved using integer valued chromosomes. They also explore the effects of changing the probabilities at which genetic operators are applied.

3   The Problem

The path planning variant we consider is distinct from those in most previous work in several significant ways. Most are required by deployment of our algorithm on a physical vehicle that operates in a real environment rather than a highly constrained, theoretical domain. The environment in which our algorithm, and subsequently the vehicle, perform is continuous, locations for starting and ending points are unconstrained, and obstacles can be arbitrary polygons that need not be axis aligned. Further, the problem may include intermediate destinations in addition to the final destination. The introduction of these intermediate destinations also allows the starting and ending points to be colocated. To this point, the algorithm is limited to planning in two dimensions. This is borne of necessity as the vehicle is intended to operate in spaces containing man-made structures. While building locations are easily found, reliable height information is much less commonly available. Thus, it is necessary to fly between obstacles at a fixed altitude rather than assume that flight over them is safely possible.

Here we formally define an instance of the path planning problem. The elements of a problem instance are: origin location s; destination location d; list of obstacles obs; and list of intermediate destinations $goals$. Both obs and $goals$ may be empty. All elements in the environment are represented by GPS coordinates which is a pair consisting of latitude and longitude. Obstacles consist of a list of GPS coordinates representing the vertices. Intermediate destinations, which do not appear in the genetic algorithms literature, represent mission-critical locations the vehicle must visit, in any order, for successful completion of the mission. These may be places from which sensor readings or photographs must be taken for security purposes. Each element of $goals$ is a circle represented by a center point and a definable radius. Some input maps on which the algorithm was tested appear in Fig. 2.

A solution to this variant of the path planning problem consists of a sequence, $S = \langle s, w_0, \dots, w_r, d \rangle$ of waypoints. Note that $r > 0$ is not a fixed value. In other words, we do not predetermine the number of waypoints a solution must contain. Waypoint w_i is represented as an ordered pair, (lat, lon), consisting of the latitude and longitude of the location. S is considered a *valid* path if it starts at s, ends at d, avoids all obstacles in obs and reaches all intermediate destinations in $goals$.

4   The Genetic Algorithm

One of the hallmarks of genetic algorithms is the breadth of different techniques, operators, and parameters that can be brought to bear on a problem. In this section, we introduce our algorithm and provide details of and design decisions for relevant features.

4.1   Chromosome Representation

The unique chromosome data structure plays a central role in genetic algorithms by both storing the information relevant to solving the problem and representing that data in such a way that the genetic operators can make randomized, yet meaningful, changes to the data. Typically, the chromosome encodes a candidate solution to the problem and therefore is highly dependent on the specifics of the problem and the requirements of a solution. For example, in discretized path planning environments like grids, which dominate the literature, the genes of a chromosome often represent row or column values (with the other value, column or row, implied by the ordering of the genes) (Hermanu et al. 2004; Sedighi et al. 2004). In some cases, position is represented as offsets from the previous gene rather than as absolute coordinates (Ahmed and Deb 2011). Although continuous environments with real valued coordinates are rare in the literature some exceptions do exist (Zheng et al. 2004; Hasircioglu et al. 2008). Zheng et al. (2004) uses a continuous environment with the addition of validity indicators for each coordinate.

The most common representation of information in genetic algorithm chromosomes is a binary string. This provides genetic operators with bit-level access to the information encoded. However, this representation may not be appropriate for some variants of path planning. Consider a GPS-specifies environment with a geo-fence, a limit on the distance traveled from the relevant area of focus, such as the one in this work. If encoded as binary, the bits in the chromosome would, represent real-valued latitudes and longitudes. Changing a low order bit in one of these values could be a very reasonable mutation. However, changing a high order bit could result in a mutation of enormous magnitude, perhaps moving a waypoint across a continent or around the globe.

Instead, we implement a chromosome that accommodates the real-world constraints that result from the environment in which our system operates. The encoding is a real-valued vector comprised of waypoint sequence $S = \langle s, w_0, \ldots, w_r, d \rangle$, where s and d are the origin and destination as described above. Genetic operators are applied to the real-valued latitudes and longitudes, providing constraints on the magnitude of waypoint mutations. We note that this is done as a concession to the limitations of the physical system within which our algorithm operates.

4.2   Multi-objective Optimization

The objectives chosen to be optimized in this work are: path length, number of obstacles hit, number of intermediate destinations reached and smoothness. Smoothness, defined as $\sum_{i=0}^{n-1} \angle w_i w_{i+1}$, is the sum of the angles between adjacent path segments. Minimizing the magnitude of changes in direction is beneficial as saves time and, as a result, battery life. Intermediate destinations reached is not an integer count of the *goal* locations visited by a path. Rather, for each goal, the algorithm calculates a value that is inversely proportional to the square of the distance of the closest approach made by the path. The objective value is the sum of these values for all intermediate destinations.

For simplicity, the algorithm works to minimize all objective values. Intuitively, larger values of the intermediate destinations objective is meant to reward the algorithm, and as such its value is negated before evaluation by the fitness function so that it too can be minimized without driving the objective values in the opposite direction. Initially, the total number of waypoints was considered as a objective to be minimized, however tests showed it provided no benefit to the quality of the solutions or the speed of the algorithm. Reducing the number of waypoints to a practical level is important however, because ultimately the path must be flown. As such, the number of waypoints in a path is used as a means of sorting paths to break ties among equally good solutions.

As stated previously, in Sect. 1, the use of many objectives provides a means to better solutions, but at the cost of complicating the search process. One method of implementing multiple objectives is to define a function f : $[o_1, \ldots, o_k] \to \mathbb{R}$. This method, known as scalarization, uses an algebraic function f over the objective variables to map them onto the real numbers. This method has the advantage of assigning each solution a single, easily comparable, fitness value. However, constructing f in a way that provides quality solutions is extremely challenging.

Many multi-objective optimization techniques make use of *Pareto optimization*, in which the result of a search is an ordered set of possibly unordered subpopulations called fronts. Assignment of population members to fronts is based on the concept of domination. Let O be the set of optimization objectives. One candidate solution X *dominates* another candidate solution Y if $\forall\, o \in O,\ X[o] \leq Y[o] \land \exists\, o\ s.t.\ X[o] < Y[o]$ (Deb et al. 2002). A solution that cannot be dominated is known as *Pareto optimal*. The set of all Pareto optimal solutions defines the *Pareto front*.

To use Pareto optimization, first assign each candidate solution to one of some number of *domination fronts* in which each candidate in front $i + 1$ is dominated by at least one candidate in front i. Thus, the fronts impose a partial order on the candidates. Front 0 contains all non-dominated candidates. Front 0 is an approximation of the Pareto front; the members of front 0 are the best found approximations of Pareto optimal solutions. Due to the random nature of genetic algorithms there is no guarantee, or assumption, that a Pareto optimal solution will be found. However, the algorithm will converge to an extremely good approximation by focusing the search around front 0 at every generation.

Pareto optimization is, in one sense, incompatible with the needs of our system. Our algorithm must identify one candidate solution, which is to say one path, preferable the "best" path, to fly. However, Pareto optimization does not identify a single best member. To identify the solution that will form the basis of our mission, a series of stable sorts is performed on the final front 0. The order of sort keys places highest priority on obstacle collisions and intermediate destinations reached to ensure that only a valid path is chosen. Among valid solutions, those with the shortest path lengths are chosen.

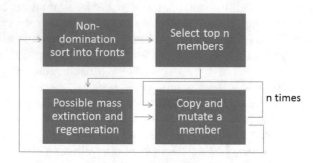

Fig. 1. Depiction of the main loop of our genetic algorithm

4.3   Population

The initial population created by the algorithm, denoted g_0, is comprised of n randomly generated members. Recall that a member of the population consists of a sequence of waypoints from s to d. The number of waypoints in each member of g_0 is in the range 1..5 and each is within a specified geo-fence distance from origin s. After creation, g_0 is sorted into domination fronts.

Illustrated in Fig. 1, the main loop of our genetic algorithm is as follows. We consider a run of the algorithm at the beginning of generation g_j. We denote the population at this point, which will serve as the parent population, p_p. Child population p_c is created by copying and mutating members of p_p. Members to mutate are chosen from p_p via a binary tournament. $|p_c| = |p_p| = n$. p_p and p_c are then merged into a combined population p with size $2n$. p is sorted into domination fronts.

As there is no domination relationship among members of a Pareto front, imposing an order, or even a partial order, on the members requires a niching measure (Mengshoel and Goldberg 2008). NSGA II (Deb et al. 2002) employs a niching measure called *crowding distance*. Crowding distance is, in effect, a uniqueness measure. More accurately, the crowding distance for a member is a measure of the dissimilarity of that member from other members of the population. To ensure diversity in the gene pool, greater crowding distance values are preferred.

From combined population p of size $2n$, the algorithm selects n members to form the next generation's parent population. Recall that the members of p have been sorted into a sequence of domination fronts (see Sect. 4.2). The *selection* operator works as follows. Let k be the largest index such that $\sum_{i=0}^{k} |f_i| = y \leq n$

where each f_i is a domination front. Then g_{j+1} consists of all members from f_0 through f_k and $n - y$ members from f_{k+1}. Crowding distances are used to select the members of f_{k+1} to include in g_{j+1}.

4.4 Mutation

Mutation consists of 4 independent operators: *add, delete, swap,* and *move.* Exactly 1 of the 4 operators, chosen at random, is applied to a member to be mutated. The probabilities with which the respective operators are applied are 0.1, 0.05, 0.1 and 0.75, though these are parameterized and easily changed. Delete, swap, and move each modify a waypoint in a genome. As the name implies, the add operator places a new waypoint in a genome. The origin and destination cannot be mutated.

Consider waypoint sequence $S = \langle s, w_0, \ldots, w_r, d \rangle$. The *delete* operator removes randomly selected waypoint w_i from S for some $0 \leq i \leq r$. The *swap* operator randomly selects a pair of adjacent waypoints w_i and w_{i+1}, $0 \leq i \leq r$, and exchanges their positions in S. The *add* operator selects a pair of adjacent waypoints w_i and w_{i+1}, $0 \leq i \leq r$ at random. The midpoint between w_i and w_{i+1} defines a new waypoint. This is inserted into S and then moved using the move operator .

The *move* operator is applied to randomly selected waypoint, w_i, $0 \leq i \leq r$. Let x represent a longitude value and y a latitude value. Then before move is invoked, $w_i = [y_i, x_i]$. After the move operator, $w_i = [y_i + c_0, x_i + c_1]$. c_0, c_1 are chosen independently at random according to a Gaussian distribution with mean 0 and standard deviation σ selected by fair coin flip from among two candidates, *small_move* and *big_move*. *small_move* is fixed at $0.00002°$, approximately 2 m. The value of *big_move* depends on the problem instance. For instances with small obstacles, it is $0.0002°$ (~ 20 m) and for instances with large obstacles, it is $0.00067°$ (~ 69 m).

4.5 Crossover

Reproduction, in which genes from two individuals are combines to create offspring, helps ensure diversity within a population. In genetic algorithms, this activity is modeled by crossover. The general template for crossover is that each of two parents is divided into some number of segments and the segments alternated to create new members. The variants of crossover differ primarily in the number of segments in to which the parents are divided, ranging from two (single-point crossover) to one per bit in a binary representation (uniform crossover). In general, a crossover event results in the creation of two children.

In this work, we employ single-point crossover. This decision was primarily a concession to the structure of the genome. In practice, the genomes are too short to reasonably support two-point or multi-point crossover. Similarly, uniform crossover is not well-suited to a real-valued, variable length genome.

Recall that member p_k is represented by its chromosome, waypoint sequence $\langle s, w_0, \ldots, w_{r_k}, d \rangle$. We define cutpoint t_i as a partition of p_k into $p_k^1 = \langle s, w_0, \ldots, w_{i-1} \rangle$ and $p_k^2 = \langle w_i, \ldots, w_{r_k}, d \rangle$. Neither subsequence can be empty. With this, given members p_1 and p_2, our algorithm chooses random cutpoint t_i in p_1 and t_j in p_2 and creates two children c_1 and c_2 such that $c_1 = p_1^1 + p_2^2$ and $c_2 = p_2^1 + p_1^2$. It is probable that $|c_1| \neq |c_2|$. After creation, the children may, with some probability, undergo mutation.

The use of crossover does not change the overall structure of the algorithm. As before, the algorithm creates a child population equal in size to the parent population. The parents and children are combined into a single population. After sorting into domination fronts, the combined population is winnowed by selection to form the parent population for the next generation.

Experiments conducted with crossover, described in Sect. 6, show that, in this form, it is at best ineffective and at worst highly negative. Thus, ultimately it is excluded from the implementation.

4.6   Mass Extinction

While evolutionary computation models many aspects of biological evolution, there is little discussion, and even less implementation, of mass extinction in the genetic algorithms literature (Jaworski et al. 2012; Lehman and Miikkulainen 2015). With respect to path planning, no previous work was found. However, there is some intuition for a positive impact due to mass extinction events in genetic algorithms. There may be support for this is in the biological record, as Lehman & Miikkulainen note that extinction events can result in a period of accelerated evolution.

In artificial evolution, a mass extinction event eliminates some large fraction of the population, significantly reducing the population size, temporarily. The motivation for doing this is to invigorate the evolutionary process which may benefit from refocusing on a smaller population and from an infusion of new genetic information. Though its biological analog in this context is questionable, *elitism* can exempt some of the best members of the population from extinction. In earlier work, (Mathias and Ragusa 2016) demonstrated that mass extinction may be beneficial for path planning.

We test three implementations of mass extinction. They differ in the following details: what fraction of the population is killed off, how is the population regenerated, and how much elitism is employed.

- Implementation 1: Random Regeneration
    - Extinction: 80% of the population chosen at random
    - Elitism: None
    - Regeneration: 100% andomly generated members
- Implementation 2: Regeneration via Crossover
    - Extinction: 80% of the population chosen at random
    - Elitism: Top 3 members

      - Regeneration: 60% of deleted members are replaced with randomly generated members; remaining 40% are created via crossover
- Implementation 3: Regeneration via Mutation
  - Extinction: All but the best 10 members of the population
  - Elitism: Top 10 members
  - Regeneration: 10 new members are randomly generated; remaining new members are mutations of the members that survived extinction

4.7 Computational Complexity

The run-time of NSGA II is dominated by the non-domination sort that allocates members of the population to domination fronts. As our algorithm adheres to the NSGA II framework, it's runtime is likewise dominated by this step in the process. For number of objectives m and population size n, the complexity is in $O(mn^2)$ for each generation. Thus, for number of generations g, the total run-time is in $O(gmn^2)$. The linear dependence on number of generations and quadratic dependence on population size suggests it is better to use longer runs with smaller populations, provided, of course, that such runs produce solutions of similar or better utility.

5 Experimental Design

The experiments described in this section, with results presented in Sect. 6, were performed only to assess the efficacy of crossover and mass extinction operators for evolving solutions to path planning instances. To achieve this, we tested the algorithm on a set of five maps (see maps 1–5 in Fig. 2). Trials were conducted for each map. A trial consists of ten runs of the algorithm with different values for extinction probability and a flag indicating whether or not to use crossover. Specifically, extinction probability $p \in \{0.0, 0.05, 0.10, 0.15, 0.20\}$. Each was tested with and without crossover. Every 50 generations, an extinction event occurs with probability p. The population contained 100 members.

    Recall that there are multiple versions of the mass extinction operator. For each, we performed 40 trials for each input. Thus, each input was run 400 times. To ensure, to the extent possible, similarity among the runs, for each input, all runs in a trial were performed with the same random seed.

    A near optimal path length, denoted len_{opt}, was determined a priori for each map. These values provide path length targets. Each trial produces three metrics measured in the mean number of generations to achieve: a valid solution, denoted mng_valid; a solution within 10% of len_{opt}, mng_10; and a solution within 5% of len_{opt}, mng_5. Runs terminate when a 5% solution is achieved but are capped at 3000 generations if no such solution is found.

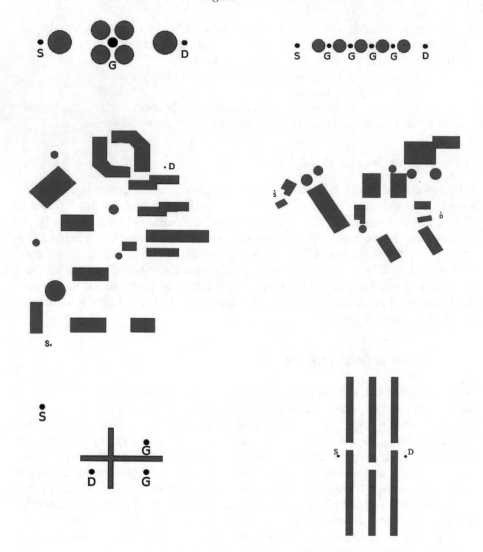

Fig. 2. The maps for which results are reported. Origin and destination points are labeled S and D, respectively. Obstacles are shown in gray. If a map includes intermediate destinations, they are shown in black and labeled G. Inputs 3 and 4 are drawn from the real world. Each represents a portion of a college campus. The remaining maps are artificial. Euclidean distances between S and D range from 62 meters for maps 1, 2, and 6 to 360 meters for map 4

6   Experimental Results

The results show that both extinction and crossover offer some benefit for pathfinding in some cases. Unfortunately, there are also instances in which these operators have a negative effect on the number of generations required,

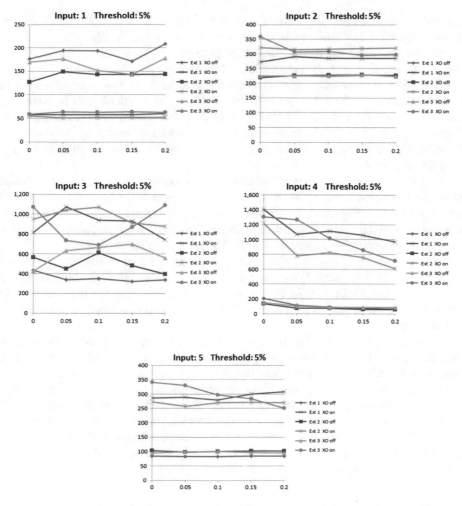

Fig. 3. *mng_5* graphs for each of the five inputs tested in this work. Each graph illustrates performance of six cases: crossover on and off for each of the three extinction implementations. Note that the scales differ

sometimes dramatically so. Crossover is predominately negative, though there are notable exceptions. For extinction, in most cases, the effect ranges from slightly negative to very positive.

The data presented are trimmed means. The initial samples of 40 runs for each experiment were trimmed by removing the top 5% and bottom 5% of values. Graphs showing *mng_5* data for each input appear in Fig. 3. Due to space considerations, we cannot present graphs for all experiments. Therefore, graphs are all for *mng_5* to allow comparison across inputs. For each mean, a confidence interval was calculated. To isolate the effect of extinction, comparisons were performed between the mean values with extinction probability $p = 0$ to a mean calculated for all probabilities $p \in \{0.05, 0.10, 0.15, 0.20\}$ (see Fig. 4). Two-sample T-tests explored the impact of crossover (see Table 1).

Crossover improved mean number of generations for measures mng_10 and mng_5 for input 1 (Fig. 3) independent of extinction implementation. For input 2, crossover was beneficial for mng_valid, with extinction implementations 1 and 2, but detrimental for both mng_10 and mng_5. For inputs 3 and 5, crossover shows a benefit for mng_valid for all extinction probabilities and all implementations. However, it is detrimental with respect to mng_10 and mng_5 thresholds; dramtically so for input 5. For input 4, crossover is extremely negative with respect to mng_5.

In the presence of crossover, extinction implementation 1 is, in general, the worst of the three versions, though in several cases it is on par with the others and for input 3 (Fig. 3) it is arguably the best implementation. On the other hand, it is the only implementation that exhibits very large negative effects, for example mng_valid for inputs 2 and 4. The overall inferiority is not surprising given the absence of elitism. Comparing extinction 2 and extinction 3, we see that extinction 2 is, with few exceptions, more effective, though the differences are small, often negligible. In the absence of crossover, there is little difference between extinction versions for inputs 2, 4, 5. For input 4, all extinction versions are beneficial.

In some cases, the impact due to extinction appears to improve with number of generations. For this reason, mng_5 is the measure most positively impacted and mng_valid least. Input 4 offers the best illustration. All extinction versions have a positive impact but the effect is strongest in the presence of crossover. The runs with crossover for that input take significantly more generations. For input 3, all extinction versions show some promise, though at varying extinction probabilities, with extinction 3 showing the greatest improvement at probability 0.1. Extinction 3 also shows the best improvement for input 5, though in this regard it benefits from an inferior baseline.

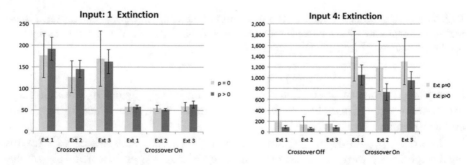

Fig. 4. The effect of extinction for inputs 1 and 4. The algorithm's performance for all non-zero extinction probabilities are aggregated into a single value. The y-axis shows number of generations

Figure 4 shows the effect of extinction for inputs 1 and 4. In each pair, the left bar is the trimmed mean for extinction probability 0 and the right bar is the trimmed mean for all other probabilities studied. 95% confidence intervals are also displayed. Again, we see that for input 4, extinction is positive in all cases, while for input 1 there is no trend.

Table 1 contains data for two sample T-tests for inputs 1 and 4. In each case, the two samples are for the same parameters except for crossover. In one sample crossover is used and in the other it is not. Instances for which the difference is statistically significant for a 95% confidence interval are in bold. For mng_10 and mng_5, crossover makes a statistically significant difference in almost all cases, beneficial for input 1 and detrimental for input 4. This trend holds for all other inputs.

For all but one of the simplest inputs tested, the effect of crossover was negative for the most important metric, mng_5. For the inputs with the highest degree of realism, crossover was negative to extremely negative for mng_5. For this reason, we have removed crossover from the algorithm. Mass extinction remains as an operator due to its promise with respect to this metric.

7   The Micro Aerial Vehicle

While path planning is of theoretical interest, it is a problem based in the physical realm. It is, therefore, important to ensure that solutions to the problem are suitable for the real world. As detailed previously, evolutionary computation has formed the basis for a number of path planning algorithms. However, few, if any, of these algorithms have been tested on physical vehicles. Such testing is significant because vehicles and the environment introduce constraints and complexities that could affect details of the algorithm. In this section, we introduce the micro aerial vehicle used in this work along with the relevant hardware and software subsystems.

The flight platform used in this work is a 3DRobotics Iris+, a 550 mm class, ready-to-fly quadcopter (Fig. 5). Navigation is supported by an internal uBlox GPS unit with integrated 3-axis compass. The integrated Pixhawk flight control board hosts the ArduPilot firmware that controls the vehicle. The Iris+ is powered by 4 950 kV motors run through a 4-in-1 electronic speed controller. Claimed flight time is in the range of 16–22 min.

The open-source Pixhawk flight control board was developed by Meier et al. at ETH Zurich. A wide array of connectivity options allow integration of numerous auxiliary devices such as range finders for collisions avoidance and accurate altitude determination, external GPS units for more precise positioning, and companion computers for running a variety of algorithms to control higher-order aspects of flight. The Pixhawk incorporates numerous sensors including a three-axis accelerometer, a three axis gyroscope, a three-axis compass and a barometer. These provide access to data concerning motion, orientation, heading and relative altitude. Pixhawk can be used in multiple vehicle types, both airborne and ground-based.

Table 1. An evaluation of the statistical significance of the effect of crossover. We use two sample T-tests for inputs 1 and 4 comparing crossover on versus crossover off. As is the case for nearly all of the inputs, we see that for *mng_10* and *mng_5* crossover does have a statistically significant effect. In nearly all cases, that effect is negative

		Input 1								
		Valid			10%			5%		
Ext Ver	Prob	Est Diff	P-value	T-value	Est Diff	P-value	T-value	Est Diff	P-value	T-value
1	0.00	5.0	0.121	1.57	88.1	**0.000**	4.85	118.1	**0.000**	4.65
	0.05	22.1	0.090	1.74	98.7	**0.000**	5.93	136.9	**0.000**	4.74
	0.10	40.8	0.051	2.02	106.4	**0.000**	5.69	135.7	**0.000**	4.70
	0.15	20.4	**0.029**	2.27	92.7	**0.000**	6.49	113.6	**0.000**	5.67
	0.20	43.8	**0.024**	2.36	108.3	**0.000**	5.57	147.8	**0.000**	4.98
2	0.00	1.7	0.599	0.53	55.5	**0.000**	4.60	72.0	**0.000**	3.87
	0.05	2.1	0.538	0.62	65.3	**0.000**	5.37	97.5	**0.000**	4.52
	0.10	2.1	0.542	0.61	62.9	**0.000**	5.31	91.4	**0.000**	4.70
	0.15	2.1	0.542	0.61	63.2	**0.000**	5.34	91.6	**0.000**	4.71
	0.20	2.1	0.542	0.61	61.2	**0.000**	5.28	91.8	**0.000**	4.34
3	0.00	2.1	0.518	0.65	74.1	**0.000**	4.08	110.4	**0.001**	3.48
	0.05	2.6	0.461	0.74	74.0	**0.000**	5.08	112.8	**0.001**	3.51
	0.10	2.6	0.466	0.73	68.7	**0.000**	5.42	88.9	**0.001**	3.70
	0.15	2.0	0.538	0.62	65.9	**0.000**	5.48	79.8	**0.001**	3.59
	0.20	2.0	0.538	0.62	69.5	**0.000**	5.37	114.6	**0.003**	3.16

		Input 4								
		Valid			10%			5%		
Ext Ver	Prob	Est Diff	P-value	T-value	Est Diff	P-value	T-value	Est Diff	P-value	T-value
1	0.00	4.8	0.308	1.03	21.1	0.539	0.62	−1195.0	**0.000**	−4.83
	0.05	−13.3	0.301	−1.05	−62.3	0.069	−1.85	−955.0	**0.000**	−4.72
	0.10	−165.9	**0.015**	−2.55	−247.8	**0.003**	−3.15	−1019.0	**0.000**	−5.16
	0.15	−288.1	**0.001**	−3.45	−385.0	**0.001**	−3.78	−981.0	**0.000**	−5.27
	0.20	−318.0	**0.008**	−2.82	−403.0	**0.002**	−3.44	−905.0	**0.000**	−4.92
2	0.00	5.8	**0.021**	2.40	−9.0	0.410	−0.83	−1082.0	**0.000**	−4.51
	0.05	5.36	**0.015**	2.53	−14.8	0.071	−1.84	−704.0	**0.000**	−3.92
	0.10	5.36	**0.015**	2.53	−16.9	**0.020**	−2.39	−753.0	**0.000**	−4.36
	0.15	5.36	**0.015**	2.53	−16.9	**0.020**	−2.39	−704.0	**0.000**	−4.71
	0.20	5.36	**0.015**	2.53	−16.9	**0.020**	−2.39	−554.0	**0.000**	−4.48
3	0.00	2.8	0.174	1.38	−15.08	**0.006**	−2.84	−1151.0	**0.000**	−5.12
	0.05	3.0	0.172	1.39	−14.64	**0.009**	−2.71	−1212.0	**0.000**	−6.11
	0.10	3.0	0.172	1.39	−12.53	**0.014**	−2.53	−930.0	**0.000**	−5.41
	0.15	3.0	0.172	1.39	−12.53	**0.014**	−2.53	−772.0	**0.000**	−5.21
	0.20	3.0	0.172	1.39	−12.56	**0.014**	−2.53	−626.0	**0.000**	−537

Fig. 5. Our micro aerial vehicle, the 3DR Iris+, shown here with the Odroid XU4 attached to the underside of the body. In the foreground is a LidarLite v2 laser range finder

While the Pixhawk firmware handles functions related to keeping the Iris+ in the air, mission applications such as path planning run on a separate computer. This can be a ground-based laptop or a smaller, lighter computer incorporated into the vehicle. Use of a ground-based laptop is the most straightforward implementation, however, communication with the vehicle then take places via a slow and potentially noisy radio link. While incorporating a single-board computer is more complex, it solves the problems of slow and unreliable communication. For these reasons, our Iris+ has been fitted with an Odroid XU4 single-board computer.

Measuring $83 \times 8 \times 20$ mm without the cooling fan, the XU4 includes a heterogenous 2.0 GHz octocore ARM processor with 2 GB of RAM. Storage is provided by a solid state drive in the form of a 32 GB eMMC 5.0 module. Connections with the Pixhawk and a WifI dongle are via USB. The board runs Ubuntu Linux 15.04.

When algorithms such as path planning are run on a companion computer, a mechanism is needed to provide communication between that computer and the flight control board. Pixhawk, and the ArduPilot firmware, us a protocol called MAVLink. MAVLink is a a message-based protocol that provides message templates for controlling vehicle position, altitude, and velocity, among many other, more complex behaviors. It does not, however, provide direct control or roll, pitch and yaw. ArduPilot controls these to achieve the navigational requests made via MAVLink. However, MAVLink is open-source and extensible, allowing users to create new messages to specific to their particular missions. Dronekit-Python, an API developed by 3D Robotics, provides a Python interface to MAVLink. The

API provides functions for connecting to the vehicle, arming, taking off, directly controlling velocity (in all three axes) and heading, and landing, in addition to GPS-based flight control.

8 Flying a Mission

Vehicle flight is autonomous though the mission must be initiated by the user. The process is outlined in Fig. 6. The first step is to power the vehicle, which in turn powers the XU4 onboard computer. Communication with the XU4 is necessary to initiate the mission and to monitor progress. Therefore, the XU4 establishes a WiFi access point and the user creates an ssh connection from a ground-based laptop. After creating a hardwired connection to the vehicle's Pixhawk flight control board, the computer then requests a series of pre-arm checks that include a GPS signal from at least 6 satellites and communication with a remote control device (as a failsafe). Subsequently, the vehicle's current position is determined by GPS and stored. This serves as the origin for path planning.

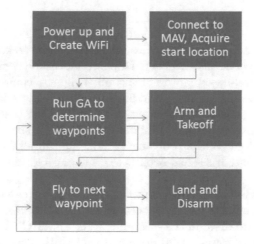

Fig. 6. Schematic depicting an abstraction of our system, including the genetic algorithm and the flight platform. At runtime, up to three attempts are made to evolve a valid path. If none is found, the program terminates. If a valid path is discovered, the vehicle executes the mission

With the vehicle ready to fly, the software runs the genetic algorithm to plan a path from the vehicle's current location to the predetermined destination. It is possible that the algorithm will fail to find an obstacle-avoiding path. In this case, the evolved path is discarded and the algorithm is called again. This process repeats up to three times. When a valid path is returned, the mission begins. The MAV is armed and takes off to a predetermined altitude. Recall that the flight path consists of a sequence of waypoints. Our software controls the vehicle through a flight that visits each of the waypoints in turn, terminating at the destination. At that point, the vehicle lands, disarms and closes the connection between the companion computer and the Pixhawk. At that point, no further autonomous flight is possible without intentional human action.

It is worth noting that US Federal Aviation Administration rules limit the format of flight testing. For example, map 4 depicts part of the Florida Southern College campus. We cannot, however, actually fly the paths calculated by our algorithm for that map. Doing so would not allow the human pilot, who must be prepared to takeover control of the vehicle in the event of an emergency, to maintain line-of-sight with the vehicle, due to the large number of buildings between origin and destination. Thus, while such real-world maps are interesting for the path planning algorithm, ironically they do not allow for real-world testing. Instead, test flights occur in a large, open field with obstacles marked by soccer training poles. GPS coordinates for environmental features are translated to that field.

One of our flight control functions is provided in Fig. 7. This function controls flight to the next waypoint in the mission. It makes use of a Dronekit-Python function, simple_goto, that flies the vehicle toward a location specified by GPS coordinates. Client code is responsible for determining when the vehicle has reached the waypoint.

```
def fly_waypoint(self, waypt, index):
    print "Flying to waypoint %s..." % index
    # create the waypoint in global_relative_frame
    dest_point = LocationGlobalRelative(waypt[0], waypt[1], self.alt_0)
    # fly to the waypoint at specified velocity
    self.vehicle.simple_goto(dest_point, groundspeed=self.v_base)

    # calculate current location and distance to waypoint so
    # that we can determine when we have gotten close enough
    current_loc = self.vehicle.location.global_relative_frame
    dist_lat, dist_lon, dist = self.distance_meters(current_loc, dest_point)
    while dist > self.dist_threshold:
        time.sleep(0.25)
        current_loc = self.vehicle.location.global_relative_frame
        dist_lat, dist_lon, dist = self.distance_meters(current_loc, dest_point)

    print "    At waypoint %s..." % index
```

Fig. 7. Example of a client function using the DroneKit-Python API. This function controls flight to the next waypoint in the mission

9   Performance

Performance in the context of this work is considered in two distinct stages: performance of the genetic algorithm and performance of the flight system, consisting of the micro aerial vehicle and devices incorporated to allow autonomous operation. Evaluating the genetic algorithm requires considering two factors: the quality of the evolved path and the efficiency with which that path was determined. Evaluating the flight system simply involves determining how effectively the system completed the mission.

The testing detailed in this section is distinct from that in Sects. 5 and 6. In this section, we use the algorithm without crossover and with a single extinction probability determined by the earlier experiments. The methods and metrics used are the same as those previously described. In addition to the five inputs used in testing for crossover and mass extinction, a sixth input (see Fig. 2) was used in this stage (Tables 2 and 3).

A standard measure of genetic algorithm efficiency is the number of generations required. Not surprisingly, this quantity was heavily dependent on the problem instance. Simple maps (see, for example, map 1 in Fig. 2) was solved much more quickly, on average, than more complex, real-world maps (see, for example, maps 3 and 4 in Fig. 2).

While the Odroid XU4 is a powerful single board computer, it cannot match a modern laptop for processing speed. Thus, algorithmic efficiency and tuning of relevant parameters in our genetic algorithm were particularly important. In use, the companion computer has proven capable of running the algorithm. Leveraging the XU4's 8 core architecture, child evaluation within the genetic algorithm is performed in parallel, typically using 4 processes. Another process, controls the graphical which allows users to monitor progress. The graphics window is transmitted to, and displayed on, a ground-based computer via ssh (Fig. 8).

Table 2. Parameter values used during testing. Values in columns 4–8 are probabilities for the various mutations and for extinction. The last column represents the minimum generations between extinction events

Parameter	Runs	Pop size	Add	Del	Swap	Move	Extinct	Period
Value	40	100	0.1	0.05	0.1	0.75	0.15	100

Table 3. Mean number of generations for a valid solution, a solution within 10% of optimal, and a solution within 5% of optimal for each of the six inputs tested

Input	Valid	10%	5%
1	20	112	170
2	61	188	234
3	537	560	572
4	22	42	135
5	45	75	109
6	99	352	674

Fig. 8. Time lapse imagery from a run of the genetic algorithm for Input 4. The first image shows the default, straight-line path from origin to destination. The last shows a solution with path length within 5% of optimal

As discussed in Sect. 4.7, genetic algorithms based on NSGA II have a quadratic dependence on population size and a linear dependence on the number of generations. Therefore, in order to speed runtimes on the companion computer, testing was performed with a population size of 50. Such runs averages roughly 35–60 s, depending on the complexity of the problem instance. More recently, the population size has been reduced further, to 30 members. To compensate for the reduction in genetic material available to evolve solutions, the number of generations in a run was extended to 1000. These changes have resulted in reduced runtimes and more reliable results.

Recall that a solution to an instance of the path planning problem is a sequence of waypoints defining an obstacle-free path from origin to destination. With such a solution in hand, the flight system must successfully execute the mission. Success is determined by completion of the mission and the accuracy with which the evolved path is flown. The Iris+ proved to be a very capable flight platform. In numerous test flights, the vehicle, flight control software, and communication links all performed well and without incident. All physical tests were successful. This is due in no small part to extensive testing on a simulator prior to testing on the MAV.

10   Conclusion

Instances of path planning occur in many real-world contexts. Consequently, many approaches have been brought to bear to find solutions to such problems. This includes, of course, genetic algorithms. In this work, we have demonstrated the first genetic algorithm for path planning to be used in a physical vehicle. Based on previous results, the algorithm does not use crossover but does use mass extinction. Mass extinction is a far less commonly used genetic operator.

Numerous areas for further investigation follow from this work. Because mass extinction has not seen broad adoption, there are many opportunities to improve its implementation. These include significant changes to the way in which members are killed off and regenerated to tuning the many parameters that control various aspects of extinction events.

Similarly, a different implementation of crossover might prove more beneficial. As discussed, other standard crossovers seem unlikely to work but perhaps there is a form of the operator specific to path planning that could provide improvement. In addition, other domain-specific genetic operators could improve performance.

Finally, experimentation has begun with a ground-based rover in place of the MAV. There are a number of benefits to this platform. These include greatly improved battery life, no need to worry about rules related to aviation, and reduced consequence of system failures. On the other hand, ground-based vehicles encounter a number of obstacle types, such as stairs, benches, and shrubs, that are of no consequence to aerial vehicles.

Acknowledgement. The authors thank Florida Southern College for financial support of this work. They also thank Annie Wu for valuable insights and feedback, Susan Serrano and Isabel Loyd for assistance with the statistical analysis, and members of the ArduPilot and Dronekit-Python development teams for valuable suggestions for developing flight control client code.

References

Ahmed, F., Deb, K.: Multi-objective optimal path planning using elitist non-dominated sorting genetic algorithms. Tech. Rep. 20111013, Kanpur Genetic Algorithms Laboratory (KanGAL), Indian Institute of Technology (2011)

Al-Sultan, K., Aliyu, M.: A new potential field-based algorithm for path planning. J. Intell. Robot. Syst. **17**(3), 265–282 (2010)

Buniyamin, N., Ngah, W.W., Shariff, N., Mohammad, Z.: A simple local path planning algorithm for autonomous mobile robots. J. Syst. Appl. Eng. Dev. **5**(2), 151–159 (2011)

Burchardt , H., Salomon, R.: Implementation of path planning using genetic algorithms on mobile robots. In: Proceedings of Congress on Evolutionary Computation, pp. 1831–1836, IEEE (2006)

Choset H., Pignon, P.: Coverage path planning: the boustrophedon decomposition. In: International Conference on Field and Service Robotics (1997)

Deb, K., Jain, H.: An evolutionary many-objective optimization algorithm using reference-point-based non-dominated sorting approach, part i: solving problems with box constraints. IEEE Trans. Evol. Comput. **18**(4), 577–601 (2014)

Deb, K., Pratap, A., Agarwal, S., Meyarivan ,T.: A fast and elitist multiobjective genetic algorithm: NSGA-II. IEEE Trans. Evol. Comput. **6**(2), 182–197 (2002)

Ferguson, D., Likachev, M., Stentz, A.: A guide to heuristic-based path planning. In: Proceedings of International Conference on Automated Planning and Scheduling (2005)

Fonseca, C., Fleming, P.: Multiobjective genetic algorithms. In: IEEE Colloquium on Genetic Algorithms for Control Systems Engineering, vol Digest No. 1993/130 (1993)

Glasius, R., Komoda, A., Gielen, S.: Neural network dynamics for path planning and obstacle avoidance. Neural Networks **8**(1), 125–133 (1995)

Goldberg, D.: Genetic algorithms for search, optimization, and machine learning. Addison-Wesley (1989)

Hasircioglu, I., Topcuoglu, H., Ermis, M.: 3-d path planning for the navigation of unmanned aerial vehicles by using evolutionary algorithms. In: Genetic Algorithms and Evolutionary Computation Conference, pp. 1499–1506. ACM (2008)

Hermanu, A., Manikas, T., Ashenayi, K., Wainwright, R.: Autonomous robot navigation using a genetic algorithm with an efficient genotype structure. In: Intelligent Engineering Systems Through Artificial Neural Networks: Smart Engineering Systems Design: Neural Networks, Fuzzy Logic, Evolutionary Programming, omplex Systems and Artificial Life. ASME Press (2004)

Jain, H., Deb, K.: An evolutionary many-objective optimization algorithm using reference-point-based non-dominated sorting approach, part ii: handling constraints and extending to an adaptive approach. IEEE Trans. Evol. Comput. **18**(4), 602–622 (2014)

Jaworski, B., Kuczkowski, L., Smierzchalski, R., Kolendo, P.: Extinction event concepts for the evolutionary algorithms. Przeglad Elektrotechniczny (Electrical Review) **88**(10b) (2012)

Jun, H., Qingbao, Z.: Multi-objective mobile robot path planning based on improved genetic algorithm. In: IEEE International Conference on Intelligent Computation Technology and Automation, pp. 752–756 (2010)

Knowles, J., Corne, D.: The pareto archived evolution strategy: a new baseline algorithm for pareto multiobjective optimization. In: Proceedings of Congress on Evolutionary Computation (1999)

Konak, A., Coit, D., Smith, A.: multi-objective optimization using genetic algorithms: a tutorial. Reliab. Eng. Syst. Saf. **91** (2006)

Lehman, J., Miikkulainen, R.: Extinction events can accelerate evolution. PLoS ONE. **10**(8) (2015)

Li, K., Deb, K., Zhang, Q., Kwong, S.: Efficient non-domination level update approach for steady-state evolutionary multiobjective optimization. Tech. Rep. 2014014, Computational Optimization and Innovation (COIN) Laboratory, Michigan State University (2014)

Lin, HS., Xiao, J., Michalewicz, Z.: Evolutionary navigator for a mobile robot. In: International Conference on Evolutionary Computation, pp. 2199–2204. IEEE (1994)

Mathias, D., Ragusa, V.: An empirical study of crossover and mass extinction in a genetic algorithm for pathfinding in a continuous environment. In: Proceedings of Congress on Evolutionary Computation, IEEE (2016)

Mengshoel, O., Goldberg, D.: The crowding approach to niching in genetic algorithms. Evol. Comput. **16**(3), 315–354 (2008)

Page, W., McDonnell, J., Anderson, B.: An evolutionary programming approach to multidimensional path planning. In: First Annual Conference on Evolutionary Programming, pp. 63–70 (1992)

Sedighi, K., Ashenayi, K., Manikas, T., Wainwright, R., Tai, H.M.: Autonomous local path planning for a mobile robot using a genetic algorithm. In: Proceedings of Congress on Evolutionary Computation, pp. 1338–1345, IEEE (2004)

Siddiqi, U., Shriraishi, Y., Sait, S.: Memory-efficient genetic algorithm for path optimization in embedded systems. IPSJ Trans. Math. Model. Appl. **6**(1) (2013)

Srinivas, N., Deb, K.: Multiobjective optimization using nondominated sorting in genetic algorithms. Evol. Comput. **2**(3), 221–248 (1994)

Tonupunuri, P.: Evolutionary based path-finding for mobile agents in sensor networks. Master's thesis, Southern Illinois University (2008)

Xiao, J., Michalewicz, Z., Zhang, L., Trojanowski, K.: Adaptive evolutionary planner/navigator for mobile robots. IEEE Trans. Evol. Comput. **1**(1) (1997)

Yun, G., Lu, H.: Dynamic multiobjective evolutionary algorithm: adaptive cell-based rank and density estimation. IEEE Trans. Evol. Comput. **7**(3), 253–274 (2003)

Zhang, Q., Li, H.: A multiobjective evolutionary algorithm based on decomposition. IEEE Trans. Evol. Comput. **11**(6) (2007)

Zheng, C., Ding, M., Zhou, C., Li, L.: Coevolving and cooperating path planner for multiple unmanned air vehicles. Eng. Appl. Artif. Intell. **17**, 887–896 (2004)

Zitzler, E., Thiele, L.: Multiobjective evolutionary algorithms: a comparative case study and the strength of the pareto approach. IEEE Trans. Evol. Comput. **3**(4), 257–271 (1999)

Zitzler, E., Laumanns, M., Thiele, L.: Spea2: improving the strength of pareto evolutionary algorithms. Tech. Rep., ETH Zurich (2001)

Mining Process Model Descriptions of Daily Life Through Event Abstraction

N. Tax[1]([✉]), N. Sidorova[1], R. Haakma[2], and W. van der Aalst[1]

[1] Technische Universiteit Eindhoven, Eindhoven, The Netherlands
{n.tax,n.sidorova,w.m.p.v.d.aalst}@tue.nl
[2] Philips Research, Eindhoven, The Netherlands
w.m.p.v.d.aalst@tue.nl

Abstract. Methods from the area of Process Mining traditionally focus on extracting insight in business processes from event logs. In this paper we explore the potential of Process Mining to provide valuable insights in (un)healthy habits and to contribute to ambient assisted living solutions when applied on data from smart home environments. Events in smart home environments are recorded at the level of sensor triggers, which is too low to mine habit-related behavioral patterns. Process discovery algorithms produce then overgeneralizing process models that allow for too much behavior and that are difficult to interpret for human experts. We show that abstracting the events to a higher-level interpretation can enable discovery of more precise and more comprehensible models. We present a framework to automatically abstract sensor-level events to their interpretation at the human activity level. Our framework is based on the XES IEEE standard for event logs. We use supervised learning techniques to train it on training data for which both the sensor and human activity events are known. We demonstrate our abstraction framework on three real-life smart home event logs and show that the process models that can be discovered after abstraction improve on precision as well as on F-score.

1 Introduction

Process mining has emerged as a research area that employs methods from data mining, business process modeling, statistics, and business process analysis in order to create different insights in processes: their structure, performance, conformance to their requirements, etc. [2]. *Process discovery* is the task of extracting *process models* from logs, which plays an important role in process mining. There are many different process discovery algorithms [3,9,23,41–43], which can discover a variety process model notations, including Business Process Model and Notation (BPMN), Workflow nets, Unified Modeling Language (UML) activity diagrams, and Statechart diagrams.

While originally the scope of process mining has been on business processes, it has broadened in recent years towards other application areas, including the

Y. Bi et al. (eds.), *Intelligent Systems and Applications*,
Studies in Computational Intelligence 751,
https://doi.org/10.1007/978-3-319-69266-1_5

analysis of human behavior [24,34,36,37]. Process model descriptions of human behavior can be used amongst others to aid lifestyle coaching for healthy living, or to assess the ability of independent living of elderly or people with illness.

Events in the event log are generated by e.g. motion sensors placed in the home, power sensors placed on appliances, open/close sensors placed on closets and cabinets, etc. This clearly distinguishes process mining for smart homes from the traditional application domain of business processes, where events in the log are logged by IT systems when an business tasks are performed.

In event logs from business processes the event labels generally have a clear semantic meaning, like *register mortgage request*. In the smart home domain the events are on the sensor level, while the human expert is interested in analyzing the behavior in terms of activities of daily life. Additionally, simply using the sensor that generated the event as the event label has been shown to result in non-informative process models that overgeneralize the event log and allow for too much behavior [39]. In the field of process mining such overgeneralizing process models are generally referred to as being *imprecise*.

In our earlier work [36] we showed how to discover more precise process models by taking the name of the sensor as a starting point for the event label and then refine the labels using the time of the day at which the event occurred. However, labels in such process models still represent sensors, and they have no direct interpretation on the level of human activity, called *activities of daily living* (*ADL*) [19]. This problem was also recognized by Leotta et al. [24] who stated the need to bridge the gap between sensor events and human activities as one of the main research challenges in applying process mining to smart spaces. In this paper we leverage diary style annotations of the activities performed on the ADL-level and use them learn a mapping from sensor-level events to ADL-level events. This enables discovery of process models that describe the ADL-level activities directly, leading to more comprehensible and more precise descriptions of human behavior. Often it is infeasible or simply too expensive to obtain such diaries for periods of time longer than a couple of weeks. However, to mine a process model of human behavior more than a couple of weeks of data is needed. Therefore, there is a need to infer human level interpretations of behavior from sensors.

Methods from the area of supervised learning can be used to learn how to transform data at the sensor-level to events at the ADL-level through examples, without providing hand-made descriptions how human activities relate to sensor events. Comparable methods have been employed in the area of *activity recognition*, in which continuous-valued time series originating from sensors are mapped to time series of ADL-level events. Sensor-level events, such as *opening the freezer*, trigger change points in these time series, and corresponding ADL events (e.g. *preparing dinner*) are generally collected by manual logging of activity diaries. However, unlike the techniques from the activity recognition field, we operate on discrete events at the sensor-level instead of continuous time series.

In this paper we extend the work started in [38]. We propose a framework for abstraction of events based on supervised learning. Our framework facilitates data preprocessing. We show that the event logs produced by our framework allow for the mining of more precise process models from smart home event logs. Additionally, the process models obtained represent ADL activities directly, thereby enabling direct analysis of human behavior itself, instead of indirect analysis through sensor-level models. In Sect. 2 we describe related work, focusing both on the process mining and the activity recognition areas. Basic concepts, notations, and definitions that we use in the remainder of this paper are introduced in Sect. 3. In Sect. 4 we explain conceptually why abstraction from sensor-level to ADL events can help to the process discovery step to find more precise process models. In Sect. 5 we describe a framework for retrieving useful features for abstraction from event logs using specific concepts of the IEEE XES standard for event logs [16]. In Sect. 6 we apply our framework to three real life smart home event logs and show that the discovered models are more precise compared to the models discovered on the unpreprocessed data at the sensor level. Section 7 concludes the paper and identifies some areas of future work.

2 Related Work

Abstraction of events to a lower-level of granularity has hardly being explored till now in the context of process mining. Tax et al. [38] proposed an initial technique in this direction, working in a supervised setting. In this paper we improve this abstraction technique by improving the proposed feature extraction steps and by performing an elaborate evaluation of the abstraction technique using additional real life case studies.

The majority of the work on abstractions from sensor-level to human activity level events can be found activity recognition research area, which addresses the challenge of recognizing different types of *activities of daily living* (ADL) [19] from e.g. in-house sensors [18,35], or on-body sensors [5,21].

Techniques for activity recognition typically operate on discretized time windows from series of continuous-valued sensor readings and aim to classify each window onto the correct ADL type, such as *showering* or *eating*. Activity recognition methods can be classified into two categories: probabilistic approaches [5,18,21,35] and ontological reasoning approaches [7,31]. The advantage of probabilistic approaches over ontological reasoning approaches is their capability to deal with sensor data that is noisy, uncertain and/or incomplete [7].

The first application of supervised learning to the problem of ADL activities detection based on the data from in-house sensors [35] was based on a naive Bayes classifier. Recently, a variety of activity recognition approaches were proposed that are based on probabilistic graphical models (PGMs) [17,18], including [18] (based on Conditional Random Fields (CRFs) [22] and Hidden Markov Models (HMMs) [30]), and [17] (based on Bayesian Networks [14]). [20] applied both HMMs and CRFs to the activity recognition problem, and observed that HMMs have difficulties to capture long-range dependencies between observations as well as transitive dependencies. The consequence of this is that HMMs have trouble to detect activities that are performed concurrently, or interleaved. contrary to HMMs, CRFs have no difficulties to capture transitive and long-term dependencies.

Our work differentiates itself from existing activity recognition work in the form of the input data on which they operate and in the goal that it aims to achieve. Activity recognition methods take multi-dimensional time series data as input; this input consists thus of consecutive sensor values. Sliding window segmentation methods are then used to identify ADL activities to each window. Choosing an appropriate size for such a window, however, requires domain expertise and is particular to the data set. This dependence on the domain specifics hinders the generality of window-based approaches. In this paper, our goal is a generic method that does not require this domain knowledge, and that works in general for any event log. Any chosen time window size t would be too small for some event logs (i.e., when the typical time between two consecutive high-level events in the event log is much larger than t, resulting in high computation time), while being too large for other event logs (i.e., when the typical time between two consecutive high-level events in the log is smaller than t, resulting in short high-level events not being detected). Therefore, time range based approaches are not suitable for supervised abstraction of events in the general case. Instead of operating on time ranges, we make a prediction of the ongoing ADL activity for each event on the low, sensor, level. Each change in the value of a binary-valued in-house sensors can be regarded as an occurrence of a low-level event. A second difference with existing activity recognition techniques is that our framework aims to find an abstraction of the data that enables discovery of more precise process models, where classical activity recognition methods do not have a link with the application of process mining.

Other related work can be found in the area of process mining, where several techniques handles the problem of dealing with low-level events (e.g. sensor-level) by abstracting them to events at a higher level [6, 12, 15, 26, 27]. Most existing event abstraction methods rely on clustering methods where each cluster of low-level events is interpreted as one single event on the higher level. Abstraction methods that are based on unsupervised learning have two fundamental limitations. First, no labels are generated for the created high-level events, and it is often not trivial to label them manually, as this requires domain knowledge that might not be available. Secondly, unsupervised learning does not provide any guidance concerning the degree to which the log should be abstracted. Normally, such unsupervised abstraction approaches are parameterized to control the size of the obtained clusters, thereby impacting the degree of abstraction. However, finding the right settings of such parameters, so that abstraction methods produce meaningful results, generally comes down to trial and error.

Two abstraction techniques from the process mining field that rely on the domain knowledge are proposed in [4, 27]. The technique from [4] requires knowledge about a single model of the overall process, and uses this model to map events in the log to the correct granularity level that is present in the model. However, for our application domain in mining on ADL activities, no overall process model on the desired granularity is available. The technique from [27] relies on domain knowledge to perform the abstract step, requiring the user to specify a low-level process model for each high-level activity. However, in the context of human behavior it is unreasonable to expect the user to provide the process model in sensor terms for each human activity. A recent work [26] attempts to replace the required domain knowledge required in [27] with automatically mined patterns in business process model notation (i.e., Local Process Models (LPMs) [37,40]), so that each mined pattern represents one high-level event and describes the behavior over the low-level events during execution of the high-level event. However, the automatic pattern mining step makes this approach unsupervised, again resulting in problems with the high-level labels and desired granularity.

3 Preliminaries

This section delineates concepts and terminology used in later sections of this paper.

3.1 Event Logs and Process Models

$X = \{a_1, a_2, \ldots, a_n\}$ denotes a finite set. $X \backslash Y$ denotes the set of elements that are in set X but not in set Y, e.g., $\{a, b, c\} \backslash \{a, c\} = \{b\}$. X^* denotes the set of all sequences over a set X and $\sigma = \langle a_1, a_2, \ldots, a_n \rangle$ denotes a sequence of length n, with $\langle \rangle$ the empty sequence. $\sigma_1 \cdot \sigma_2$ denotes the concatenation of sequences σ_1 and σ_2, e.g., $\langle a, b, c \rangle \cdot \langle d, e \rangle = \langle a, b, c, d, e \rangle$. A multiset (or bag) over X is a function $B : X \rightarrow \mathbb{N}$ which we write as $[a_1^{w_1}, a_2^{w_2}, \ldots, a_n^{w_n}]$, where for $1 \leq i \leq n$ we have $a_i \in X$ and $w_i \in \mathbb{N}^+$. The set of all bags over X is denoted $\mathscr{B}(X)$.

In the context of process mining, we assume the set of all *process activities* Σ to be given. Event logs consist of sequences of events where each event represents a process activity.

Definition 1 (*Event, Trace, and Event Log*) An *event* e in an event log is the occurrence of an activity $e \in \Sigma$. We call a sequence of events $\sigma \in \Sigma^*$ a *trace*. An *event log* $L \in \mathscr{B}(\Sigma^*)$ is a finite multiset of traces.

One process model notation that is commonly used in the area of process mining is a workflow net [1]. Workflow nets can be converted into process model notations that are more commonly found in business environments, such as BPMN and BPEL [25]. A workflow net is a directed bipartite graph consisting of places (depicted as circles) and transitions (depicted as rectangles), connected by arcs. A transition describes an activity, while places represent the enabling conditions of transitions. Labels of transitions indicate the type of activity that they represent. Unlabeled transitions (τ-transitions) represent invisible transitions (depicted as gray rectangles), which are only used for routing purposes and are not recorded in the event log. Examples of workflow nets are shown in Fig. 1.

Definition 2 (*Labeled Workflow Net*) A *labeled workflow net* $N = \langle P, T, F, \ell \rangle$ is a tuple where P is a finite set of places, T is a finite set of transitions such that $P \cap T = \emptyset$, $F \subseteq (P \times T) \cup (T \times P)$ is a set of directed arcs, called the flow relation, and $\ell: T \nrightarrow \Sigma$ is a partial labeling function that assigns a label to a transition, or leaves it unlabeled (the τ-transitions).

Workflow nets are a special type of Petri nets [29]. We write $^\bullet n$ and $n^\bullet$ for the input and output nodes of $n \in P \cup T$ (according to F). A state of a workflow net is defined by its *marking* $m \in \mathscr{B}(P)$ being a multiset of places. A marking is graphically denoted by putting $m(p)$ tokens on each place $p \in P$. State changes occur through transition firings. A transition t is enabled (can fire) in a given marking m if each input place $p \in {}^\bullet t$ contains at least one token. Once t fires, one token is removed from each input place $p \in {}^\bullet t$ and one token is added to each output place $p' \in t^\bullet$, leading to a new marking $m' = m - {}^\bullet t + t^\bullet$.

A firing of a transition t leading from marking m to marking m' is denoted as step $m \xrightarrow{t} m'$. Steps are lifted to sequences of firing enabled transitions, written $m \xrightarrow{\gamma} m'$ and $\gamma \in T^*$ is a *firing sequence*.

A partial function $f \in X \nrightarrow Y$ with domain $dom(f)$ can be lifted to sequences over X using the following recursive definition: (1) $f(\langle \rangle) = \langle \rangle$; (2) for any $\sigma \in X^*$ and $x \in X$:

$$f(\sigma \cdot \langle x \rangle) = \begin{cases} f(\sigma) & \text{if } x \notin dom(f), \\ f(\sigma) \cdot \langle f(x) \rangle & \text{if } x \in dom(f). \end{cases}$$

Defining an *initial* and *final* markings allows to define the *language* accepted by a workflow net as a set of finite sequences of activities.

Definition 3 (*Accepting Workflow Net*) An *accepting workflow net* is a triplet $AWN = (N, m_0, m_f)$, where N is a labeled workflow net, $m_0 \in P$ is its input (source) place, and $m_f \in P$ is its output (sink) place. A sequence $\sigma \in \Sigma^*$ is a *trace* of an accepting workflow net AWN if there exists a firing sequence $m_0 \xrightarrow{\gamma} m_f$ such that $\gamma \in T^*$ and $\ell(\gamma) = \sigma$. Every node must be on some path from m_0 to m_f. Furthermore, an accepting workflow net is required to be *sound*. The *soundness* property requires that each transition can be potentially executed, and that the process can always terminate properly, i.e., finish with only one token in m_f.

Visually, the source place contains a token and the sink place is marked as $\bigcirc$.

The *language* $\mathscr{L}(AWN)$ is the set of all its traces, i.e., $\mathscr{L}(AWN) = \{l(\gamma) | \gamma \in T^* \wedge m_0 \xrightarrow{\gamma} m_f\}$, which can be of infinite size when AWN contains loops. Even though we define language for accepting workflow nets, in theory $\mathscr{L}(M)$ can be defined for any process model M with formal semantics. We denote the universe of process models as $\mathscr{M}$. For each $M \in \mathscr{M}$, $\mathscr{L}(M)$ is defined.

One of the main challenges in process mining is process discovery, i.e., the task of extracting a process model from an event log that accurately describes the behavior seen in that log. A process discovery method is a function $PD : \mathscr{B}(\Sigma^*) \rightarrow \mathscr{M}$ that obtains a process model for a given event log, aiming at the discovery of a process model that is a good description of the process from which the event log was obtained, i.e., it allows for all the behavior seen in the event log (*fitness*) while not allowing for too much behavior that was not seen in the event log *precision*. For an event log L and a model M we say that L is *fitting* on model M if $L \subseteq \mathfrak{L}(M)$. Precision is related to the behavior that is allowed by a model M that was not observed in the event log L, i.e., $\mathfrak{L}(M) \backslash L$.

3.2   Conditional Random Fields

We consider the detection of human activity (ADL) events from sensor-level events as a sequence labeling problem where each event on the sensor-level is classified into one of the human activity level events. Linear-chain Conditional Random Fields (CRFs) [22] are a special type of Probabilistic Graphical Model (PGM) that has proven to be suitable for a variety of sequence labeling tasks in natural language processing and biological sequences. Conceptually, CRFs can be regarded as a sequential version of multiclass logistic regression, i.e., the predictions in the prediction sequence are dependent on each other. CRFs model the probability distribution over the possible labellings of an input sequence, conditional to that input sequence. Formally, linear-chain CRFs are formulated as follows:

$$p(y|x) = Z(x)^{-1} e^{\sum_{t=1} \sum_k \lambda_k f_k(t, y_{t-1}, y_t, x)} \tag{1}$$

in which $Z(x)$ is a normalization term which makes sure that the resulting function is a valid probability distribution (i.e., it sums to one). $X = \langle x_1, \ldots, x_n \rangle$ is an input sequence (the sensor-level events), $Y = \langle y_1, \ldots, y_n \rangle$ is the associated sequence of label value (the human activity (ADL) level events), f_k are feature functions that describe the sensor-level events and λ_k represents the weights of those feature functions. The values of λ_k are optimized during model training, such that the cross-entropy error of the predicted labels for the input sequences on the training set is minimized. Where Hidden Markov Models [30] assume the feature functions to be uncorrelated, CRFs do not state this assumption.

4   The Discrepancy Between Low-Level and High-Level Structure

Figure 1 demonstrates through a simplistic example that a process can seem unstructured at the sensor level of events, while being structured at a human behavior (ADL) level. The workflow net in Fig. 1c depicts the actual process at the ADL level. The *Taking medication* ADL activity is itself represented by a process, which is shown in Fig. 1a. *Eating* is also defined as a process, which is shown in Fig. 1b. The Inductive Miner (IM) [23] applied on traces of events on the sensor-level, simulated from the process of Fig. 1c, results in the workflow net shown in Fig. 1d. This process model allows for almost all possible traces over alphabet $\{WT, PMC, PC, D, CD\}$, with the only behavioral restriction specified by the workflow net being that if a W occurs, then it has to be preceded by a PMC event. The gray τ-transitions allow skipping the execution of the other

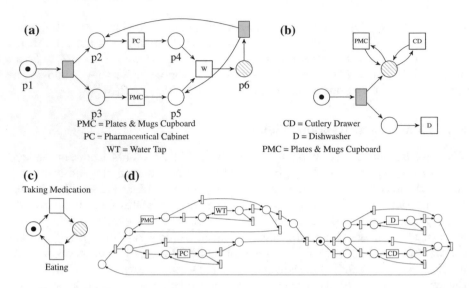

Fig. 1. A human activity level process model (**c**) where the two transitions themselves are defined as process models (shown in **a** and **b**), and the Inductive Miner (IM) result on the sensor-level traces generated from this model (**d**)

transitions in the workflow net. The model from Fig. 1d has a very low precision with respect to the simulated log, as it overgeneralizes, i.e., it allows for too much behavior that is not seen in the event log. Behaviorally this model is almost equivalent to the *flower model* [2], i.e., the model that does not specify any behavioral restrictions over the activities. The alternating pattern consisting of *taking medication* and *eating* in the ADL level process of Fig. 1c cannot be found in Fig. 1d. The high variance in the *start events* and *end events* of the sensor-level subprocesses of *taking medication* and *eating* as well as by the overlap in types of activities between the subprocesses are the cause of this effect. Both subprocesses contain *PMC*, and the miner cannot see that there are actually two different contexts for the *PMC* activity to split the label in the model. Abstracting the sensor-level events to their respective human activity level events before applying process discovery to the resulting human activity log unveils the alternating structure between *eating* and *taking medication* as shown in Fig. 1c. The resulting model after abstraction is both *fitting*, as it allows for all the behavior that is should allow, and *precise*, as it does not allow for any more behavior than it should allow for.

5   Abstracting Events Through Sequence Labeling

This section presents the framework for supervised abstraction of events based on Conditional Random Fields (CRFs). Furthermore, we introduce feature functions that can be extracted from event logs and that are applicable to any log that conforms to the IEEE XES standard [16]. XES, which is an abbreviation for *eXtensible Event Stream*, is the IEEE standard for process mining event logs. Figure 2 provides an overview of the IEEE XES file structure. Like the formal definition of event logs that we provided in Sect. 3, the IEEE XES standard defines an event *log* as a set of *traces*. A trace is defined as sequences of *events*. Logs, traces and events can contain *attributes*, containing a *key* and *value*. *Global* attributes are a set of attribute keys for traces or events, which indicate respectively that all traces respectively all events in the log contain an attribute with that key. *Classifiers* define viewing perspectives on logs, and can be used to transform traces from complex sequences of events to simple sequences of labels, taking the labels from one or more global event attributes that are defined by the classifier. A special collection of global attributes for logs, traces, or events are provided by *extensions*, which have clear, globally defined, semantics for some attribute keys. The IEEE XES standard [16] lists the following set of extensions, and their semantic interpretation:

Concept   stores a generally understood name for any hierarchy element. For logs, the name attribute may store the name of the process having been executed. For traces, the name attribute usually stores the case ID. For events, the name attribute represents the name of the event, e.g. the name of the executed activity represented by the event (*key:* **concept:name**).

Lifecycle   specifies the lifecycle phase (*key:* **lifecycle:transition**) that the event represents in a transactional model of its producing activity. The lifecycle extension also specifies a standard transactional model for activities.

Organizational   the name, or identifier, of the resource having triggered the event (*key:* **org:resource**), the role of the resource having triggered the event, within the organizational structure (*key:* **org:role**), and the group within the organizational structure, of which the resource having triggered the event is a member. (*key:* **org:group**).

Time   The date and time, specifying the point in time when the event has taken place (*key:* **time:timestamp**).

In addition to the existing *concept:name* attribute key, we propose attribute key *label*, representing the activity at the human activity (ADL). We use the existing key *concept:name* sto represent sensor-level activity. Each event then contains a low-level *concept:name* attribute and a high-level *label* attribute, linking a sensor event (e.g., *plates & mugs cupboard*) to an ADL event (e.g., *taking medication*), and potentially linking another sensor event of the same type to a different ADL event (e.g. *eating*). Some traces can consist of events with empty *label* attributes, in case ADL level annotations are not available for the trace. We aim to provide accurate estimations of activities on the ADL level for traces with missing *label* attributes, thereby replacing the missing *label* value with an estimated value, by using a model learned from traces where *label* attributes are available. This allows traces with missing *label* attributes to be used for process mining applications.

A schema of the abstraction approach is provided in Fig. 3. The method uses as input a collection of traces with *label* attribute (i.e., the ADL activity for the

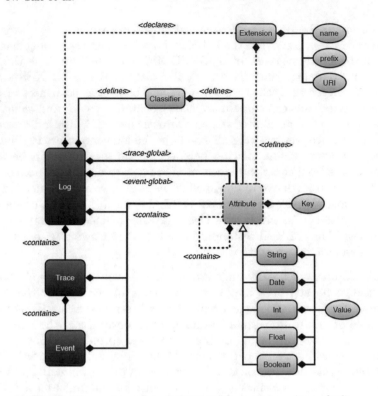

Fig. 2. The metamodel of IEEE XES event logs (from [16])

sensor events are known) and a collection of traces without *label* attribute (i.e., only the sensor activity, but no ADL activity, is known). A CRF is trained on the traces with *label* attributes, resulting in a function from sensor-level events to human activity level events. The trained CRF model can be used to estimate the *label* attributes for traces where these are missing. Generally, multiple consecutive sensor-level events will have identical *label* attribute values. We assume that multiple human activity level events cannot occur concurrently. This assumption allows merging of a sequence of events with identical *label* values into a two ADL events, where the first event has value *start* and the second has value *complete* for attribute *lifecycle:transition*. Tables 1 and 2 illustrate this collapsing procedure with an example.

The CRF-based abstraction approach is implemented as part of the process mining framework ProM [13] and available as package *AbstractEventsSupervised*.

Now we specify how each of the XES extensions can be used to extract helpful feature functions for event abstractions. In the training phase we search for values of weight vector λ that minimize the cross entropy between the ground truth label and the predicted label on the training data. The method to extract features from IEEE XES logs generates a feature space consisting of an unknown number of features, which could cause problems with overfitting for high numbers of features. To address this we apply ℓ_1 regularization in the training of the CRFs:

Table 1. A case consisting of events on the sensor level and their respective predicted label (i.e., human activity annotations)

Trace	Time:timestamp	Concept:name	Label
23	04/12/2016 08:46:24	Pharmaceutical cabinet	Taking medication
23	04/12/2016 08:47:12	Plates and mugs cupboard	Taking medication
23	04/12/2016 08:47:46	Water tap	Taking medication
23	04/12/2016 08:48:60	Plates and mugs cupboard	Eating
23	04/12/2016 08:48:90	Dishwasher	Eating
23	04/12/2016 17:11:59	Plates and mugs cupboard	Taking medication
23	04/12/2016 17:11:70	Pharmaceutical cabinet	Taking medication
23	04/12/2016 17:12:19	Water tap	Taking medication

Table 2. The resulting ADL-level log, obtained after merging in Table 1 consecutive events in a trace with identical values for the label attribute. Lifecycle:transition s = 'start' and c = 'complete'

Trace	Time:timestamp	Concept:name	Lifecycle:transition
23	04/12/2016 08:46:24	Taking medication	s
23	04/12/2016 08:47:46	Taking medication	c
23	04/12/2016 08:48:60	Eating	s
23	04/12/2016 08:48:90	Eating	c
23	04/12/2016 17:11:59	Taking medication	s
23	04/12/2016 17:12:19	Taking medication	c

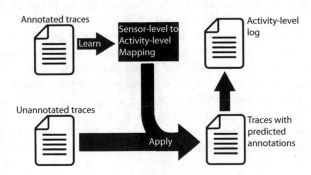

Fig. 3. An overview of the supervised event abstraction technique

a penalty term is added to the cross entropy loss that is proportional to the size of weight vector λ. This gives the model an incentive to use only a subset of the available features (i.e., setting some features to zero weight), thereby resulting in

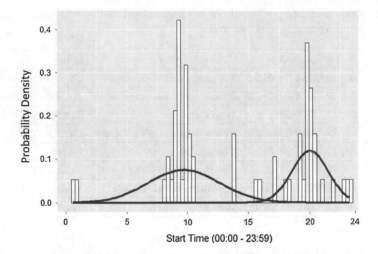

Fig. 4. The histogram representation and a Gaussian Mixture Model fitted on the timestamp values of the plates cupboard sensor of a real-life smart home environment sensor log

prediction models that are sparse and therefore simpler, which helps to prevent overfitting.

5.1 Extracting a Feature Vector from an IEEE XES Event Log

We now discuss per XES extension how feature functions can be obtained.

Concept Based on the k-gram $\langle a_1, a_2, \ldots, a_k \rangle$ consisting of the sensor-level activities (*concept:name*) of the k last-seen events, we estimate a categorical probability distribution over the classes of human activity level activities from the training log, such that the probability of class l is equal to the number of times that the n-gram was observed while the k-th event was annotated with class l, divided by the total number of times that the n-gram was observed. A feature function based on the concept extension has two parameters, k and l, and is valued with the estimated categorical probability density of the current sensor-level event having human activity level label l given the n-gram with the last k sensor-level event labels. It can be useful to combine multiple features that are based on the concept extension, where the features have different values for k and l.

Organizational Analogous to the concept extension, a categorical probability distribution over the ADL activities can be estimated on the training data for *resource*, *role*, or *group* attribute k-grams. Similarly, a three parameter organizational extension based feature function, with k-gram size parameter k, parameter $o \in \{group, role, resource\}$, and label parameter l, is valued with the estimated probability density according of the categorical probability distribution for ADL activity l given the last k values of o.

Time Certain type of ADL activities could have particularly high probability density in certain parts of the day (e.g. *sleeping* at night), week (e.g. *working* during business days), month, etc. A naive attempt to model the probability density of ADL activities over the time-of-the-day, time-of-the-week, or time-of-the-month could be based on a Gaussian Mixture Model (GMM), where each component of the mixture is defined by a normal distribution. The circular, non-Euclidean, nature of the data space of time-of-the-day, time-of-the-week, or time-of-the-month however introduces problems for the GMM, as, using time-of-the-day as an example, 00:00 is actually very close to 23:59. Figure 4 illustrates this problem. The Gaussian component with a mean around 10 o'clock has a standard deviation that is much higher than what one would expect when looking at the histogram, as the GMM tries to explain the data points just after midnight with this component. These data points just after midnight would however have been much better explained with the Gaussian component with the mean around 20 o'clock, which is much closer in time. Alternatively, we use a mixture model with components of the von Mises distribution, which is a close approximation of a normal distribution wrapped around the circle. To determine the correct number of components of such a von Mises Mixture Model (VMMM) we use Bayesian Information Criterion (BIC) [33], choses the number of components which explains the data with the highest likelihood, while adding a penalty for the number of model parameters. A VMMM is learned from training data and models the probabilities of each type of human activity based on the amount of time that has passed since the day start, week start, month start. A feature function for the time extension with parameters $t \in \{day, week, month, \dots\}$ and label l, is valued with the probability density of label l according to the learned VMMM model given the view on the timestamp of the event according to t. An alternative approach to estimate the probability density on data that lies on a manifold, such as a circle, is described by Cohen and Welling [8].

Lifecycle and Time The IEEE XES standard [16] describes lifecycle phases of activities. Lifecycle values that are commonly found in real life logs are *start* and *complete* which respectively represent when this activity started and ended However, a larger set of lifecycle values is defined in the IEEE XES standard, including *schedule*, *suspend*, and *resume*. The time differences between different stages of an activity lifecycle can be calculated for event logs that contain the lifecycle extension as well as and the time extension. For example, when observing the *complete* of an activity, the time between this *complete* and the corresponding *start* of this activity can contain useful information for predicting the correct human activity label. Identifying the corresponding *start* event is non-trivial when multiple instances of the same activity run concurrently. We make the assumption that only one sensor-level event of the same sensor type can be on-going at any point in time.

The IEEE XES standard defines an ordering over the lifecycle values. For each type of human activity, we fit a GMM to the set of time differences between each two consecutive lifecycle steps. A feature based on both the combination of the lifecycle and the time extension with activity label parameter l and lifecycle c is valued the probability density of activity l as estimated by the GMM given the time between the current event and lifecycle value c. We decide on the number of components of the GMM using BIC [33]. Note that while these features are time-based, regular GMMs can be used instead of VMMMs since time duration is a Euclidean, non-circular, space.

5.2 Evaluating the Estimated Human Activity Events for Discovering ADL Models

A well-known metric for the distance of two sequences is the Levenshtein distance. However, Levenshtein distance is not suitable to compare sequences of human actions, as human behavior sometimes includes branches in which it does not matter in which order two activities are performed. For example, most people *shower* and *have breakfast* after waking up, but people do not necessarily always perform the two in the same order. Indeed, when $\langle a, b \rangle$ is the sequence of predicted human activities, and $\langle b, a \rangle$ is the actual sequence of human activities, we consider this to be only a minor error, since it is often not relevant in which order two parallel activities are executed. Levenshtein distance would assign a cost of 2 to this abstraction, as transforming the predicted sequence into the ground truth sequence would require one deletion and one insertion operation. For example, most people *shower* and *have breakfast* after waking up, but people do not necessarily always perform the two in the same order. An evaluation measure that better reflects the prediction quality of event abstraction is the Damerau-Levenstein distance [10], which adds a swapping operation to the set of operations used by Levenshtein distance. Damerau-Levenshtein distance would assign a cost of 1 to transform $\langle a, b \rangle$ into $\langle b, a \rangle$. To obtain comparable numbers for different numbers of predicted events we normalize the Damerau-Levenshtein distance by the maximum of the length of the ground truth trace and the length of the predicted trace and subtract the normalized Damerau-Levenshtein distance from 1 to obtain Damerau-Levenshtein Similarity (DLS).

6   Case Studies

In this section we evaluate the supervised event abstraction framework on three case studies on real life smart home data sets.

6.1   Experimental Setup

We include three real life smart home event logs in the evaluation: the Van Kasteren event log [18], and two event logs from a smart home experiment conducted by MIT [35]. All three event logs used in for the evaluation consist of time series of multiple dimensions, in which every time series contains the binary value that characterizes the state of one sensor over time. These data sets include motion sensors, open/close sensors, and power sensors (discretized to binary states). We have preprocessed sensor data into events, such that each change point in the value of a sensor results in an event. Events that have occurred in the same day are grouped together to form a case. Annotations at the ADL level are included in all three logs. The following XES extensions can be used for these event logs:

Concept   Specifies which sensor in the smart home environment caused this event.

Time   Specifies the timestamp at which the changepoint in the sensor occurred.

Lifecycle   Has the value *Start* when this event represents change in sensor value from 0 to 1 and *Complete* otherwise.

For evaluation of the abstraction quality, we split the data into separate parts for training and testing. The predicted ADL labels are evaluated by comparing them the actual ADL labels in a Leave One Trace Out Cross Validation (LOTOCV) setup, i.e., we iteratively leave out one trace from the data set to evaluate on, while training on the other traces of the data set. We measure the accuracy of the ADL level traces compared to the ground truth ADL traces in terms of Damerau-Levenshtein similarity [10].

Additionally, we evaluate the quality of the process model that can be discovered from the ADL level traces. We use the Inductive Miner (IM) [23] to mine a workflow net from the estimated ADL level log. There are several criteria to express the fit between a process model and an event log in the area of process mining. Two of those criteria are *fitness* [32], which measures the degree to which the behavior that is observed in the event log can be replayed on the process model, and *precision* [28], which measures the degree to which the behavior that was never observed in the event log cannot be replayed on the process model. Low precision typically is indicates an overly general process model, that allows for too much behavior. We compare the *fitness* and *precision* of the models produced by the IM algorithm on the sensor-level log and the ADL level log. Finally, we investigate the effect on *F-score* [11], which is the harmonic mean between fitness and precision.

6.2   Case Study 1: Van Kasteren Event Log

For the first case study we use an event log from a smart home setting with in-house sensors [18]. This Van Kasteren log contains 1285 events divided over fourteen different sensors. The log contains 23 days of data. The average Damerau-Levenshtein similarity between (1) the ADL level traces that were predicted in the Leave One Trace Out Cross Validation (LOTOCV) experimental setup and (2) the actual ADL level traces is 0.7516, which shows that the estimated ADL level traces as produced by the approach are reasonably similar to the ground truth.

Figure 5 depicts the resulting workflow net obtained with the IM algorithm [23] from the sensor-level events. The workflow net starts with a choice between four activities: *hall-toilet door*, *hall-bedroom door*, *hall-bathroom door*, and *front-door*. After this choice the model branches into three parallel blocks, where the

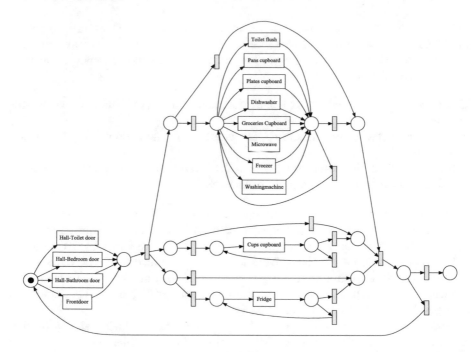

Fig. 5. Inductive Miner output on the sensor-level events of the Van Kasteren log

upper block consists of a large choice between eight different activities. The other two parallel blocks respectively contain a loop of the *cups cupboard* and the *fridge*. This model allows for many traces, and it has high similarity to the flower model. The data seems to be unstructured at the sensor level granularity events, or, at least the IM algorithm seems to be unable to find such structure.

Figure 6 depicts the resulting workflow net obtained by applying the IM algorithm on the aggregated set of predicted test sequences. The workflow net shows a clear daily routine, which starts with having *breakfast*, followed by *leaves the house*, presumably to go to work. When back from work, the person *prepares dinner* and *goes to bed*. Activities *use toilet* and *take shower* are put in parallel to this sequence of activities, indicating that they occur at different places in the sequences of activities.

Table 3 shows the effect of the abstraction on the fitness, precision, and F-score of the models discovered by the IM algorithm of the three case studies shown in this section and the following sections. It shows that the precision of the model discovered on the abstracted log is much higher than the precision of the model discovered on the sensor data, indicating that the abstraction helps discovering a model that is more behaviorally constrained and more specific. At the same time, the drop in fitness as a result of abstracting events is limited, indicating that the obtained models on the higher level still allow for most of the behavior seen in the log. The F-score values show that the trade-off between

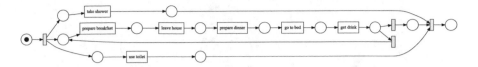

Fig. 6. Inductive Miner output on the human activity level Van Kasteren log

Table 3. Effect of abstraction on fitness, precision, and F-score of the process model discovered by the Inductive Miner

Event log	Abstraction	Fitness	Precision	F-score
Van Kasteren	No (Fig. 5)	0.9111	0.3308	0.4854
Van Kasteren	Yes (Fig. 6)	0.7918	0.7804	0.7861
MIT household A	No (Fig. 7)	0.9916	0.2289	0.3719
MIT household A	Yes (Fig. 8)	0.9880	0.3711	0.5395
MIT household B	No (Fig. 9)	1.0	0.2389	0.3857
MIT household B	Yes (Fig. 10)	0.9305	0.4319	0.5900

fitness and precision improves as a result of abstraction of events for all three smart home event logs.

6.3   Case Study 2: MIT Household A Event Log

For the second case study we use the data of *household A* of a smart home experiment conducted by MIT [35]. Household A contains data of 16 days of living, 2701 sensor-level events registered by 26 different sensors. The human level activities are provided in the form of a taxonomy of activities on three levels, called *heading*, *category* and *subcategory*. On the *heading* level the human activities are very general in nature, such as the activity *personal needs*. The eight different activities on the *heading* level branch into 19 different activities on the *category* level, where *personal needs* branches into e.g. *eating*, *sleeping*, and *personal hygiene*. The 19 *categories* are divided over 34 *subcategories*, which contain very specific human activities. At the *subcategory* level the *category meal cleanup* is for example divided into *washing dishes* and *putting away dishes*. At the *subcategory* level there are more types of human activities than there are sensors-level activities, which makes the abstraction task very hard. Therefore, we set the target label to the *category* level.

The model that is discovered with the IM algorithm on the sensor events in the MIT household A log is shown in Fig. 7. The model obtained allows for too much behavior, as it contains two large choice blocks. We found a Damerau-Levenshtein similarity of 0.6348 in the LOTOCV experiment. Note that the abstraction accuracy on this log is lower than the abstraction accuracy on the Van Kasteren event log. However, the MIT household A log contains more different types of human activity, resulting in a more difficult prediction task with

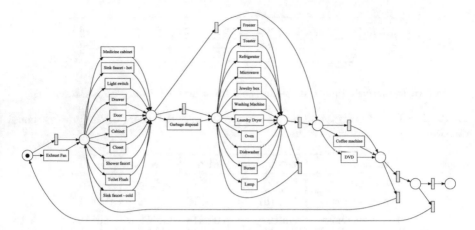

Fig. 7. The resulting process model of the Inductive Miner applied on the sensor-level MIT household A event log

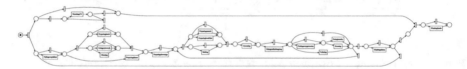

Fig. 8. The resulting process model of the Inductive Miner applied on the discovered human activity level events on the MIT household A log

a higher number of possible target classes. Figure 10 shows the process model discovered with the IM algorithm from the ADL level traces that we predicted from the sensor-level events. Even though the model is too large to print in a readable way, from its shape it is clear that the abstracted model is much more behaviorally constrained than the sensor-level model. The precision and fitness values in Table 3 show that indeed the process model after abstraction has become behaviorally more specific while the portion of behavior of the data that fits the process model remains more or less the same.

6.4 Case Study 3: MIT Household B Event Log

For the third case study we use the data of *household B* of the MIT smart home experiment [35]. Household B contains data of 17 days of living, 1962 sensor-level events registered by 20 different sensors. Identically to MIT household A the human-level activities are provided as a three-level taxonomy. Again, we use the *subcategory* level of this taxonomy as target activity label.

The model discovered with the IM algorithm [23] from the sensor events is shown in Fig. 9. The model obtained allows for too much behavior, as it contains two large choice blocks. We found a Damerau-Levenshtein similarity of 0.5865 in the LOTOCV experiment, which is lower than the similarity found on the

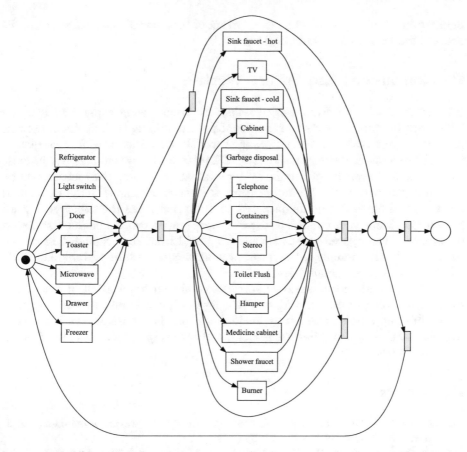

Fig. 9. The resulting process model of the Inductive Miner applied on the sensor-level MIT household B event log

Fig. 10. The resulting process model of the Inductive Miner applied on the discovered human activity level events on the MIT household B log

MIT A data set while the target classes of the abstraction are the same for the two data sets. This can be explained by the fact that there is less training data for this event log, as household B contains 1932 sensor-level events where household A contains 2701 sensor-level events. Figure 10 shows the process model discovered from abstracted log. Again this model is not readable due to its size, but its shape shows it to be behaviorally quite specific. The precision and fitness values in Table 3 also that process model after abstraction has indeed become behaviorally more specific while the portion of behavior of the data that fits the process model decreased only slightly.

7   Conclusions and Future Work

We presented a novel framework to abstract events using supervised learning which has been implemented in the ProM process mining toolkit. An important part of the framework is a generic way to extract useful features for abstraction from the extensions defined in the XES IEEE standard for event logs. We propose the Damerau-Levenshtein Similarity for evaluation of the abstraction results, and motivate why it fits the application of process mining. Finally, we showed on three real life smart home data sets that application of the supervised event abstraction framework enables us to mine more precise process model description of human life compared to what could be mined from the original data on the sensor-level. Additionally, these process models contain interpretable labels on the human behavior activity level.

In future work, we plan to explore the combined application of techniques for preprocessing of and mining on event logs of human behavior event logs on a real life smart home case study, including supervised abstraction, refinement of event labels based on timestamps [36], and discovery of local instead of global process models [37, 40].

References

1. van der Aalst, W.M.P.: The application of petri nets to workflow management. J. Circuits Syst. Comput. **8**(01), 21–66 (1998)
2. van der Aalst, W.M.P.: Process Mining: Data Science in Action. Springer (2016)
3. van der Aalst, W.M.P., Weijters, T., Maruster, L.: Workflow mining: discovering process models from event logs. IEEE Trans. Knowl. Data Eng. **16**(9), 1128–1142 (2004)
4. Baier, T., Mendling, J., Weske, M.: Bridging abstraction layers in process mining. Inf. Syst. **46**, 123–139 (2014)
5. Bao, L., Intille, S.S.: Activity recognition from user-annotated acceleration data. In: Pervasive Computing, pp. 1–17. Springer (2004)
6. Bose, R.P.J.C., van der Aalst, W.M.P.: Abstractions in process mining: a taxonomy of patterns. In: Proceedings of International Conference on Business Process Management, pp. 159–175. Springer (2009)
7. Chen, L., Nugent, C.: Ontology-based activity recognition in intelligent pervasive environments. Int. J. Web Inf. Syst. **5**(4), 410–430 (2009)
8. Cohen, T., Welling, M.: Harmonic exponential families on manifolds. In: Proceedings of 32nd International Conference on Machine Learning, JMLR Workshop and Conference Proceedings, pp. 1757–1765 (2015)
9. Conforti, R., Dumas, M., García-Bañuelos, L., La Rosa, M.: BPMN miner: automated discovery of BPMN process models with hierarchical structure. Inf. Syst. **56**, 284–303 (2016)
10. Damerau, F.J.: A technique for computer detection and correction of spelling errors. Commun. ACM **7**(3), 171–176 (1964)

11. De Weerdt, J., De Backer, M., Vanthienen, J., Baesens, B.: A robust F-measure for evaluating discovered process models. In: Proceedings of Symposium Series on Computational Intelligence (SSCI), pp. 148–155. IEEE (2011)
12. van Dongen, B.F., Adriansyah, A.: Process mining: fuzzy clustering and performance visualization. In: Proceedings of International Conference on Business Process Management Workshop, pp. 158–169. Springer (2010)
13. van Dongen, B.F., de Medeiros, A.K.A., Verbeek, H.M.W., Weijters, A.J.M.M., van der Aalst, W.M.P.: The ProM framework: a new era in process mining tool support. In: International Conference on Applications and Theory of Petri Nets, pp. 444–454. Springer (2005)
14. Friedman, N., Geiger, D., Goldszmidt, M.: Bayesian network classifiers. Mach. Learn. **29**(2–3), 131–163 (1997)
15. Günther, C.W., Rozinat, A., van der Aalst, W.M.P.: Activity mining by global trace segmentation. In: Proceedings of International Conference on Business Process Management Workshop, pp. 128–139. Springer (2010)
16. IEEE: IEEE standard for eXtensible Event Stream (XES) for achieving interoperability in event logs and event streams. IEEE Std 1849-2016, pp. 1–50. https://doi.org/10.1109/IEEESTD.2016.7740858 (2016)
17. van Kasteren, T., Kröse, B.: Bayesian activity recognition in residence for elders. In: Proceedings of 3rd IET International Conference on Intelligent Environments, pp. 209–212. IEEE (2007)
18. van Kasteren, T., Noulas, A., Englebienne, G., Kröse, B.: Accurate activity recognition in a home setting. In: Proceedings of 10th International Conference on Ubiquitous Computing, pp. 1–9. ACM (2008)
19. Katz, S.: Assessing self-maintenance: activities of daily living, mobility, and instrumental activities of daily living. J. Am. Geriatrics Soc. **31**(12), 721–727 (1983)
20. Kim, E., Helal, S., Cook, D.: Human activity recognition and pattern discovery. Pervasive Comput. **9**(1), 48–53 (2010)
21. Kwapisz, J.R., Weiss, G.M., Moore, S.A.: Activity recognition using cell phone accelerometers. ACM SIGKDD Explor. Newslett. **12**(2), 74–82 (2011)
22. Lafferty, J., McCallum, A., Pereira, F.C.N.: Conditional random fields: probabilistic models for segmenting and labeling sequence data. In: Proceedings of 18th International Conference on Machine Learning. Morgan Kaufmann (2001)
23. Leemans, S.J.J.: Robust process mining with guarantees. Ph.D. thesis, Eindhoven University of Technology (2017)
24. Leotta, F., Mecella, M., Mendling, J.: Applying process mining to smart spaces: perspectives and research challenges. In: International Conference on Advanced Information Systems Engineering, pp. 298–304. Springer International Publishing (2015)
25. Lohmann, N., Verbeek, E., Dijkman, R.: Petri net transformations for business processes—a survey. In: Transactions on Petri Nets and Other Models of Concurrency II, pp. 46–63. Springer (2009)
26. Mannhardt, F., Tax, N.: Unsupervised event abstraction using pattern abstraction and local process models. In: Proceedings of International Workshop on Business Process Modeling, Development and Support, CEUR (2017)
27. Mannhardt, F., de Leoni, M., Reijers, H.A., van der Aalst, W.M.P., Toussaint, P.J.: From low-level events to activities—a pattern-based approach. In: Proceedings of International Conference on Business Process Management, pp. 125–141. Springer (2016)

28. Munoz-Gama, J., Carmona, J.: A fresh look at precision in process conformance. In: Proceedings of International Conference on Business Process Management, pp. 211–226. Springer (2010)
29. Peterson, J.L.: Petri Net Theory and the Modeling of Systems. Prentice Hall PTR, Upper Saddle River, NJ, USA (1981)
30. Rabiner, L.R., Juang, B.H.: An introduction to hidden Markov models. ASSP Mag. **3**(1), 4–16 (1986)
31. Riboni, D., Bettini, C.: OWL 2 modeling and reasoning with complex human activities. Pervasive Mobile Comput. **7**(3), 379–395 (2011)
32. Rozinat, A., van der Aalst, W.M.P.: Conformance checking of processes based on monitoring real behavior. Inf. Syst. **33**(1), 64–95 (2008)
33. Schwarz, G.: Estimating the dimension of a model. Ann. Stat. **6**(2), 461–464 (1978)
34. Sztyler, T., Carmona, J., Völker, J., Stuckenschmidt, H.: Self-tracking reloaded: applying process mining to personalized health care from labeled sensor data. In: Transactions on Petri Nets and Other Models of Concurrency XI, pp. 160–180. Springer, Berlin, Heidelberg (2016)
35. Tapia, E.M., Intille, S.S., Larson, K.: Activity recognition in the home using simple and ubiquitous sensors. In: Pervasive Computing, pp. 158–175. Springer (2004)
36. Tax, N., Alasgarov, E., Sidorova, N., Haakma, R.: On generation of time-based label refinements. In: Proceedings of the 25th International Workshop on Concurrency, Specification and Programming, CEUR, pp. 25–36 (2016a)
37. Tax, N., Sidorova, N., van der Aalst, W.M.P., Haakma, R.: Heuristic approaches for generating local process models through log projections. In: Proceedings of Symposium Series on Computational Intelligence (SSCI), pp. 1–8. IEEE (2016b)
38. Tax, N., Sidorova, N., Haakma, R., van der Aalst, W.M.P.: Event abstraction for process mining using supervised learning techniques. In: Proceedings of the SAI Intelligent Systems Conference, pp. 161–170. Springer (2016c)
39. Tax, N., Sidorova, N., Haakma, R., van der Aalst, W.M.P.: Log-based evaluation of label splits for process models. Proc. Comput. Sci. **96**, 63–72 (2016d)
40. Tax, N., Sidorova, N., Haakma, R., van der Aalst, W.M.P.: Mining local process models. J. Innov. Digital Ecosyst. **3**(2), 183–196 (2016e)
41. Weijters, A.J.M.M., Ribeiro, J.T.S.: Flexible heuristics miner (FHM). In: IEEE Symposium Series on Computational Intelligence (SSCI), pp. 310–317. IEEE (2011)
42. Wen, L., van der Aalst, W.M.P., Wang, J., Sun, J.: Mining process models with non-free-choice constructs. Data Mining Knowl. Discov. **15**(2), 145–180 (2007)
43. van Zelst, S.J., van Dongen, B.F., van der Aalst, W.M.P.: Avoiding over-fitting in ILP-based process discovery. In: Proceedings of International Conference on Business Process Management, pp. 163–171. Springer International Publishing (2015)

A Novel Approach for Time Series Forecasting with Multiobjective Clonal Selection Optimization and Modeling

N. N. Astakhova[1(✉)], L. A. Demidova[1,2], and E. V. Nikulchev[2,3]

[1] Ryazan State Radio Engineering University, Gagarin Str., 59/1, Ryazan 390005, Russia
asnadya@yandex.ru, liliya.demidova@rambler.ru
[2] Moscow Technological Institute, Leninskiy pr., 38A, Moscow 119334, Russia
nikulchev@mail.ru
[3] Moscow Technological University MIREA, 78, Vernadskogo pr., Moscow 119454, Russia

Abstract. In this paper a novel approach for time series forecasting with multiobjective clonal selection optimization and modeling has been considered. At first, the main principals of the forecasting models (FM) on the base of the strictly binary trees (SBT) and the modified clonal selection algorithm (MCSA) have been discussed. Herewith, it is suggested, that the principles of the FMs on the base of the SBT can be applied to creation the multi-factor FMs, if we are aware of the presence of the several interrelated time series (TS). It will allow increasing the forecasting accuracy of the main factor (the forecasting TS) on the base of the additional information on the auxiliary factors (the auxiliary TS). Then, it is offered to develop the multiobjective MCSA (MMCSA) on the base of the notion of the "Pareto dominance", and use the affinity indicator (AI) based on the average forecasting error rate (*AFER*), and the tendencies discrepancy indicator (TDI) in the role of the objective functions in this algorithm. It will allow to improve the results of the solution of a problem of the short-term forecasting and to receive the adequate results of the middle-term forecasting. This MMCSA can be applied for solving problems of individual and group forecasting. Also, the application of the principles of the attractors' forming on the base of the long TSs to creation of the training data sequence (TDS) with the adequate length for the FM on the base of the SBT has been discussed. aBesides, the possibilities of the FMs on the base of the SBT and the MMCSA in the problem of the TS restoration with aim of the fractal dimension definition have been discussed. It is offered to carry out restoration of the TS elements' values as for the timepoints in the past as for the timepoints in the future simultaneously, using two FMs of the middle-term forecasting. The experimental results which confirm the efficiency of the offered novel approach for time series forecasting with multiobjective clonal selection optimization and modeling have been given.

© Springer International Publishing AG 2018
Y. Bi et al. (eds.), *Intelligent Systems and Applications*,
Studies in Computational Intelligence 751,
https://doi.org/10.1007/978-3-319-69266-1_6

1 Introduction

The problem of the right choice of the best forecasting model is the main problem of forecasting. In particular, in the FM on the base of the SBT and the basic MCSA this model is presented in the form of antibodies [1–3]. The MCSA allows generating the different variants of the forecasting models to find among them the best forecasting model according with the chosen quality indicator.

The MCSA possesses a number of benefits in comparison with analogic optimization algorithms. It allows weakening the requirements to data representation and establishing the optimum settings. Also, the MCSA provides high speed and high efficiency of calculations. This algorithm applies the evolutionary principles to modification and selection of antibodies constructed on the base of the SBT. As a result, it is possible to find the best antibody and the best forecasting model corresponding to this antibody. The MCSA is applicable for forecasting of TSs with a short length of an actual part which is characteristic for a large number of real processes and statistical indicators. It allows receiving the adequate results of the short-term forecasting.

Nowadays, the one-factor FM and multi-factor FM on the base of the SBT and the MCSA which use the elements' values of the TSs and the values of the elements' increments for creation of the training data sequence are developed by authors of this paper [1]. Also, we have offered the forecasting method for the groups of the TSs which uses the MCSA to create the general FM for each cluster of the TSs [2].

Usually, the *AFER*, calculated for the training data sequence, is used as the main quality indicator (QI) of the FM. Herewith, the *AFER* should be minimized [1, 2]. The MCSA steadily shows the low value of this QI for the TSs of various subject domains.

However, the use of the *AFER* as the single QI of the FM is not always sufficient to determine the best FM. Often it is required to consider the additional QI of the FM, for example, the compliance to the trend of the TS, the emissions' lack, the FM complexity, etc. [3]. But it isn't possible to choose the QI unambiguously. Therefore, the problem of the QI choice is very actually. It is expedient to use the additional QI which will allow estimating the general tendency of the values' change of the known elements of the TS (for example, the TDI) along with the *AFER* [3].

Now, the problem of the simultaneous accounting of the several QIs can be successfully solved using the multiobjective optimization algorithms, including, multiobjective evolutionary algorithms (MOEAs). Herewith, it is necessary to say about the multiobjective genetic algorithms (MOGA) [4–13] and the multiobjective clonal selection algorithms (MCSA) [14–20]. These algorithms allow taking into consideration the several objective functions (QI, criteria).

The analysis of advantages and shortcomings of the MOEAs shows that such the MOGA as the NSGA-II is significantly better than others because it can successfully solve more difficult multiobjective optimization problems [11–13]. Besides, the MOGAs have more successful development and application in the solution of many real difficult problems. Herewith, the MOCSAs borrows some ideas of the MOGAs. In this regard, the decision on expediency of the adaptation of the ideas put in the NSGA-II at the realization of the multiobjective MCSA (MMCSA) which is applied for the selection of the FMs on the base of the SBT had been made. In this case, it is necessary to understand the FM as the decision, and the QI of the FM as the objective function at the realization of the MMCSA. All FMs with use of the notion of "Pareto dominance" can be divided to dominated and non-dominated FMs. Obviously, it is possible to increase the efficiency of the FMs on the base of the SBT, using the multiobjective MCSA (MMCSA) at the solution of the problem of the middle-term forecasting. It is suggested to use in this algorithm the affinity indicator (AI) based on the *AFER* and the TDI in the role of the objective functions.

The FMs on the base of the SBT use the short time series (about 10–30 elements) as the TDS [1–3]. It is offered in [21] to use the attractor-based approach to find the adequate length of the TDS for creation of the FMs on the base of the SBT. As a result, it is possible along with the reduction of the time expenditures on the FMs creation to minimize the values of the forecasting errors thanks to refusal from the attempt to pick up a FM based on the original data sequence of the big length, which may fail due to the specifics of the applied mathematical tools.

Very often the TSs are so short that it is impossible to use them for solving many important problems, in particular, the problem of the fractal dimension definition [22, 23]. It is offered to carry out restoration of the values of the TS elements as for the timepoints in the past as for the timepoints in the future simultaneously, using two FMs of the middle-term forecasting. Use of such FMs will allow increasing the length of the TS on 6–10 elements and will provide the possibility of adequate application of the corresponding methods of the fractal dimension definition. As a result, it is possible to assess more precisely the state of these or those processes, forecast their future development, analyzing the flags of accidents, etc., and take the relevant managerial decisions. Nevertheless, the problem of the analysis of the extremely short TSs remains and requires the solution as by means of adaptation of the corresponding FMs, as by modification of the methods of the fractal dimension definition.

The rest of this paper is structured as follows. Section 2 describes the main principles of the FMs on the base of the SBT and the MCSA. Section 3 devoted to forecasting for the grouped time series. Section 4 details the attractor-based approach to the choice of the TDS for the FM on the base of the SBT. Section 5 describes the possibilities of the FMs on the base of the SBT and the MMCSA in the problem of the TS restoration with aim of the fractal dimension. Also, the improved method of the fractal dimension definition has been considered. The experimental results which clearly highlight the efficiency of a novel approach for time series forecasting with multiobjective clonal selection optimization and modeling follow in Sect. 5. Finally, conclusions are drawn in Sect. 6.

2   Forecasting Models on the Base of the Strictly Binary Trees and the Modified Clonal Selection Algorithm

The principles of the forecasting models on the base of the one-objective MCSA (the basic MCSA) were investigated in [1]. The main ideas of the multiobjective MCSA are described in [2]. The MCSA allows forming an analytical dependence on the base of the SBT at an acceptable time expenses. This analytical dependence in the one-objective MCSA describes the TS values and provides a minimum value of the *AFER*:

$$AFER = \frac{1}{n-k} \sum_{j=k+1}^{n} \left| \frac{f^j - d^j}{d^j} \right| \cdot 100\%, \tag{1}$$

where d^j and f^j are respectively the actual (fact) and forecasted values for the j-th element of the TS; n is the number of TS elements; k is the model order. Also the *AFER* (1) can be named as the affinity indicator *Aff* [3].

The multiobjective MCSA (MMCSA) uses two QIs: the AI *Aff* and the TDI *Tendency*:

$$Tendency = \frac{h}{n-k-1}, \tag{2}$$

where h is the number of negative multiplications $(f^{j-1} - f^j) \cdot (d^{j-1} - d^j)$; $j = \overline{k+2, n}$; d^j and f^j are respectively the actual (fact) and forecasted values for the j-th element of TS; n is the number of TS elements; k is the model order; $n - k - 1$ is the total number of multiplications $(f^{j-1} - f^j) \cdot (d^{j-1} - d^j)$.

This QI describes the rate of discrepancy between the tendencies of two TSs (the real TS and the model TS). It allows adapting the FMs on the base of the SBT and the MCSA for the middle-term forecasting.

Both QI (1) and (2) must be minimized for receiving the required FM.

2.1   Forecasting Models on the Base of the Strictly Binary Trees

Possible options for analytical dependences are presented in the form of antibodies *Ab* which recognize antigens *Ag* (certain TS values). An antibody *Ab* is selected as "the best one". It provides the minimum value of the AI *Aff* [1]. Coding of an antibody *Ab* is carried out by recording signs in a line. The signs are selected from three alphabets [1–3]: the alphabet of arithmetic operations (addition, subtraction, multiplication and division) – *Operation* = {'+', '−', '•', '/'}; the functional alphabet, where letters 'S', 'CH', 'Q', 'L', 'E' define mathematical functions "sine", "cosine", "square root", "natural logarithm", "exhibitor", and the sign '_' means the absence of any mathematical function, – *Functional* = {'S', 'C', 'Q', 'L', 'E', '_'}; the alphabet of terminals, where letters 'a', 'b', ..., 'z' define the arguments required analytical dependence and the sign '?' defines a constant, *Terminal* = {'a', 'b', ..., 'z', '?'}. The use of these three signs alphabets provides a correct conversion of randomly generated antibodies into the analytical dependence. The structure of such antibodies can be described with the help of the SBT [1]. Figure 1 shows the example of the SBT, which is created on the base of these three signs alphabets.

It is possible to use such SBT to create as the one-factor FM as the multi-factor FM (for example, two-factor FM). In the one-factor FM the forecasting TS with the real (fact) values d^j of the j-th element $(j = \overline{1, n})$ is the main factor (the forecasting factor). In this case the number of signs in the alphabet of terminals in the antibody *Ab* determines the maximal possible order K of the FM with $K \geq k$, where k is the real model order), i.e. having the value of the element d^j in the forecasting TS at the j-th moment of time, K values of TS elements can be used as: $d^{j-K}, \ldots, d^{j-2}, d^{j-1}$ [1–3].

In the two-factor FM the main factor is forecasted on the base of the elements' values of two factors (two TSs) such as the main factor and the auxiliary factor. The main factor corresponds to the forecasting TS with the real (fact) values d_1^j of the j-th element $(j = \overline{1, n})$. The auxiliary factor corresponds to the auxiliary TS with the real (fact) values d_2^j of the j-th element $(j = \overline{1, n})$. In this case the number of signs in the alphabet of terminals has to be halved. Therefore, it is more convenient that this number was even. Then, the maximal possible order of the FM is equals to the half of the number K of elements in the alphabet of terminals: $K/2$. Hence, it is necessary to work with two TSs with the following elements: $d_1^{j-K/2}, \ldots, d_1^{j-2}, d_1^{j-1}$, and $d_2^{j-K/2}$, $\ldots, d_2^{j-2}, d_2^{j-1}$ to make the forecast to the j-th element d_1^j $(j = \overline{1, n})$ of the main TS. The real order of this FM is equal to the maximal real number of elements of the main TS and the auxiliary TS in the code of antibody.

The ideas mentioned above can be applied to create the multi-factor FMs with the much bigger number of factors. But these models will have very long trees.

The use of the SBT, illustrated in Fig. 1, allows building the complex analytical dependence and provides high accuracy of the TS forecasting [1–3].

For example, the antibody formed on the base of the SBT as shown in Fig. 1 is coded by the line of signs: $L \cdot S/SeSdC - S + EcCbEa$, which can be transformed into an analytical dependence:

$$f(a, b, c, d) = \ln\left[\cos(\sin(\exp(a) + \cos(b)) - \exp(c)) \cdot \sin\left(\frac{\sin(d)}{\sin(a)}\right)\right].$$

For the one-factor forecasting model with $k = 4$ this analytical dependence can be written as:

$$f(d^{j-1}, d^{j-2}, d^{j-3}, d^{j-4}) = \ln\left[\cos(\sin(e^{d^{j-1}} + \cos(d^{j-2})) - e^{d^{j-3}}) \cdot \sin\left(\frac{\sin(d^{j-4})}{\sin(d^{j-1})}\right)\right].$$

For the two-factor forecasting model with $k = 2$ this analytical dependence can be written as:

$$f(d_1^{j-1}, d_1^{j-2}, d_2^{j-1}, d_2^{j-2}) = \ln\left[\cos(\sin(e^{d_1^{j-1}} + \cos(d_1^{j-2})) - e^{d_2^{j-1}}) \cdot \sin\left(\frac{\sin(d_2^{j-2})}{\sin(d_1^{j-1})}\right)\right].$$

The FM on the base of the SBT can be applied for the individual and groups' forecasting of the TSs [2].

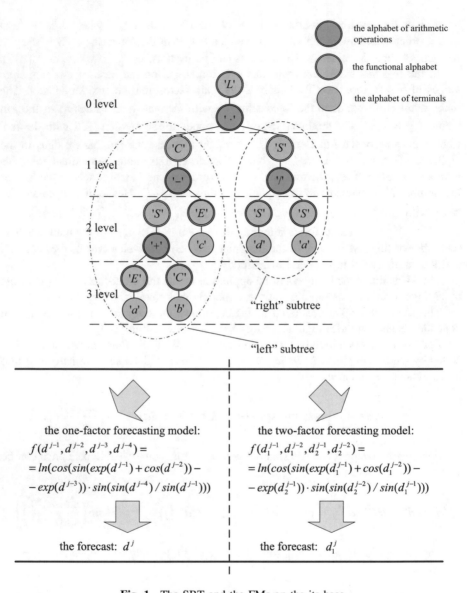

Fig. 1. The SBT and the FMs on the its base

2.2 The Modified Clonal Selection Algorithm

The MCSA based on the application of one QI (the AI *Aff*) is used to the searching for "the best" antibody defining "the best" analytic dependence includes the preparatory part (realizes the formation of the initial antibody population) and iterative part (presupposes the ascending antibodies ordering of affinity *Aff* the selection and cloning the part of "the best" antibodies, that are characterized by the least affine value *Aff* of the hyper mutation of the antibodies clones; self-destruction of the antibodies clones "similar" to the other clones and antibodies of the current population; calculating the affinity of the antibodies clones and forming the new antibodies population; suppression of the population received; generation of the new antibodies and adding them to the current population until the ingoing size; the conditional test of the MCSA completion [1, 2].

The FMs on the base of the SBT and the MCSA have well proven themselves in the short-term forecasting [1, 2].

2.3   The Multiobjective Optimization

The main common features of the MOEAs are the following: (i) assigning fitness to the population members based on the non-dominated sorting and (ii) preserving the diversity among the solutions of the same non-dominated front.

The multiobjective optimization problem in case of the FMs on the base of the SBT is very urgent. Two QIs (the AI *Aff* and the TDI *Tendency*) must be used simultaneously at the quality assessment of these FMs to solve the problem of the middle-term forecasting [3]. Herewith, both QIs must be minimized (the smaller the values of the QIs, the better the FM).

The rank R_s must be calculated for every s-th FM ($s = \overline{1, S}$). The rank R_s is equal to the number of the FMs in the population which dominate over the s-th FM. The rank R_s of the s-th non-dominated FM is equal to zero [11, 12].

Let $V = 2$, $Q_{s,1} = Aff_s$, $Q_{s,2} = Tendency_s$, where Aff_s and $Tendency_s$ are the values of the AI (1) and the TDI (2) for the s-th FM ($s = \overline{1, S}$) accordingly. Then the s-th FM is dominated by the z-th FM ($s = \overline{1, S}$); $z = \overline{1, S}$); if the following conditions are satisfied:

$$(Q_{s,1} \geq Q_{z,1} \text{ and } Q_{s,2} > Q_{z,2}) \text{ or } (Q_{s,1} > Q_{z,1} \text{ and } Q_{s,2} \geq Q_{z,2}),$$

that is

$$(Aff_s \geq Aff_z \text{ and } Tendency_s > Tendency_z) \text{ or } (Aff_s > Aff_z \text{ and } Tendency_s \geq Tendency_z).$$

The crowing distance (CD) τ_s ($s = \overline{1, S}$) can be calculated using the following algorithm [11, 12].

Step 1   To calculate ranks for all FMs in the population. To unite the FMs with identical values of the rank into one group.

Step 2   For every group of the FMs:

- to sort the FMs according to each QI value in ascending order of magnitude;
- to assign infinite distance to boundary values of the FMs in the group, i.e. $\tau_1 = \infty$ and $\tau_{G_w} = \infty$, where G_w is the number of the FMs in the w-th group ($w = \overline{1, W}$); W is the groups' number; to assign $\tau_s = 0$ for $s = \overline{2, G_w - 1}$;
- to calculate the CD τ_s as:

$$\tau_s = \sum_{v=1}^{V} \frac{Q_{s-1,v} - Q_{s+1,v}}{Q_v^{\max} - Q_v^{\min}}, \qquad (3)$$

where $Q_{s-1,v}$ and $Q_{s+1,v}$ are the values of the v-th QI ($v = \overline{1,V}$) for the FMs with the numbers $(s-1)$ and $(s+1)$, which are the nearest "neighbors" for the s-th FM; $Q_v^{\min}$ and $Q_v^{\max}$ are the minimum and maximum values of the v-th QI ($v = \overline{1,V}$) accordingly.

Figure 2 shows, how we can calculate the CD on the base of two QIs. The points, marked with solid circles, correspond to the FMs with the minimum (zero) value of the rank (i.e. these points correspond to the Pareto front with the zero rank).

To calculate the CD for the s-th FM it is required to define values of both QIs for the $(s-1)$-th and the $(s+1)$-th FMs, which are the nearest "neighbors" for the s-th FM and have the same rank (Fig. 2). Also, it is necessary to define the best and worst values of each QI. The CD τ_s ($s = \overline{1,S}$) for the s-th FM on the base of two QIs can be calculated as:

$$\tau_s = \frac{Q_{s-1,1} - Q_{s+1,1}}{Q_1^{\max} - Q_1^{\min}} + \frac{Q_{s-1,2} - Q_{s+1,2}}{Q_2^{\max} - Q_2^{\min}}. \qquad (4)$$

At the realization of the MMCSA the s-th FM will be better than the z-th FM, if: $(R_s < R_z)$ or $(R_s = R_z$ and $\tau_s > \tau_z)$. If the s-th FM is better, than the z-th FM, the s-th FM is the candidate for transfer into the new generation.

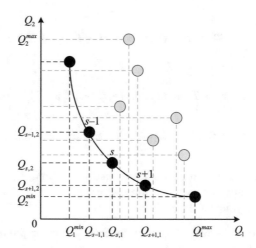

Fig. 2. The points using for calculation of the CD

The Pareto set of the non-dominated FMs will be received as a result of performance of the offered MMCSA. The received FMs can be applied for the middle-term forecasting.

3   Forecasting for the Grouped Time Series

It is possible to apply the FM on the base of the SBT and MCSA as the general FM for describing the clusters' centroids, when we must to make forecast for the grouped TSs. In this case, we must, firstly, cluster the group of the TSs by means of some clustering algorithm (for example, by means of the k-means algorithm or the FCM algorithm), and, secondly, to create the general FM for each centroid of cluster. Such general FM can be applied to forecast every private TS in the corresponding cluster [2]. Herewith, the general FM can be specified for some individual TS during the additional iterations of the MCSA (MMCSA).

It is necessary to say, that we offer to modify the Euclidean metric, which is usually used in the mentioned above clustering algorithms. In the context of solving the problem of the TS clustering on the base of similarity of the mathematical laws of the TSs values changing, it is suggested to do modification to perform this metric so as to take into account various relevance of the TS elements (most distant in time from the time of forecasting) and larger relevance of other TS elements (closest in time to the moment of forecasting) [2]:

$$
dist(v_r, d_i) = \left[\sum_{j=1}^{n} \frac{j}{n} \cdot \left(v_r^j - d_i^j \right)^2 \right]^{0,5},
\tag{5}
$$

where v_r is the r-th cluster center with the coordinates v_r^j and d_i is the i-th TS with the elements d_i^j; $i = \overline{1, m}$; $j = \overline{1, n}$; m is the number of the TSs; n is the number of the TS elements.

The utility of such a modification can be explained so that with the passage of time the dependencies between socio-economic indicators change. So, during the forming of the FMs, larger preference should be given to the closest TS elements in time of forecasting. The use of the weighting coefficients j/n suggests the most significant differences between the values of the most relevant TS elements (for example, if $j = n$ then the weight value is 1, and if $j = 1$ then it is $1/n$). The use of formula (5) will not only take into account the relevance of the TS elements, but also in a view of high sensitivity to the divergent TS trends, it will provide unification of clusters on the base of the trends' similarity. The method of the groups' forecasting with the use of the FMs on the base of the SBT is described in [2].

4   Attractors-Based Forecasting Models

The attractor is a set of numerical values toward which a system tends to evolve, for a wide variety of starting conditions of the system. The attractor can be applied for the forecasting problems' solution [24–27].

One of approaches to identification of the phase curve with aim of the attractors' detection uses the difference equations and assumes that the element's value d^j ($j = \overline{2, n}$) of the analyzed TS is postponed on the abscissa axis in each concrete timepoint, and the corresponding chain pure gain:

$$\Delta d^j = d^j - d^{j-1} \; (j = \overline{2, n}) \tag{6}$$

is postponed on the ordinate axis.

The attractors can be used to form reasonably the TDS for the FM on the base of the SBT. In this case, it is necessary to analyze the last attractor found in data.

The technique of the attractors' use for development of the FMs on the base of the SBT is described in details in [21]. This technique can be applied as for the individual forecasting as for the group's one.

5   Forecasting Models on the Base of the Strictly Binary Trees and the Fractal Dimension of Time Series

The FM on the base of the SBT and MMCSA can be used for the length "restoration" of the short TS, namely for "accumulation" of the values of its elements with the aim of application of methods of the fractal dimension definition of the TS. Herewith, it is offered to carry out "accumulation" of the TS elements' values as for the timepoints in the past (that is to build the retrospective forecast) as for the timepoints in the future (that is to build the perspective forecast). As a result, it will be possible to assess more precisely the state of various processes, forecast their future development, analyzing the flags of accidents, etc., and take the relevant managerial decisions.

As the analysis shows, the use of the FMs on the base of the SBT and MMCSA allows obtaining the acceptable results of the medium-term forecasting (that is the forecasting results for 3–5 steps forward).

Therefore, if we are able to realize the forecasting for 5 steps forward in the future and for 5 steps backwards in the past, it will be possible to increase the number of the values of the TS elements on 10.

For example, if we possess the initial TS of 15 elements as a result we will be able to receive the new TS of 25 elements to which it is already possible to apply this or that method of the fractal dimension definition of the TS intended for the analysis of the short TS.

Herewith, it will be necessary to construct 2 FMs on the base of the SBT and MMCSA: the first FM will be constructed on the base of the known TS (without any additional manipulations with the TS), and for creation of the second FM it will be necessary "to turn" the initial TS, that is the first element of the TS must become the last, the second element of the TS must become the last penultimate, etc.

The new TS will contain in the structure 5 values of the retrospective forecast, the initial TS and 5 values of the perspective forecast (in the specified sequence of elements).

The example of the restored TS "Gold prices in Russian Rubles" (from 16.02.2016 to 29.03.2016) is shown in Fig. 3. The real data are represented in black, the retrospective forecast is represented in blue, the perspective forecast is represented in red. The retrospective and the perspective forecasts were received by application of two FM on the base of the SBT.

Further, the method of the fractal dimension definition of the TS, which is offered in the work [22, 23], can be applied to this TS. In our researches the improved version of this method in which the additional steps to provide more exact conclusions when calculating the fractal dimension of the random TS of the same length of n are taken (where n is the number of elements of the restored TS, for example, equal to 25 (5 + 15 + 5 = 25)) has been applied.

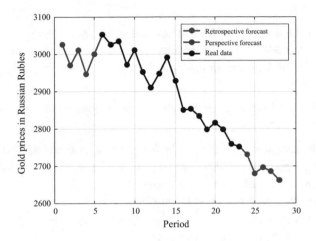

Fig. 3. The example of the restored TS

The length L_0 of the broken line of the TS can be found as the sum of the lengths l_j^0 of the j-th sections $(j = \overline{1, J_0})$, that make up it:

$$L_0 = \sum_{j=1}^{J_0} l_0^j, \tag{7}$$

where $l_0^j = \sqrt{\left(d_0^j - d_0^{j+1}\right)^2 + \left(t_0^j - t_0^{j+1}\right)^2}$, $d_0^j = d^j$ is the value of the current element of the TS; $d_0^{j+1} = d^{j+1}$ is the value of the next element of the TS; $t_0^j = t^j$ is the current value of the time scale axis; $t_0^{j+1} = t^{j+1}$ is the next value of the time scale axis.

In general, the length L_q of the q-th $(q = \overline{0, Q})$ averaged broken line of the TS can be found as the sum of the lengths l_q^j of the j-th sections $(j = \overline{1, J_q})$, that make up it:

$$L_q = \sum_{j=1}^{J_q} l_q^j, \tag{8}$$

where $l_q^j = \sqrt{\left(d_q^j - d_q^{j+1}\right)^2 + \left(t_q^j - t_q^{j+1}\right)^2}$, d_q^j, d_q^{j+1} are the consecutive average values calculated on the base of the elements' values of the TS, which formulas for calculation are various for various values q and will be defined below; t_q^j, t_q^{j+1} are the consecutive values of the time scale axis.

At realization of this method of the fractal dimension definition it is necessary to calculate 7 lengths of the average broken lines (sometimes it is enough to consider 5 lengths of the average broken lines). Herewith all of them will be more smoothed in comparison with the broken line with the length L_0.

The method of the fractal dimension definition can be described by the following sequence of steps.

1. To calculate the length L_q of the average broken line according to the formula (8). It is necessary to consider the consistently numbers from 0 to 6 as the number q (it is possible to be limited also to numbers from 0 to 4), that is $q = \overline{0, Q}$, where Q is equal to 6 or 4.

The length L_q of the average broken line at various values of the number q is calculated as follows:

- if $q = 1$ the length of the average broken line L_1 is calculated according to the formula (8), where $d_1^j = \frac{d^j + d^{j+1}}{2}$; $d_1^{j+1} = \frac{d^{j+1} + d^{j+2}}{2}$; $d_1^{j+2} = \frac{d^{j+2} + d^{j+3}}{2}$, etc. (that is the averaging is carried out for two points);
- if $q = 2$ the length of the average broken line L_2 is calculated according to the formula (8), where $d_2^j = \frac{d^j + d^{j+1}}{2}$; $d_2^{j+1} = \frac{d^{j+2} + d^{j+3}}{2}$; $d_2^{j+2} = \frac{d^{j+4} + d^{j+5}}{2}$, etc. (that is the averaging is carried out for two points);
- if $q = 3$ the length of the average broken line L_3 is calculated according to the formula (8), where $d_3^j = \frac{d^{j+1} + d^{j+2}}{2}$; $d_3^{j+1} = \frac{d^{j+3} + d^{j+4}}{2}$; $d_3^{j+2} = \frac{d^{j+5} + d^{j+6}}{2}$, etc. (that is the averaging is carried out for two points);
- if $q = 4$ the length of the average broken line L_4 is calculated according to the formula   (8),   where $d_4^j = \frac{d^j + d^{j+1} + d^{j+2}}{3}$; $d_4^{j+1} = \frac{d^{j+3} + d^{j+4} + d^{j+5}}{3}$; $d_4^{j+2} = \frac{d^{j+6} + d^{j+7} + d^{j+8}}{3}$, etc. (that is the averaging is carried out for three points);
- if $q = 5$ the length of the average broken line L_5 is calculated according to the formula   (8),   where $d_5^j = \frac{d^{j+1} + d^{j+2} + d^{j+3}}{3}$; $d_5^{j+1} = \frac{d^{j+4} + d^{j+5} + d^{j+6}}{3}$; $d_5^{j+1} = \frac{d^{j+7} + d^{j+8} + d^{j+9}}{3}$, etc. (that is the averaging is carried out for three points);
- if $q = 6$ the length of the average broken line L_6 is calculated according to the formula   (8),   where $d_6^j = \frac{d^{j+2} + d^{j+3} + d^{j+4}}{3}$; $d_6^{j+1} = \frac{d^{j+5} + d^{j+6} + d^{j+7}}{3}$; $d_6^{j+2} = \frac{d^{j+8} + d^{j+9} + d^{j+10}}{3}$, etc. (that is the averaging is carried out for three points).

2. To calculate the natural logarithms of integers from 1 to $Q + 1$ (that is numbers from 1 to 7 or from 1 to 5 (that is depends on the value of Q) and natural logarithms of the series of the average broken lines lengths L_q ($q = \overline{0, Q}$). To execute the linear approximation of the $(\ln(q + 1), \ln(L_q))$ points and to define the tangent of the inclination angle α of the straight line. To calculate the value $FD = 2 - tg(\alpha)$, which represents the assessment of the fractal dimension of the TS.

3. To execute the correction of the number FD calculated for the TS taking into account that theoretical value of the fractal dimension of the stochastic time series is equal to 1.5.

3.1 To create in a random way the TS of the same length as the analyzed TS, calculating its elements in accordance with the formula:

$$\tilde{d}^j = rnd \cdot \left(\max_j (d^j) - \min_j (d^j) \right) + \min_j (d^j),$$

where $j = \overline{1, J_0 + 1}$ (herewith, the numerical values $\tilde{d}^j$, simulating the elements' values of the TS in a random way will keep within the interval $\left[\min_j (d^j), \max_j d^j \right]$; rnd is the uniform distribution sensor in [0, 1].

3.2 To calculate the lengths of the average broken line $L_q^{stochastic}$ $(q = \overline{0, Q})$ for the stochastic TS according to the formula (8).

3.3 To fulfill the items 3.1 and 3.2 repeatedly, that is (G time), (for example, 5000 times), summarizing the lengths $L_q^{stochastic}$ of the average broken lines: $S_q = S_q + L_q^{stochastic}$, where the initial value of S_q is equals to 0 $(S_q = 0, q = \overline{0, Q})$.

3.4 To calculate the value: $averageL_q^{stochastic} = S_q/G$ $(q = \overline{0, Q})$.

3.5 To calculate the natural logarithms of integers from 1 to $Q + 1$ (that is numbers from 1 to 7 or from 1 to 5 (that is depends on the value of Q) and natural logarithms of the series of the average broken lines lengths $averageL_q^{stochastic}$ $(q = \overline{0, Q})$. To execute the linear approximation of the $\left(\ln(q + 1), \ln \left(averageL_q^{stochastic} \right) \right)$, points and to define the tangent of the inclination angle $\alpha^{stochastic}$ of the straight line. To calculate the value $FD^{stochastic} = 2 - tg \left(\alpha^{stochastic} \right)$, which represents the assessment of the fractal dimension of the stochastic TS.

3.6 To calculate the deviation value of $FD^{stochastic}$ from theoretical value of the fractal dimension of the stochastic time series: $\Delta FD^{stochastic} = FD^{stochastic} - 1.5$, which will differ from zero owing to errors of the measurements and the heuristic of the formula (9) for the assessment of the average length of the broken line;

3.7 To carry out the correction of the number FD calculated for the TS at the size $\Delta FD^{stochastic}$ (with the accounting of a sign of $\Delta FD^{stochastic}$). The obtained number will be the assessment of the fractal dimension of the analyzed TS.

The improvement of the basic method of the fractal dimension definition takes place in items 3.1–3.4, where we calculate the average lengths of the average broken lines.

6 Experimental Studies

A novel approach for time series forecasting with multiobjective clonal selection optimization and modeling assumes the accomplishment of the following steps: to reveal the problem type (individual or groups' forecasting); to analyze the length of the

TSs and apply the technique of the attractors' forming, if it is necessary; to apply the multiobjective optimization to the FMs on the base of the SBT.

The approbation of the offered approach for time series forecasting with multiobjective clonal selection optimization and modeling was executed with the use of the several TSs (in particular, the socio-economic indexes) with the different length, the various laws of behavior, etc. Therefore, we can consider the received conclusions as the generalizing ones.

At first, we have examined the TS "The number of unemployed officially registered" in Russia (in millions of people) with data from 01-2004 to 06-2005 (www. mzsrrf.ru): T1 = [1.64; 1.67; 1.66; 1.65; 1.59; 1.56; 1.57; 1.59; 1.63; 1.71; 1.84; 1.92; 1.94; 2.02; 2.00; 1.97; 1.88; 1.82]. We have used this TS to develop the one-factor FM and the two-factor one for the short-term forecasting for 3 steps forward. In the last case we have considered this TS as the main factor. Herewith, the TS "The number of unemployed by the technology of the International Labour Organization (ILO)" in Russia (in millions of people) with the same data period was used as the auxiliary factor: T2 = [6.62; 6.92; 6.48; 6.03; 5.58; 5.53; 5.47; 5.42; 5.67; 5.90; 6.14; 6.11; 6.08; 6.00; 5.76; 5.64; 5.55]. It is obvious, that these TSs are interrelated. The length of these TSs is equal to 18 elements. We have used the first 15 elements of TSs to create the FMs on the base of the SBT and two variants of optimization (with the basic MCSA and the MMCSA), and the last 3 elements of the TSs—to forecast.

Herewith, we have obtained the 6-order one-factor FM:

$$f = \ln\left(\sin(f_1 \cdot e^{d^{i-5}}) + e^{d^{i-1}}\right),$$

where

$$f_1 = \cos\left(\frac{\cos\left(\sin\left(\sin\left(\sqrt{d^{i-4}} - e^{d^{i-1}}\right) - \sin(d^{i-6})\right) \cdot \ln(d^{i-3})\right)}{\sin(d^{i-2})}\right),$$

and the 5-order two-factor FM:

$$f = \ln\left(f_2 + \ln(d_2^{t-1}) - \sqrt{d_1^{t-3}} + e^{d_1^{t-1}}\right),$$

where

$$f_2 = \ln\left((\exp\{e^{f_1} \cdot \sin(d_2^{t-4})\} \cdot \ln(d_2^{t-2})) \cdot d_2^{t-4} + \ln(d_1^{t-3})\right) \cdot \cos(d_1^{t-5}),$$

$$f_1 = \sin\left(\exp\left\{\frac{(e^{d_1^{t-2}} - \sin(d_2^{t-3})) \cdot d_1^{t-4}}{\ln(d_1^{t-4})}\right\} - e^{d_2^{t-2}}\right) \cdot \sqrt{d_2^{t-5}}.$$

The average value of the affinity indicator *Aff* in the first and second cases have constituted 0.852% and 0.693% respectively. The average value of the forecasting error on 3 steps forward in the first and second cases have constituted 2.415% and 0.298%

respectively. In case of the MMCSA we have received the same forecasting results as on 3 steps forward as for 5 steps forward. It can be explained with the fact that for this example the FMs on the base of the basic MCSA already have a small value of the TDI *Tendency*, which failed to reduce using the MMCSA. Figure 4 shows the obtained forecasting results with the basic MCSA.

Then, we have examined the group of the TSs for 19 macroeconomic indicators of Russian Federation taken from the site World DataBank from 1999 to 2014 (http://databank.worldbank.org/data/views/reports/tableview.aspx?isshared=true#). Data from 1999 to 2009 were used for creation of the FMs. Data from 2010 to 2014 were used for forecasting of the private TSs on 5 steps forward. All indicators were divided into 6 clusters (subgroups) with the use of the k-means algorithm. Information on the clusters' contents is given in the second column of Table 1.

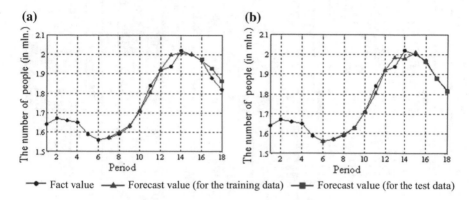

(a)   **(b)**

—●— Fact value —▲— Forecast value (for the training data) —■— Forecast value (for the test data)

Fig. 4. The forecasting results: **a** The one-factor FM; **b** The two-factor FM

For each cluster of the TSs the general FM on the base of one and two QIss have been constructed. Herewith, by results of forecasting for 5 steps forward the average forecasting error *Error* was calculated as:

$$Error = \frac{100\%}{5} \cdot \sum_{j=n+1}^{n+5} \left| \frac{f^j - d^j}{d^j} \right|. \tag{9}$$

In this formula $n = 11$ and the values f^j and d^j with $j = 12, j = 13, j = 14, j = 15$ and $j = 16$ correspond to forecasting and real elements' values of the TSs for 2010, 2011, 2012, 2013 and 2014 years accordingly.

The averaged values of the relative forecasting errors at the 5 steps, the averaged values of the AI and the averaged values of the TDI received by the results of 10 runs of the MCSA for each TS are presented in Table 2.

Tables 1 and 2 show the values of *Aff* (*AFER*) and *Tendency* for data from 1999 to 2009; the values of *Error* for data from 2010 to 2014; the average values of *Aff* (*AFER*), *Tendency* and *Error* for every cluster and for all clusters in general according to results of forecasting on the base of optimization with one and two QIs correspondingly.

Table 1. The forecasting results on the base of one QI

The 1-st cluster

No	Indicator name	Measurement unit	AFER, %	Steps					Error	Tendency
				1	2	3	4	5		
1	The expected life expectancy at the birth	Years' quantity	1.46	5.11	3.42	4.47	4.48	5.11	4.52	0.3
2	The improved sanitation means	%	2.45	0.38	0.92	7	7.03	7.03	4.47	0.3
3	Value added in a services sector	% of GDP[a]	0.28	6.34	2.18	3.99	0.64	2.25	30.8	0.5
	The average value of the cluster		1.40	3.94	2.17	5.16	4.05	4.80	4.02	0.37

The 2-nd cluster

No	Indicator name	Measurement unit	AFER, %	Steps					Error	Tendency
				1	2	3	4	5		
4	Coefficient of teenage fertility	Births' quantity/1000 women	0.8	1.36	0.74	2.49	5.1	7.56	3.45	0.1
5	Export of goods and services	% of GDP[a]	1.35	4.19	9.03	1.86	11.5	0.97	5.51	0.2
6	Income (except for grants)	% of GDP[a]	1.32	4.46	2.91	1.55	5.17	4.51	3.72	0.3
7	Mortality aged till 5 years	%	1.43	5.14	2.52	4.04	1.48	0.59	2.75	0.2
8	Value added in the industry	% of GDP[a]	1.17	7.14	6.96	7.91	2.82	2.46	5.46	0.5
	The average value of the cluster		1.21	4.46	4.43	3.57	5.21	3.22	4.18	0.26

The 3-rd cluster

No	Indicator name	Measurement unit	AFER, %	Steps					Error	Tendency
				1	2	3	4	5		
9	Fertility rate	Births' quantity/1000 women	2.81	0.63	5.63	10.93	4.63	5.79	5.52	0.1
10	Military expenditure	% of GDP[a]	1.08	5.14	2	4.16	2.28	5.79	3,87	0.5
	The average value of the cluster		1.95	2.88	3.81	7.54	3.45	5.79	4.70	0.30

The 4-th cluster

No	Indicator name	Measurement unit	AFER, %	Steps					Error	Tendency
				1	2	3	4	5		
11	Export of high technologies	%	1.11	4.25	8.33	6.53	6.15	3.5	5.75	0.56
12	Procedures' start of for business registration	Quantity	0.29	0	0	0.63	14.29	8.06	4.59	0
	The average value of the cluster		0.70	2.13	4.17	3.58	10.22	5.78	5.17	0.28

(continued)

Table 1. (*continued*)

The 5-th cluster

No	Indicator name	Measurement unit	AFER, %	Steps					Error	Tendency
				1	2	3	4	5		

The 5-th cluster

No	Indicator name	Measurement unit	AFER, %	Steps					Error	Tendency
				1	2	3	4	5		
13	Gross accumulation of capital	% of GDP[a]	1.57	1.08	0.98	0.48	0.48	1.01	0.81	0.2
14	Import of goods and services	% of GDP[a]	0.47	4.24	5.56	4.61	5.21	4.66	4.86	0.45
	The average value of the cluster		1.02	2.66	3.27	2.55	2.85	2.83	2.83	0.33

The 6-th cluster

No	Indicator name	Measurement unit	AFER, %	Steps					Error	Tendency
				1	2	3	4	5		
15	Immunization against measles	%	2.01	4.08	4.08	3.77	3.77	3.11	3.76	0
16	Quantity of childbirth by means of qualified medical personnel	%	0.01	0	0	0.16	0.21	0.28	0.13	0
17	Ratio of girls and boys in system of primary and secondary education	%	0.05	0.2	0.3	0.43	0.01	1.02	0.39	0.4
18	The improved water sources	%	0.04	1.24	2.06	1.26	1.8	1.5	1.57	0.1
19	The population percent with primary education	%	0.17	1.56	4.74	0	0.01	1.27	1.51	0.56
	The average value of the cluster		0.46	1.42	2.24	1.12	1.16	1.44	1.47	0.22

[a]GDP gross domestic product

Table 2. The forecasting results on the base of two QIs

The 1-st cluster

No	Indicator name	Measurement unit	AFER, %	Steps					Error	Tendency
				1	2	3	4	5		
1	The expected life expectancy at the birth	Years' quantity	0.09	5.55	3.89	4.40	4.20	3.35	4.28	0.1
2	The improved sanitation means	%	0.24	2.32	0.67	2.36	2.03	3.54	2.18	0.2
3	Value added in a services sector	% of GDP[a]	0.15	1.89	0.98	1.58	0.79	1.88	1.42	0.5
	The average value of the cluster		0.16	3.25	1.84	2.78	2.34	2.92	2.63	0.27

The 2-rd cluster

No	Indicator name	Measurement unit	AFER, %	Steps					Error	Tendency
				1	2	3	4	5		
4	Coefficient of teenage fertility	Births' quantity/1000 women	0.18	0.11	3.04	1.45	4.86	1.93	2.28	0
5	Export of goods and services	% of GDP[a]	0.8	1.66	0.34	0.03	2.51	1.93	1.29	0
6	Income (except for grants)	% of GDP[a]	1.15	5.21	3.72	4.68	5.17	5.89	4.93	0
7	Mortality aged till 5 years	%	0.47	2.08	0.00	5.88	3.19	2.52	2.74	0.2
8	Value added in the industry	% of GDP[a]	0.18	3.64	5.57	0.35	4.78	4.59	3.79	0
	The average value of the cluster		0.63	2.54	2.53	2.48	4.10	3.37	3.01	0.04

The 3-rd cluster

No	Indicator name	Measurement unit	AFER, %	Steps					Error	Tendency
				1	2	3	4	5		
9	Fertility rate	Births' quantity/1000 women	0.72	0.63	1.88	9.29	0.00	0.83	2.52	0.2
10	Military expenditure	% of GDP	1.42	2.97	5.50	1.47	3.58	0.76	2,86	0.1
	The average value of the cluster		1.07	1.80	3.69	5.38	1.79	0.79	2.69	0.05

The 4-th cluster

No	Indicator name	Measurement unit	AFER, %	Steps					Error	Tendency
				1	2	3	4	5		
11	Export of high technologies	%	0.51	2.63	2.62	0.48	1.56	3.85	2.23	0.2
12	Procedures' start of for business registration	Quantity	0.67	1.88	1.88	1.26	0.71	0.38	1.22	0.1
	The average value of the cluster		0.59	2.25	2.25	0.87	1.14	2.12	1.72	0.15

(continued)

Table 2. (continued)

The 5-th cluster

No	Indicator name	Measurement unit	AFER, %	Steps					Error	Tendency
				1	2	3	4	5		

The 5-th cluster

No	Indicator name	Measurement unit	AFER, %	Steps					Error	Tendency
				1	2	3	4	5		
13	Gross accumulation of capital	% of GDP[a]	1.09	0.40	1.63	3.01	2.85	2.66	2.11	0.2
14	Import of goods and services	% of GDP[a]	1.04	2.30	3.59	4.11	1.69	3.61	3.06	0.1
	The average value of the cluster		1.07	1.35	2.61	3.56	2.27	3.13	2.58	0.15

The 6-th cluster

No	Indicator name	Measurement unit	AFER, %	Steps					Error	Tendency
				1	2	3	4	5		
15	Immunization against measles	%	0.04	0.51	0.51	0.84	0.84	0.12	0.56	0
16	Quantity of childbirth by means of qualified medical personnel	%	0.01	0.15	0.10	0.11	0.16	0.20	0.14	0.1
17	Ratio of girls and boys in system of primary and secondary education	%	0.04	0.05	0.25	0.17	0.45	1.91	0.57	0.1
18	The improved water sources	%	0.04	0.62	0.41	0.51	0.06	0.17	0.35	0.2
19	The population percent with primary education	%	1.22	1.24	4.74	3.01	2.05	2.06	2.62	0.2
	The average value of the cluster		0.27	0.51	1.20	0.93	0.71	0.89	0.85	0.12

On the base of the analysis of the values of the average forecasting error *Error* for 5 steps forward (Fig. 5), the AI *Aff* (Fig. 6) and the TDI *Tendency* (Fig. 7), it is possible to draw the following conclusions:

- the accounting of several QIs leads to reduction of the forecasting error both in case of middle-term forecasting, and in case of short-term forecasting;
- the accounting of the additional QI of the FM—the TDI—allows keeping the low level of the values of the forecasting error in case of increase in the forecasting horizon.

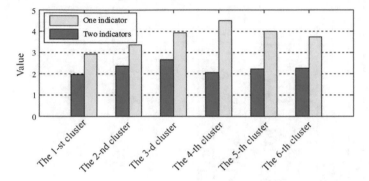

Fig. 5. The comparison of the values of the average forecasting error *Error*

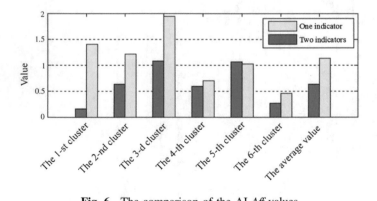

Fig. 6. The comparison of the AI *Aff* values

In this example 400 iterations of optimization algorithm (the basic MCSA or the MMCSA) for population of 20 antibodies were executed. 77 s were spent for creation of one forecasting model. Thus, 462 s (7 min 42 s) are necessary for creation of 6 models, and 1463 s (24 min 23 s) are necessary for creation of 19 models, that in 3.2 times more.

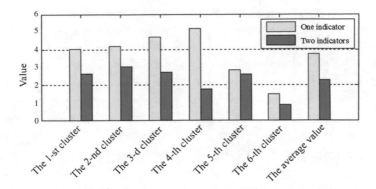

Fig. 7. The comparison of the TDI *Tendency* values

The attractor-based approach to creation of the TDS was applied for the individual and groups' forecasting of the TSs. These TSs describe the references' number of the E-Commerce systems in the requirements to vacancies posted on the websites of 2 recruiting network services (HeadHunter.ru (Russia) and Indeed.com (USA)). Figure 8, b shows the results of the attractors' identification for the TS "Magento (USA)" (Fig. 8a). The TS "Magento (USA)" contains the information on the number of vacancies which include a specific keyword "Magento", defining the name of E-Commerce system for development of online stores. The obtained forecasting results can be used for the analysis of tendencies of the labor market. The use of the second attractor (Fig. 8b) allows reducing the time expenditures for creation of the FM more, than by 1.2 times as for one-objective MCSA as for two-objective MCSA. Also, the application of the attractor-based approach allowed reducing both the value of the *AFER* (1) and the value of the *Error* (9) on 5 steps forward about 0.2%.

The attractor-based approach was applied for groups' forecasting on the base of the model data, which were generated with the use of the same data as in the case of the

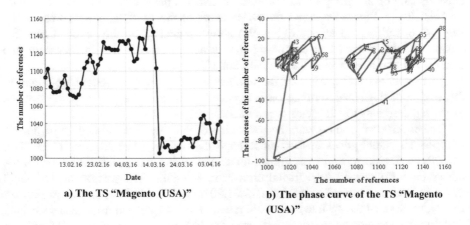

a) The TS "Magento (USA)"
b) The phase curve of the TS "Magento (USA)"

Fig. 8. The identification of attractors

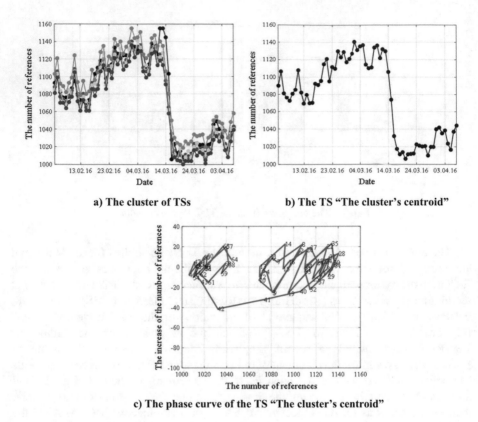

a) The cluster of TSs

b) The TS "The cluster's centroid"

c) The phase curve of the TS "The cluster's centroid"

Fig. 9. Identification of attractors for the cluster of TSs

individual forecasting of the TSs. Figure 9a shows one of clusters, received during the clusterization of the group of TSs. Figure 9b presents the TS "The cluster's centroid" for this cluster. Figure 9c shows the phase curve of the TS "The cluster's centroid". The use of the second attractor (Fig. 9c) allows reducing the time expenditures for creation of the general FM describing the cluster's centroid. If it is necessary, the general FM can be specified for the individual TS.

More detailed information on the application of the attractor-based technique to forecasting of the TSs describing the references' number of the E-Commerce systems in the requirements to vacancies is provided in [21].

The experiment dealing with the restoration of the TSs were fulfilled, in particular, with the TS "Gold prices in Russian Rubles" dated from 12.01.2016 to 07.03.2017. This TS has 8 attractors with the known fractal dimension, calculated on the base of the corresponding method of the fractal dimension definition. The part of the TS "Gold prices in Russian Rubles" dated from 12.01.2016 to 03.08.2016 is shown in Fig. 10. Firstly, we divided the TS into 8 more short TSs. Then, we deleted from each such TS 5 first and 5 last elements. Further, we used the FM on the base of the SBT and the

MMCSA to create 2 FM for each shortened TS to restore the data at the begin and at the end of this TS: we have executed the forecast for 5 steps backwards (the retrospective forecast) and for 5 steps forward (the perspective forecast). At last, we applied the method described above to the fractal dimension definition. In all cases we received the results comparable to the assessments of the fractal dimension received earlier. In the worst case, the deviation from the previously calculated assessments of the fractal dimension was less than 3%. Herewith, in all cases, it was correctly determined in what state the process is located. Hence, it is possible to use the FM on the base of the SBT and the MMCSA to the restoration of the TSs, when it is necessary to define the fractal dimension of the short TS.

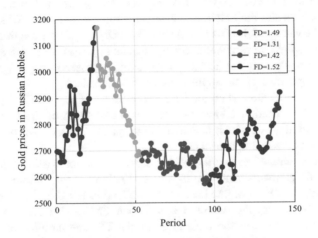

Fig. 10. The TS "Gold prices in Russian Rubles"

During the experiments the system of engineering and scientific calculations MATLAB R2016a had been used.

7 Conclusions

The experimental results confirm the expediency and prospects of the novel approach for time series forecasting with multiobjective clonal selection optimization and modeling.

Initially, the basic MCSA was developed for solving the short-term forecasting problems. However, the fulfilled researches showed the possibility of application of the MMCSA for solving the middle-term forecasting problems. It has been proved during the experiments on forecasting of the macroeconomic indicators of the Russian Federation and others, that the application of the principles of the Pareto dominance is the best solution of the account's problem of the several QIs for the FMs, which represent the analytical dependences on the base of the SBT. Ideas of the multiobjective clonal selection optimization and modeling can be applied for individual and groups' forecasting. The application of the MMCSA will allow giving the more exact description of the clusters' centroids, and, as a result, to increase the forecasting accuracy. Herewith, any general FM can be specified for some private TS in the cluster (if it is necessary) with the use of the same MMCSA.

The principles of the FMs on the base of the SBT can be applied to creation of the multi-factor FMs (for example, two-factor FMs), if we are aware of the presence of the several interrelated TSs. It will allow increasing the forecasting accuracy of the main factor (the forecasting TS) on the base of the additional information on the auxiliary factor (the auxiliary TS).

The use of the attractors-based FMs allows reducing the time expenditures on the creation of the FMs and the obtaining of the forecasting results. Herewith, it is possible to minimize the values of the QIs (the AI and the TDI) and the values of the forecasting errors on 1–5 steps forward simultaneously, thanks to refusal from the attempt to pick up a FM based on the original data sequence of the excessively big length, which may fail due to the specifics of the applied mathematical tools.

The application of the FM on the base of the SBT and the MMCSA to the problem of the short TS restoration allows to solve urgent problem of the fractal dimension definition, the solution of which is demanded in many vital spheres such as economy, ecology, medicine, social sphere, etc.

Further, it is planned to use the offered approach to forecasting the values increments of the TS elements with aim to create a technique for individual and groups' forecasting using the FMs on the base of the SBT and the MMCSA. Also, we plan to create the database with the several TSs and their FMs and use this database in the analysis of the new TSs to choose the most similar TSs and the FMs corresponding to them for the purpose of their further faster specification under the concrete forecasting TS. Furthermore, we plan to develop intensively the FM and to improve the methods of the fractal dimension definition with the aim of their application to the analysis of the extremely short TSs.

References

1. Demidova, L.A.: Time series forecasting models on the base of modified clonal selection algorithm. In: 2014 International Conference on Computer Technologies in Physical and Engineering Applications ICCTPEA 2014, pp. 33–34. IEEE Press (2014)
2. Astakhova, N.N., Demidova, L.A., Nikulchev, E.V.: Forecasting method for grouped time series with the use of K-Means algorithm. Appl. Math. Sci. **9**(97), 4813–4830 (2015)
3. Astakhova, N., Demidova, L., Nikulchev, E.: Multiobjective optimization for the forecasting models on the base of the strictly binary trees. Int. J. Adv. Comput. Sci. Appl. **7**(11), 171–179 (2016)
4. Sivanandam, S.N., Deepa, S.N.: Introduction to Genetic Algorithms, 442 pp. Springer (2008)
5. Reeves, C.R.: Genetic Algorithms, pp. 109–139. Springer, Handbook of Metaheuristics (2010)

6. Fonseca, C.M., Fleming, P.J.: Multiobjective optimization and multiple constraint handling with evolutionary algorithms—Part I: A unified formulation. Technical report 564, University of Sheffield (1995)
7. Horn, J., Nafpliotis, N., Goldberg, D.E.: A niched Pareto genetic algorithm for multiobjective optimization. In: Proceedings of the First IEEE Conference on Evolutionary Computation, vol. 1 (USA: Piscataway), pp. 82–87 (1994)
8. Zitzler, E., Thiele, L.: Multiobjective optimization using evolutionary algorithms—a comparative case study. In: Eiben, A.E., Back, T., Schoenauer, M., Schwefel, H.-P. (eds.) Parallel Problem Solving from Nature—PPSN V. LNCS, vol. 1498, pp. 292–301. Springer, Heidelberg (1998)
9. Knowles, J., Corne, D.: The Pareto archived evolution strategy: a new baseline algorithm for multiobjective optimization. In: Proceedings of the 1999 Congress on Evolutionary Computation, pp. 98–105. IEEE Service Center (1999)
10. Rudolph, G.: Evolutionary search under partially ordered sets, Department Computer Science/LS11. Technical report CI-67/99, Dortmund (1999)
11. Deb, K., Pratap, A., Agarwal, S., Meyarivan, T.: A fast and elitist multiobjective genetic algorithm: NSGA II. KanGAL Report No 200001, Indian Institute of Technology, pp. 182–197 (2000)
12. Deb, K.: Multi-objective Optimization using Evolutionary Algorithms. Wiley, Chichester (2001)
13. Seada, H., Deb, K.: U-NSGA-III: a unified evolutionary optimization procedure for single, multiple, and many objectives: proof-of-principle results. In: Evolutionary Multi-Criterion Optimization. LNCS, vol. 9019, pp. 34–49. Springer, Heidelberg (2015)
14. Coello, P., Coello, C.A., Cruz Cortés, N.: An approach to solve multiobjective optimization problems based on an artificial immune system. In: Proceedings of the First International Conference on Artificial Immune Systems, Kent, pp. 212–21 (2012)
15. Luh, G.C., Chueh, C.H.: Multi-objective optimal design of truss structure with immune algorithm. Comput. Struct. **82**(11), 829–844 (2004)
16. Campelo, F., Guimarães, F.G., Saldanha, R.R., Igarashi, H., Noguchi, S., Lowther, D.A., Ramirez, J.A.: A novel multiobjective immune algorithm using nondominated sorting. In: 11th International IGTE Symposium on Numerical Field Calculation in Electrical Engineering (2004)
17. Jiao, J., Gong, M., Shang, R., Du, H., Lu, B.: Clonal selection with immune dominance and anergy based multiobjective optimization. In: 3rd International Conference on Evolutionary Multi-Criterion Optimization, pp. 474–89 (2005)
18. Wang, X.L., Mahfouf, M.: ACSAMO: an adaptive multiobjective optimization algorithm using the clonal selection principle. In 2nd European Symposium on Nature-Inspired Smart Information Systems, pp. 959–971 (2006)
19. Zhang, Z.: Constrained multiobjective optimization immune algorithm: convergence and application. Comput. Math Appl. **52**(5), 791–808 (2006)
20. Jiao, L., Gong, M., Du, H., Bo, L.: Multiobjective immune algorithm with nondominated neighbor-based selection. Evol. Comput. **16**(2), 225–255 (2008)
21. Astakhova, N.N., Demidova, L.A., Kuzovnikov, A.V., Tishkin, R.V.: Attractor-based models for individual and groups' forecasting. In: IOP Conference Series: Materials Science and Engineering (2017)
22. Kudinov, A.N., Lebedev, D.Y., Tsvetkov, V.P., Tsvetkov, I.V.: Mathematical model of the multifractal dynamics and analysis of heart rates. Math. Models Comput. Simul. **7**(3), 214–221 (2015)

23. Ivanov, A.P., Kudinov, A.N., Lebedev, D.Y., Tsvetkov, V.P., Tsvetkov, I.V.: Analysis of instantaneous cardiac rhythm in a model of multifractal dynamics based on holter monitoring. Math. Models Comput. Simul. **8**(1), 7–18 (2016)
24. Huffaker, R.: Phase space reconstruction from time series data: where history meets theory. In: Proceedings in System Dynamics and Innovation in Food Networks, pp. 1–9 (2010)
25. Chen, M., Fang, Y., Zheng, X.: Phase space reconstruction for improving the classification of single trial EEG. Biomed. Signal Process. Control **11**, 10–16 (2014)
26. Stakhovsky, I.R.: Attractor reconstruction from the time series of information entropy of seismic kinetics process. Izvestiya Phys. Solid Earth **52**(5), 740–753 (2016)
27. Kozma, R., Wang, J., Zeng, Z.: Neurodynamics. In: Handbook of Computational Intelligence, 1633 pp. Springer (2015)

ARTool—Augmented Reality Human-Machine Interface for Machining Setup and Maintenance

Amedeo Setti[1], Paolo Bosetti[2(✉)], and Matteo Ragni[2]

[1] Pro-M Facility, Trentino Sviluppo S.P.A., Rovereto, TN, Italy
amedeo.setti@trentinosviluppo.it
[2] Department of Industrial Engineering, University of Trento, Trento, Italy
{paolo.bosetti,matteo.ragni}@unitn.it

Abstract. In modern production lines, smaller batches to be produced and higher customization level of a single component bring to higher cost, related especially to setup and preparation of machines. The setup of a milling machine is an operation that requires time and may bring to errors that can be catastrophic. In this Chapter, the **AR**Tool Augmented Reality framework for machine tool operations is presented. The framework permits to write and debug part-code in an augmented environment, to identify quicker misalignments and errors in fixing of new blank material, and to support maintenance operations. The ego-localization of the handheld device that depicts the augmented scene in machine work-area is based upon markers. The library that performs marker identification is brand-new and it is benchmarked throughout the Chapter against a state-of-the-art solution (`ARUCO`) and a ground truth (multi-stereoscopic motion capture). The Chapter also describes the general information flow and the context that brought to the conception of the **AR**Tool framework, and presents a series of applications developed using the framework.

1 Machining Economics and Augmented Reality

The economics of machining operations considers different cost authorities that should be minimized to achieve an efficient process. For each machined product, the main factors to consider are [13,17]:

- the cost of the effective machining operation, alongside with maintenance and man-hours costs;
- the cost for preparing the machine, which comprises testing of the part-program, fixing and aligning the blank material in the working area, and mounting the tools and the cutters on tool holders;

The original version of this chapter was revised: The author name and references were corrected. The erratum to this chapter is available at https://doi.org/10.1007/978-3-319-69266-1_23

© Springer International Publishing AG 2018
Y. Bi et al. (eds.), *Intelligent Systems and Applications*,
Studies in Computational Intelligence 751,
https://doi.org/10.1007/978-3-319-69266-1_7

- the costs for loading the raw material and unloading the finished part;
- the cost of tooling.

One of the cost of greater impact is due to maintenance and inactivity that directly correlates time and machining costs. In case of human operator involved in the process of loading and unloading material—e.g. in case case of shop-floor with limited automation and with small batches to be produced—optimizing the maintenance and the alignments procedures permits to reduce dramatically the costs.

The Chapter describes a framework that exploits Virtual and Augmented Reality technologies to reduce unproductive times. The platform, namely **AR**Tool, reduces errors induced by operators during procedures such as alignments of blank material. In common practice, for avoiding collisions that may result in extended damages for both machine and work-piece, in-air test are performed—i.e. a complete execution of the part program with a constant safety offset between the tool and the raw material.

The Augmented Reality (AR) component of the **AR**Tool frameworks uses the reference systems stored inside the machine controller to overlaid a properly oriented simulation of the workpiece blank, alongside with fixtures, and machine moving peripherals on the scene of the working area captured by a camera. The simulation reflects exactly what the machine is programmed to perform, thus in-air test, which may require hours to be fully executed, is substituted by an augmented simulation with time scaled. The operator concentrates the attention only on the complicated passages, and effectively identify visually evident mistakes, in less time and with an higher accuracy.

The augmented component of the framework is built to run on a personal device, and throughout the Chapter the considered device is a tablet which is relatively low cost with respect to more exotic hardware—e.g. head mounted displays. With a tablet, the operator explores the simulated scene from different perspectives. Moreover, the same framework can be easily employed to enhance the maintenance operations on a machine, and inexperienced operators largely benefit from the usage of augmented schematics and manuals.

1.1   Envisioning AR Technologies

The manufacturing industry has always envisioned the application of AR related technologies, and the strong interest is underlined in the results of the survey conducted by the *Deutsche Forschungszentrum für Künstliche Intelligenz* during the Hannover Messe of 2010. On a total of 54 industrial rappresentative, the 77.8% have every intention of deploying augmented solutions in their production lines [28].

In literature, Architecture is the first field that embraced the AR, enlarging the Building Information Modeling schemes in order to accommodate a data infrastructure for the Augmented Reality technologies [35].

Also the Cognitive Sciences inspected the application of AR technologies, evaluating the benefit from a cognitive workload point of view [15, 30].

1.2   Manufacturing and Augmented Reality

In general, the proofs-of-benefit for AR as alternative training method, described in [12,27], make educational and informational applications, such as augmented manuals and operators training, literally mainstream. In [21], the authors use a marker solution to build interactive lectures on machinery handling for completely inexperienced students, revealing once again the high acceptance of the methodology, and allowing a faster comprehension of programming caveats for complex paths [7]. Syberfeldt et al. [32] pushes towards the integration of AR for training and expert systems to support decision making for inexperienced operators.

The costs of integrating such a new technology in the process is not an easy decision. Few studies started to develop decision supporting tools ex-ante [9], for evaluating the effectiveness of the approach for a specific manufacturing process. Both Product Design and Planning (PDP) and Workplace Design and Planning (WDP) benefit from an AR developing environment [24], that aid designers and engineers in making better decision while designing new assembly lines. Lines include AR interfaces [3,5] that guide the operator in the execution of a specific task—i.e. projecting welding spots on work-piece in [8]. The ergonomy of the technology is also evaluated in literature [34].

Papers [15,36] present first implementations of virtual assembly interfaces. Cameras are used to detect position of operator hands, that are the Human Computer Interface (HCI) for the augmented renderer. Systems are desktop static prototypes, but usability is validated with respect to non-augmented real-case-scenario. Evidence of cognitive workload reduction for the operator are underlined, as also reduced time to complete tasks and reduced mean error rate.

For what concerns application on process machines, the manipulators programming and collision avoidance is for sure the most prolific field. And in fact the complex kinematic configurations during a program execution results more intuitive—e.g. programming [10] or visualizing [6,11] end-effector pose and trajectory—by the mean of different user interface—e.g. mobile, projection on half silvered glasses or head displays [16]. General survey can be found in [22,25].

The applications of AR on machine tools are limited and may be referred as proof-of-concept prototypes rather than proof-of-benefit ones. In [31], an AR application is used to help operators during manual alignment in a pipe manufacturing machine. In [20], AR is used to develop a framework for dimensional validation of finished parts. The framework is marker based, one of the more reliable solution that guarantees enough precision for manufacturing applications. The works also illustrates evidences of advantages, both economical and practical, induced by the use of AR applications in manufacturing. Another approach typically discussed in literature, is the use of super-imposition of virtual image on work-space video recording for validation of complex paths [37]. Virtual images contain augmented information about the process, and are visualized through the use of different device, such as stereo-projector [23] or mobile devices. In general, the idea is to use the augmented visualization to give more insight to the operators about the process, usually before performing the actual machining operation [39]. Other applications focused instead on active maintenance, using OCR (Optical Character Recognition) in combination with localization markers [19], but real benefits of such implementations to users were not assessed. In [38], it is worth noting the use of handheld devices, with respect to the typical static desktop setups seen in previous works.

1.3   Chapter Summary

The complete concept for the full **AR**Tool framework is deeply analyzed in Sect. 2: starting from a broader view, the single elements of the approach are described and motivated. The device layer description is the pretext to introduce the `ARSceneDetector`, in Sect. 3.2: the library is portrayed extensively and benchmarked against to the state-of-the-art equivalent library `ARUCO`, and results are illustrated. Section 4 is an application showcase that presents some of the developed applications that use **AR**Tool for data interchange.

2   The ARTool Platform

The **AR**Tool Framework is conceived to support machine manufacturers, technical offices, and machine operators in bringing augmented reality information on the machine and in the production lines.

The main objective of the framework is the optimization of the machining processes by tackling two major shortfalls:

- reducing the unproductive time between production batches, allowing the operators to test quickly the newer part-program and eventually correct misalignment of blank material with respect to reference systems saved in numerical control;
- supporting the maintenance procedures through augmented manuals that facilitate remote assistance from technical support. Failure diagnosis can be improved highlighting failing components directly on machine chassis.

2.1   From Authors to Consumers: The Flow of Data

The main source of information are the technical offices, that provide tasks to shop-floor. Tasks data include:

- part-programs;
- fixtures list and fixture sequences;
- tooling information.

The technical office stores the authored data in SCADA (Supervisory Control and Data Acquisition) servers: this permits the centralization and distribution to data consumers.

The second authoring agent of the network is the machine manufacturer that through a Content Delivery Network distributes assets for augmented manuals that the different SCADA servers of the different industries that acquired the machine download.

The SCADA server act as a gateway for delivering update data to local machine and shop-floor operators.

For both technical offices and machine manufacturers, tools for authoring information are developed as plugins for commercially available Computer Aided Engineering (CAE) software [26,33]. For technical offices, this means expand the capabilities of common Computer Aided Manufacturing software, while, for manufacturers, the plugins are related to Computer Aided Design (CAD) and Product Life-cycle Management (PLM) software. Optionally, manufacturers can exploit the framework for marketing opportunities, such as ticketing services and web store for spare parts.

The main information consumer are the machine tool and the operator device. Both consumer download data from SCADA servers. The computer numerical control (CNC) communicates using a client that can be software service, for newer machines, or a embedded computer, for older machines. The client requires an implementation of the proprietary communication protocol of the machine, while the communication with the SCADA is performed through standard protocols. The client broadcasts to the SCADA server all relevant information for diagnostic and simulation purposes, such as system states, tools table, etc.

Machine operators carry a personal device that has the hardware necessary to perform the ego-localization task—i.e. camera and inertial sensors—that is the most prominent feature of the **AR**Tool framework. Currently, **AR**Tool has been tested only on tablet devices, which are relatively low-cost and reliable with respect to other solutions.

2.2 Operator Device

Operators are equipped with personal devices that have the minimum hardware requirements to perform the ego-localization. The current release of **AR**Tool framework requires an high definition camera for gathering the scene on which assets are overlaid, an inertial measurement unit to filter the ego-localization state and a GPU for rendering the virtual scene (Figs. 1 and 2).

Localization is performed through markers that characterize a scene (cfr. Sect. 3.2 for a description of *scene* in detail). Once a scene is identified, a query to the SCADA server permits to populate the camera feed with virtual assets.

The framework eases the presentation of different information, that are contextualized with respect to a scene and a *operation mode*, or scope. When the current scope is to setup a new process for a machine, the main assets considered are:

- blank material and possibly the fixing for the bulk;
- tool and optionally machine head;
- mechanical axes simulacra;

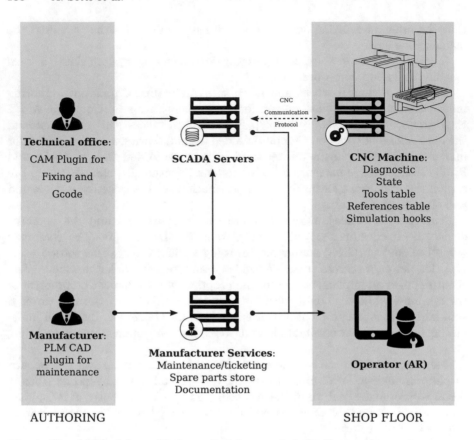

Fig. 1. The **AR**Tool flow of information, from technical offices and manufacturer, to machines and operators

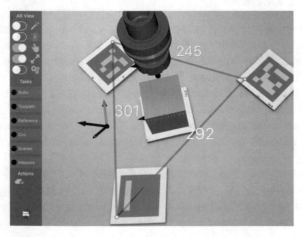

Fig. 2. Screenshot of a very first prototype **AR**Tool iPad app, showing the setup-mode augmented reality view. In this case, marker distances are measured. Camera images are localized in the working area: the application shows a bulk, a trajectory and a tool oriented with machine reference frames

- coordinate systems and oriented trajectory;
- marker anchoring elements (cfr. Sect. 2.4);
- auxiliary descriptive text.

When the intended scope is maintenance, the framework is designed to stage:

- machine contextual information;
- mask for component of the machine;
- contextual manual web pages;
- geometric primitive shapes—e.g. arrows—that can be used to draw operator attention.

The device selected as prototype is an Apple iPad 2 Air Tablet, with iOS 9.3 operating system. The framework is a C++ library that exposes Swift and Objective-C bindings. The rendering operation are handled by the Apple Framework SceneKit [1].

2.3 The SCADA and Per-Machine Server

The SCADA server is responsible for storage and distribution of augmented assets. It also challenges machine clients for information necessary to present simulation and localized elements:

- the current state of machine, that includes the current position of axes, the active coordinate system, the loaded tool on the spindle and the active part-program;
- part-program simulation hooks, that comes from the numerical control parser/interpolator. If this information is available, **AR**Tool shows the exact tool trajectory as interpolated by the numerical controller. If this information is not actually available, the framework exposes a fallback interpolator, that will generates trajectory with minimal differences;
- coordinate systems table and tool table. The tool table relates the currently loaded tool with a solid model counterpart for rendering. The reference systems table permits to project machine simulacra within the AR view, alongside the correct origins;
- optionally, diagnostic information that guides inexperienced operators in unusual situation and training.

In the experimental system, the server is a Ruby and C++ software on a separated machine, with database composed by a sequence of YAML files—i.e. a format that simplify inspection and debugging. The server provides a HTML5 web application for authoring, which exploits the C++ component of **AR**Tool framework for the creation of scenes from static images (Fig. 3).

2.4 ARTool as Input

Capabilities of **AR**Tool can be enriched from a simple output interface, to a novel, input/output human machine interface, providing functionalities for identifying exterior points and geometric features in space.

Fig. 3. An example application created using **AR**Tool framework, that helps operator in bulk alignment operations. This application is described in Sect. 4.4

The screen of a mobile device can be used to capture a bi-dimensional input. As already discussed, depth can be reliably reconstructed by using structured elements (markers). Each marker defines a *virtual plane*. Indeed, the area of the screen can be projected on this plane, associating each bi-dimensional screen coordinate to a tri-dimensional point that lies on the virtual plane. In other words, that point is the projection of the 2-D point on screen along the line of view on the virtual plane.

The procedure is explained in Fig. 4. Each machine has a *fixed* origin, which is hard-coded in machine's controller. Then, the machine may define an *active* reference frame (in this case **0**), that is used for defining the coordinates in the part-program. The transformation matrix from *fixed* to *active* reference frame is known. In the figure, reference system **A** is defined by a *machine marker*, whose position is well known with respect to the machine *fixed* origin. Through a simple coordinate transformation, the vector from reference **0** to reference **A** is known. The marker in **A** is used by **AR**Tool library to ego-localize the mobile device, so that the vector from **0** to tablet internal reference is known. **AR**Tool also reconstructs the vector pointing to the marker reference **B**, which is the movable virtual plane, closing the chain between **0** and **B**. When the user taps the mobile screen, the 2D coordinates of the tap on screen are transformed in the coordinates of a 3D point projected on the plane of **B**.

There is no need to keep both *machine marker* and *moving marker* framed at all times: indeed, once the position of the free marker is set, it can be anchored in software while framing both, then anchoring allows **AR**Tool to use the *free marker* as a *machine marker*, thus ego-localizing the device relatively to any marker in the markers chain. This opens to the possibility to create *chains* of markers, although the reliability of the ego-localization decreases exponentially at each hop.

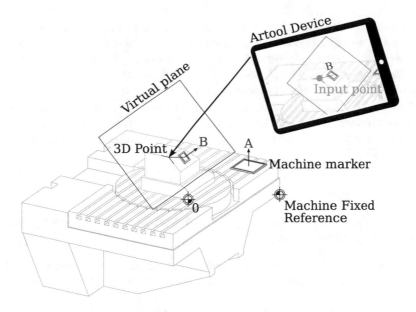

Fig. 4. Using the mobile device as a 3D input system, through a mobile marker

Figure 3 shows a practical application. In common practice, part-programs contain axes motion coordinates relative to a point in space, which is the *workpiece origin*. One of the very first operations is to identify the position of *workpiece origin* on the workpiece in the working space. This requires to approach the object with a touching probe—i.e. the tool in the figure—that returns a feedback to the machine controller upon achieving contact. The **AR**Tool application acts as a virtual touching probe, that identifies a point that lies on the virtual plane described by the marker attached on the workpiece. In the figure, the identified point is the upper-left corner of the gray cube overlaid on the workpiece, visible on the screen of the tablet.

3 ARSceneDetector: The Core of ARTool

One of the critical requirements for an augmented application is a reliable and precise localization of the device with respect to the scene observed. The library ARSceneDetector is the software component that fulfill this task.

During the early development stage, the **AR**Tool framework included the open source library ARUCO, currently distributed with the OpenCV suite [4]. ARUCO is a localization library which takes advantage of the presence of structured markers in scene for reconstruction. ARUCO was chosen after a comparison with the ArtoolKit platform: it provided a better responsiveness at the cost of a lower accuracy, on the prototype device.

In a later development stage, in order to tackle the accuracy issues and to get a more stable and reliable localization through sensor fusion, the designed from scratch `ARSceneDetector` library has been introduced as core component of the **AR**Tool framework. The library is strongly device dependent (ARM-processor) and uses specific hardware instructions to speed-up its performances. This allows to squeeze the computational power of the device, attaining a precise and yet responsive placement of virtual assets on the framed scene.

The next section describes the internal logical structure od `ARSceneDetector`, while the Sect. 3.2 is devoted to a comparison against `ARUCO`. In particular, ego-localization accuracy and computational efficiency are evaluated carefully.

3.1 `ARSceneDetector` Library Details

The `ARSceneDetector` library is logically divided in three different layers, from perception to scene rendering.

- The Sensor Acquisition and information gathering layer is written in `Swift` language. This is required by the platform and uses the current operating system API.
- The Marker Handler layer is written in `C++` and is linked to the OpenCV library. This layer handles the identification of the marker in the scene, the inter-frame tracking and the image stabilization.
- The very last layer is the Scene Detector, a classifier that extract more information based upon the relative position of the marker in the scene.

The three layers are presented in Fig. 5.

As with other computer vision algorithms, `ARSceneDetector` requires a calibration of the camera [14] which results in a camera matrix. Light parameters and thresholds are automatically evaluated through normalization procedures: each frame is enhanced and the edge detection is extracted from the frame in GPU.

Using the internal camera of the prototype device, it is possible to collect frame with 720p and 1080p resolution. The bigger the frame, the lower the update frequency guaranteed for the localization—i.e. 120 Hz and 30 Hz respectively.

Beside the camera frame, accelerations and angular ratios of the device are measured by the on-board IMU sensor. This information permits to stabilize the rendered scene [2]. The combination of the frame and IMU data are passed through the bridge `Swift`/`C++` and enters the marker handler layer, as Scene container.

The Marker Detector is the implementation od a classical one-frame-at-the-time algorithm which, for each camera frame, extracts convex quadrilateral shapes as marker candidates. The candidate are then reoriented and checked for squareness. The pose of each square element is reconstructed using different well-known algorithms [18, 29]. The algorithms return a reference system that is oriented through an asymmetrical pattern drawn on the marker itself. The pattern can be a number encoded in a binary form—e.g. the `ARUCO` encoding—or

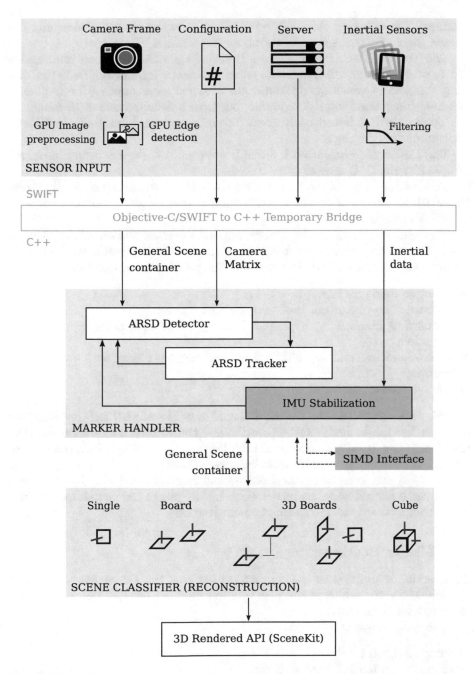

Fig. 5. Library structure. ARSD stands for `ARSceneDetector`. In gray, plugins that are disabled during benchmark

a image. The reference system is relative to the camera point-of-view and has always the $\hat{z}$ axis perpendicular to the marker surface.

The Tracker is an extension of the Detector algorithm that uses information of the previous frame to reduce the computational efforts of the Detector, limiting the area in which quadrilateral are searched, and lowering the frequency of whole-frame scanning (configurable, but with a default value of 10 frame). It can be disabled. To improve efficiency, Single Instruction Multiple Data (SIMD) instructions are employed.

The IMU Stabilization block filters the state of the device, fusing the signal sampled by the IMU sensor.

The result of the Marker Handler is a General Scene Container, a data structure with all the information about identified marker and their position with respect to the device.

The very last layer of the library performs a classification of the General Scene Container. Using a combination of markers it is possible to drastically improve the accuracy of the localization. The possible scenes contain:

- a simple single marker;
- a board of co-planar markers, with parallel $\hat{z}$ axes;
- a board of markers, with parallel $\hat{z}$ axes, and known, non-zero offset in $\hat{z}$ direction;
- a board of three markers with mutually orthogonal $\hat{z}$ axes, with known offset vectors;
- a solid cube of markers.

The SCADA server provides the list of scenes to be classified. The Scene Detector matches the most similar one. Nevertheless, the library may enrich SCADA definitions: this particular feature is used for marker chaining which consents to expand the rendering volume, reaching area in which marker are not currently visible. Once the scene has been classified and reconstructed, the General Scene Container is shared with the render engine, that places the virtual models in a virtual world that is aligned with the perceived one.

3.2 Library Benchmarking

This section is devoted to the comparison between the `ARSceneDetector` and the `ARUCO` library, which is the first solution adopted by **AR**Tool, in the very early developing stages.

The test focuses on:

- computational time;
- reliability in marker identification;
- accuracy in ego-localization.

3.2.1   Methodology

For the localization, the ground-truth is provided by a professional level Motion Capture System (MoCAP—*OptiTrack*, equipped with 8 *Prime13* cameras running at 120 fps). For the localization test, a MoCAP 3D reference is attached on the iPad that records a video of a board of 4 ARUCO markers. At least one marker is always framed during the video (see Fig. 6). The MoCAP reference frame is placed on the coordinate system of one of the corner of one of the marker—i.e. the origin have a known offset.

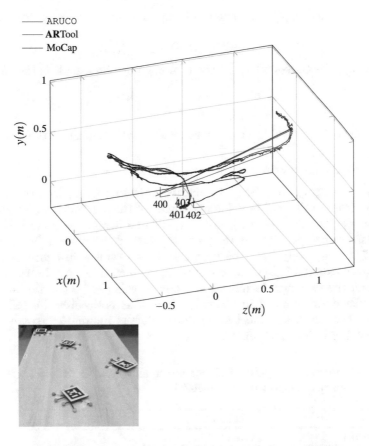

Fig. 6. A frame of the video used for bench-marking

The recorded video is than used to run a testing application with both libraries in profiling mode. Setup parameters are fine tuned to crunch the maximum performances without compromising too much reliability, but some of the very advanced feature of the ARSceneDetector—i.e. the GPU usage and the SIMD operations—are disabled for a fairer comparison. This effects the real performances of ARSceneDetector, but allows to limit the comparison only on the algorithmic level, rather than on differences in filtering and input data processing.

Since the signal length are different for MoCAP and iPad, localization data are synchronized minimizing the variance of positions with respect to time. Given the signals:

- $x_0(t)$ the x coordinate returned by the motion capture at frame t
- $x_A(t)$ the x coordinate returned by the **AR**Tool library at frame t
- $x_B(t)$ the x coordinate returned by the ARUCO library at frame t

the distance $\varepsilon_x(t, \delta)$ is evaluated as:

$$\varepsilon_x(t, \delta) = 2x_0(t) - ((x_A(t + \delta) + x_B(t + \delta))) \tag{1}$$

while the variance $\sigma_x(\delta)$ with respect to the shift δ on the x signal is obtained as:

$$\sigma_x(\delta) = E\left[\varepsilon_x(t, \delta) - E\left[\varepsilon_x(t, \delta)\right]\right] \tag{2}$$

consequently, the time-shift to be used for aligning the signals is the result of:

$$\delta^* = \arg\min_{\delta} \sum_{i=\{x,y,z\}} \sigma_i(\delta) \tag{3}$$

Position signals are used because more reliable with respect to the others.

3.2.2 Result Analysis

The benchmark trajectory in space and its projection along the principal direction is depicted in Fig. 7. The reference frames of the markers are also presented.

ARUCO localization presents instability, and in different occasions it is not able to reconstruct the pose of the markers. In particular, between the frame 5672 and 5807 it completely loses the tracking—i.e. the spikes in figure. For further analysis, the ARUCO missing trajectory is approximated linearly between the last known and the first new localization. However this segment is the main cause of the differences reported in Table 1, where it is noticeable the reliability of ARSceneDetector, that almost never drops track of the marker, scoring a quite high reliability index (98.9%).

Table 1. Comparison of speed (in frame per seconds) and reliability (percentage of frame identified with respect to total—21599)

	ARTool	ARUCO
Speed (fps)	114.5	94.3
Reliability	98.9% (21 380)	86.8% (18 739)

Figure 8 shows a comparison of the three trajectory and the error of the markers detected trajectories with respect to the MoCAP one. In Fig. 9, histograms report the probability distribution for errors. For what concern positions, the error distribution of ARUCO tends to be larger with a mode that diverges slightly from zero. Numerical analysis is reported in Table 2. Regarding attitude estimation, the performance can be considered comparable.

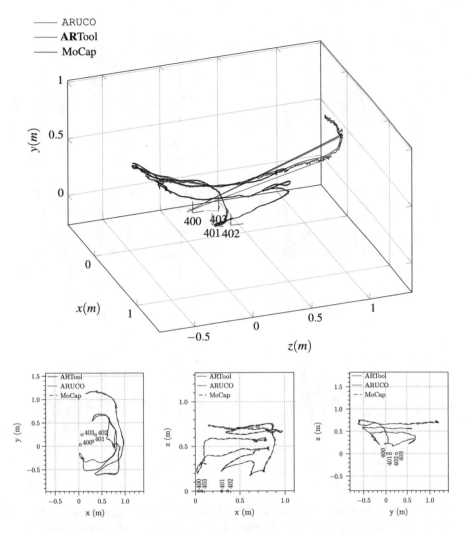

Fig. 7. On top of the image the 3D representation of the trajectories in the video. The reference frames of each marker are also reported. The spikes in the ARUCO trajectory are due to missed identification of markers

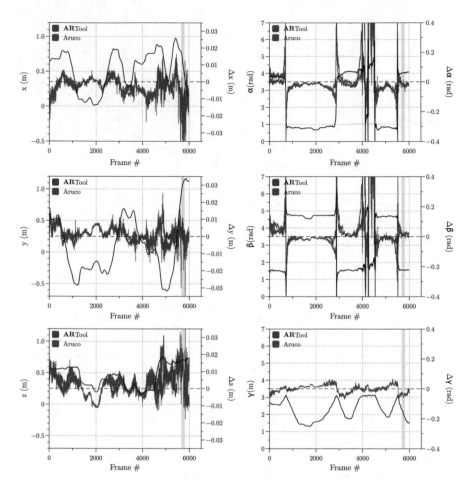

Fig. 8. Ego-localization errors. On the left, there are position plot and errors between marker libraries and MoCAP. On the right Euler's angles and their errors are plotted. ARUCO fails the identification between frames 5672 and 5807 (vertical hatched band)

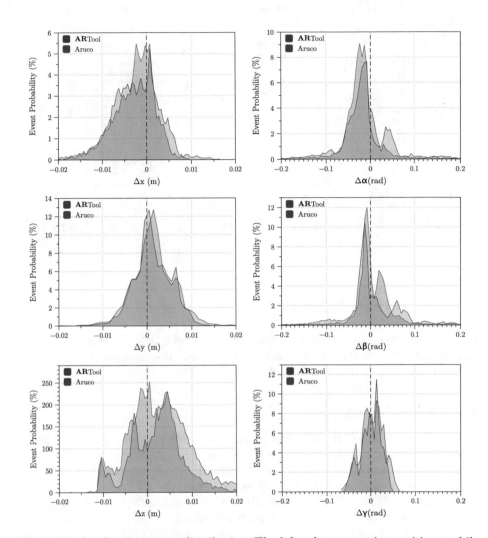

Fig. 9. Ego-localization errors distribution. The left column contains positions, while the right column contains Euler's angle

Table 2. Statistical indicators for errors distribution (mean μ, standard deviation σ and kurtosis k)

	Artool			ARUCO		
	μ	σ	k	μ	σ	k
x (mm)	-3.10	5.38	9.72	-6.04	2.93×10^1	7.58×10^1
y (mm)	1.22	4.33	8.75	4.05	2.05×10^1	5.85×10^1
z (mm)	9.37×10^{-1}	5.66	3.59	4.72	8.20	5.67
α (rad)	1.92×10^{-2}	2.99×10^{-1}	1.33×10^2	4.07×10^{-2}	3.09×10^{-1}	1.09×10^2
β (rad)	-9.07×10^{-4}	2.23×10^{-2}	2.38	-4.20×10^{-3}	5.17×10^{-2}	4.75×10^1
γ (rad)	2.13×10^{-2}	3.84×10^{-1}	1.52×10^2	1.67×10^{-2}	3.39×10^{-1}	1.53×10^2

4 Applications Showcase

The section presents a series of applications designed to test the most prominent capabilities and features of the **AR**Tool framework, leaving aside the authoring tools for machine manufacturer and technical offices.

4.1 Origin Debugger

The application visualizes the origins and the coordinate systems of both markers and numerical controller. The machine client is connected to the Heidenhain iTNC 530 of a Deckel Mori DMU-60T (5-axis milling machine), and takes advantages of an FTP connection for data exchange. From the FTP, the machine client downloads the iTNC file that stores the reference table. The file is queried at constant interval and parsed only if modification time changes.

The application permits to see selected origins of the table projected on the screen, overlaid on the frame captured by the camera. Operators can inspect the scene from different orientations. The applications shows also distances between origins for debugging (see Fig. 10).

4.2 Trajectory Inspector

The application is built upon the capabilities of the previous application. The machine client queries the controller for the currently active part-program and download it through the FTP connection, alongside origin and tool table.

The tool table is parsed, and the name is used as identifier for the digital model to render, distributed through SCADA server. Since there is no communication channel for the numerical interpolator of this particular machine, it is

Fig. 10. The Origin Debugger application: on the left, the visualization of the reference frame obtained from the machine client, while on the right a measurement between different origin is performed

the fallback **AR**Tool interpolator that parses the part-program source file and generates the tool trajectory for the simulations.

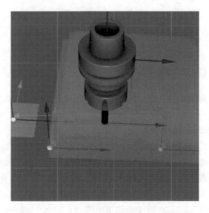

Fig. 11. The Trajectory Inspector: operator can navigate the virtual environment or fix it through a marker

Simulations are projected in a virtual environment that can be navigated by a user, exactly like a common CAE environment. It is also possible to fix the virtual environment through a marker and explore the simulation by moving and reorienting the tablet, as depicted in the screenshot of Fig. 11.

4.3   Trajectory Simulator

This application acts exactly like the Trajectory Inspector, and uses machine client and SCADA server to collect data and generate a virtually simulated environment that, in this case, is projected upon the camera feed. Operators can inspect directly the simulation in the working area, against real objects, the result of the interpolated trajectory and intercept collisions, programming errors, and misalignments (see Fig. 12).

Fig. 12. The trajectory simulator: operator can navigate the virtual environment or fix it through a marker

4.4   ARTool Zero (Concept)

ARTool Zero is the concept of an Augmented Reality application that allows operator to select directly some geometric features as reference through the touching probe of a machine tool. Leveraging the input capabilities described in Sect. 2.4, the approximated feature information input through the augmented interface is transformed on-the-fly in a part-program that allows the touching probe to precisely identify the geometry.

Before performing the actual machine movements, a simulation of the trajectory of the touching probe is presented on the device screen, in Trajectory Simulation mode, so that operator can check for collisions, with respect to different point-of-view. The part-program generated is then loaded in the machine tool controller for the actual execution. A sequence that exemplify the usage is depicted in Fig. 13.

4.5   Maintenance Mode

The maintenance mode is at an early developing stage. The application requires a series of marker installed in the different parts of the machine to allow contextualized information gathering. In this case, the placement of assets on the screen does not require the same accuracy as in simulation, and a single marker covers quite a big area of the machine.

Simulation   Execution

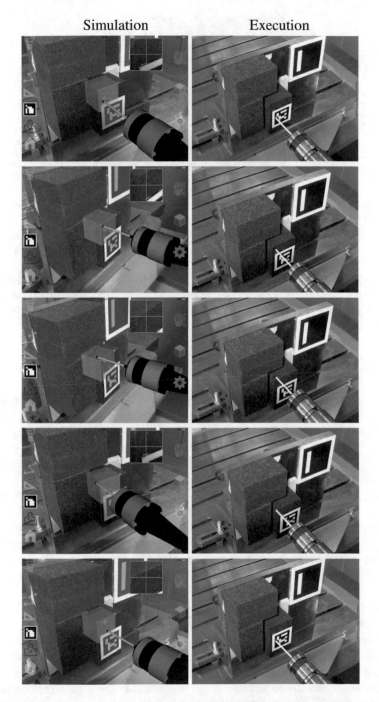

Fig. 13. Artool zero concept: the sequence on the left shows the simulated part program on the display of the tablet: users can frame the scene from different directions to check for collisions; on the right, the actually sequence of operations are depicted

Fig. 14. Maintenance mode: a failed component highlighted

If a component fails the diagnostic, it is highlighted (see Fig. 14) and it is made evident to the operators. At the same time, an operator recall the manual page of a particular component by framing and taping it on the screen (using the input capabilities described in Sect. 2.4).

5   Conclusions

The work presents an Augmented Reality software framework for supporting CNC machine tool operations, such as setting up and checking for errors in part-programs or remotely guided maintenance operations. The systems uses a portable device (an Apple iPad) that overlays information to camera images by the mean of solid models and localized text. The system can work according to two prominent scenarios: *setup-mode* and *maintenance-mode*.

In *setup-mode*, the system shows workpiece shape and position, part-program simulated trajectory, and CNC setup data (reference systems, toolpaths, etc.). The mismatch between 3-D scene and real image are easily perceived by the user, that can *quickly* and *reliably* identify (and then correct) misalignments, collisions, and other errors in part-programs. But the framework does not act only as output interface. Indeed, leveraging the communication with the machine, the mobile device can act as measuring instrument, that can identify workspace coordinates—for example, as shown in Sect. 2.4, **AR**Tool can be used to define the *workpiece origin* by the mean of a *free marker*.

The framework is also used for another scenario, namely *maintenance-mode*, that feeds the operator with service information from machine manufacturer. Visualized data include position of failing components and service operation sequences—e.g. the manufacturer may request the operator to check the axes lubricant reservoir: instructing the portable device to draw a red 3-D model of the tank, localized in space and overlaid on the real object, the operator can quickly locate it without checking machine schematics.

The framework is characterized by three layers, developed for testing purposes. The augmented interface layer comprises iOS applications, that uses camera and inertial sensors to perform ego-localization of the mobile device. In particular, camera images are processed by the custom made library **AR**Tool, designed for high performance and high reliability in manufacturing environment, tested against state of the art competitor ARUCO. **AR**Tool proved faster and more reliable when comparing the two libraries against a motion capture ground-truth. The device communicates with the machine client through a server that queries system status, positions, reference systems and part-program to be presented on the augmented application. The server also acts as information exchange systems (SCADA server). On the upper layers there are technical offices, that provide part-program to be executed and models, and machine manufacturers, that provide augmented documentations and operation sequences. Information are authored through plugin for CAE software.

References

1. Apple Inc.: SceneKit Framework. https://developer.apple.com/scenekit/ (2016)
2. Bleser, G., Stricker, D.: Advanced tracking through efficient image processing and visual-inertial sensor fusion. Comput. Graph. **33**(1), 59–72 (2009)
3. Bondrea, I., Petruse, R.: Augmented reality—an improvement for computer integrated manufacturing. Adv. Mater. Res. **628**, 330–336 (2013). https://doi.org/10.4028/www.scientific.net/AMR.628.330
4. Bradski, G., et al.: The opencv library. Doct. Dobbs J. **25**(11), 120–126 (2000)
5. Büttner, S., Sand, O., Röcker, C.: Extending the design space in industrial manufacturing through mobile projection. In: MobileHCI 2015—Proceedings of the 17th International Conference on Human-Computer Interaction with Mobile Devices and Services Adjunct, pp. 1130–1133 (2015). https://doi.org/10.1145/2786567.2794342
6. Chong, J., Ong, S., Nee, A.C., Youcef-Youmi, K.B.: Robot programming using augmented reality: an interactive method for planning collision-free paths. Robot. Comput.-Integr. Manuf. **25**(3), 689–701 (2009). https://doi.org/10.1016/j.rcim.2008.05.002
7. Ćuković, S., Devedžić, G., Pankratz, F., Baizid, K., Ghionea, I., Kostić, A.: Augmented reality simulation of cam spatial tool paths in prismatic milling sequences. IFIP Adv. Inf. Commun. Technol. **467**, 516–525 (2015). https://doi.org/10.1007/978-3-319-33111-9_47
8. Doshi, A., Smith, R., Thomas, B., Bouras, C.: Use of projector based augmented reality to improve manual spot-welding precision and accuracy for automotive manufacturing. Int. J. Adv. Manuf. Technol. 1–15 (2016). https://doi.org/10.1007/s00170-016-9164-5
9. Elia, V., Gnoni, M., Lanzilotto, A.: Evaluating the application of augmented reality devices in manufacturing from a process point of view: An ahp based model. Expert Syst. Appl. **63**, 187–197 (2016). https://doi.org/10.1016/j.eswa.2016.07.006
10. Fang, H., Ong, S., Nee, A.: Interactive robot trajectory planning and simulation using augmented reality. Robot. Comput.-Integr. Manuf. **28**(2), 227–237 (2012). https://doi.org/10.1016/j.rcim.2011.09.003

11. Fang, H., Ong, S., Nee, A.: Robot path and end-effector orientation planning using augmented reality. Proc. CIRP **3**, 191–196 (2012). https://doi.org/10.1016/j.procir.2012.07.034
12. Fiorentino, M., Uva, A., Gattullo, M., Debernardis, S., Monno, G.: Augmented reality on large screen for interactive maintenance instructions. Comput. Ind. **65**(2), 270–278 (2014). https://doi.org/10.1016/j.compind.2013.11.004
13. Gilbert, W.: Economics of Machining. Machining Theory and Practice, pp. 465–485 (1950)
14. Hartley, R., Zisserman, A.: Multiple View Geometry in Computer Vision. Cambridge University Press (2003)
15. Hou, L., Wang, X., Bernold, L., Love, P.E.: Using animated augmented reality to cognitively guide assembly. J. Comput. Civ. Eng. **27**(5), 439–451 (2013)
16. Jozef, N.M., Miroslav, J., Ludmila, N.M.: Augmented reality aided control of industrial robots. Adv. Mater. Res. **1025–1026**, 1145–1149 (2014). https://doi.org/10.4028/www.scientific.net/AMR.1025-1026.1145
17. Kalpakjian, S., Schmid, S.R., Kok, C.W.: Manufacturing Processes for Engineering Materials. Pearson-Prentice Hall (2008)
18. Lowe, D.G.: Fitting parameterized three-dimensional models to images. IEEE Trans. Pattern Anal. Mach. Intell. **5**, 441–450 (1991)
19. Martínez, H., Laukkanen, S., Mattila, J.: A new hybrid approach for augmented reality maintenance in scientific facilities. Int. J. Adv. Robot. Syst. **1729**, 8806 (2013). Ferre, M., Mattila, J., Siciliano, B., Bonnal, P. (eds.)
20. Meden, B., Knodel, S., Bourgeois, S.: Markerless augmented reality solution for industrial manufacturing. In: ISMAR 2014—IEEE International Symposium on Mixed and Augmented Reality—Science and Technology 2014, Proceedings, pp. 359–360 (2014). https://doi.org/10.1109/ISMAR.2014.6948488
21. Monroy Reyes, A., Vergara Villegas, O., Miranda Bojrquez, E., Cruz Snchez, V., Nandayapa, M.: A mobile augmented reality system to support machinery operations in scholar environments. Comput. Appl. Eng. Educ. **24**(6), 967–981 (2016). https://doi.org/10.1002/cae.21772
22. Nee, A., Ong, S.: Virtual and augmented reality applications in manufacturing. In: IFAC Proceedings Volumes (IFAC-PapersOnline), pp. 15–26 (2013). https://doi.org/10.3182/20130619-3-RU-3018.00637
23. Olwal, A., Gustafsson, J., Lindfors, C.: Spatial augmented reality on industrial cnc-machines. In: Proc. SPIE **6804**, 680, 409–680, 409–9 (2008). https://doi.org/10.1117/12.760960
24. Ong, S., Pang, Y., Nee, A.B.: Augmented reality aided assembly design and planning. CIRP Ann.—Manuf. Technol. **56**(1), 49–52 (2007). https://doi.org/10.1016/j.cirp.2007.05.014
25. Ong, S., Yuan, M., Nee, A.: Augmented reality applications in manufacturing: a survey. Int. J. Prod. Res. **46**(10), 2707–2742 (2008). https://doi.org/10.1080/00207540601064773
26. PTC Inc.: Product Lifecycle Management (PLM) Software. http://www.ptc.com/product-lifecycle-management (2016). Accessed 05 Oct 2017, 15:40:45
27. Ramrez, H.B., Mendoza, E., Mendoza, M., Gonzlez, E.: Application of augmented reality in statistical process control, to increment the productivity in manufacture. Proc. Comput. Sci. **75**, 213–220 (2015). https://doi.org/10.1016/j.procs.2015.12.240
28. Schaumlöffel, P., Talha, M., Gorecky, D., Meixner, G.: Augmented reality applications for future manufacturing. In: Proceedings of the 5th Manufacturing Science and Education-MSE, vol. 1, no. 5, pp. 2–5 (2011)

29. Schweighofer, G., Pinz, A.: Robust pose estimation from a planar target. IEEE Trans. Pattern Anal. Mach. Intell. **28**(12), 2024–2030 (2006)
30. Setti, A., Bosetti, P., Ragni, M.: ARTool—Augmented reality platform for machining setup and maintenance. In: Bi, Y., Kapoor, S., Bhatia, R. (eds.) Proceedings of SAI Intelligent Systems Conference (IntelliSys) 2016. IntelliSys 2016. Lecture Notes in Networks and Systems, vol. 15, pp. 457–475, Springer, Cham (2018)
31. Surez-Warden, F., Mendvil, E., Ramrez, H., Garza Njera, L., Pantoja, G.: Mill setup manual aided by augmented reality. In: Mechanisms and Machine Science, vol. 25, pp. 433–441 (2015). https://doi.org/10.1007/978-3-319-09858-6_41
32. Syberfeldt, A., Danielsson, O., Holm, M., Wang, L.b.: Dynamic operator instructions based on augmented reality and rule-based expert systems. In: Procedia CIRP, vol. 41, pp. 346–351 (2016). https://doi.org/10.1016/j.procir.2015.12.113
33. Systems, D.: Solidworks Model Based Definitions. http://www.solidworks.it/sw/products/technical-communication/packages.htm (2016). Accessed 26 June 2017
34. Vignais, N., Miezal, M., Bleser, G., Mura, K., Gorecky, D., Marin, F.: Innovative system for real-time ergonomic feedback in industrial manufacturing. Appl. Ergon. **44**(4), 566–574 (2013). https://doi.org/10.1016/j.apergo.2012.11.008
35. Wang, X., Love, P.E., Kim, M.J., Park, C.S., Sing, C.P., Hou, L.: A conceptual framework for integrating building information modeling with augmented reality. Autom. Constr. **34**, 37–44 (2013)
36. Wang, Z., Ong, S., Nee, A.: Augmented reality aided interactive manual assembly design. Int. J. Adv. Manuf. Technol. **69**(5–8), 1311–1321 (2013)
37. Weinert, K., Zabel, A., Ungemach, E., Odendahl, S.: Improved nc path validation and manipulation with augmented reality methods. Prod. Eng. **2**(4), 371–376 (2008). https://doi.org/10.1007/s11740-008-0115-3
38. Wójcicki, T.: Supporting the diagnostics and the maintenance of technical devices with augmented reality. Diagnostyka **15**(1), 43–47 (2014)
39. Zhang, J., Ong, S., Nee, A.: A multi-regional computation scheme in an ar-assisted in situ cnc simulation environment. CAD Comput. Aided Des. **42**(12), 1167–1177 (2010). https://doi.org/10.1016/j.cad.2010.06.007

Some Properties of Gyrostats Dynamical Regimes Close to New Strange Attractors of the Newton-Leipnik Type

Anton V. Doroshin[✉]

Space Engineering Department (Division of Flight Dynamics and Control Systems), Samara National Research University, Samara, Russia
doran@inbox.ru; doroshin@ssau.ru

Abstract. New dynamical systems with strange attractors are numerically investigated in the article. These dynamical systems correspond to the main mathematical model describing the attitude dynamics of multi-spin spacecraft and gyrostat-satellites. The considering dynamical systems are structurally related to the well-known Newton-Leipnik system. Properties of the strange attractors arising inside the phase spaces of the dynamical systems are examined with the help of the numerical modelling.

1   Introduction

The investigation of dynamical systems with strange attractors is one of the important problems of the modern nonlinear dynamics [1–19]. Especially interesting cases of such systems represent the dynamical systems describing the natural behavior of mechanical, electrodynamical, biological, meteorological and other systems.

As the important part of such mechanical systems it is possible to indicate the multi-body systems, which dynamics is described by the ordinal differential equations. As one of the partial cases of such multi-body systems, in this paper we consider the mechanical model of the multi-spin spacecraft (MSSC), also called as the gyrostat-satellite. As it was shown in previous works [e.g. 1–5], the corresponding phase space of differential equations of the MSSC attitude dynamics can contain different forms of strange attractors, including cases of Newton-Leipnik-like two-scroll strange attractors [6]. Moreover, in the framework of MSSC dynamics the strange chaotic attractors can be defined as one additional dynamical opportunity, which allows to solve the task of the spacecraft spatial reorientation using chaotic properties of its angular motion (that is quite actual in some nontrivial cases of spacecraft motion, including accidents/failures in main control systems) [3–5]. Therefore, the problem of the strange attractors examination is important not only from the mathematical point of view, but also from the side of possible technical applications.

In this work we use the MSSC equations of motion as a mathematical basis which allows to write and to investigate the dynamical systems in the form of three ordinal differential equations containing strange chaotic attractors of the Newton-Leipnik type.

© Springer International Publishing AG 2018
Y. Bi et al. (eds.), *Intelligent Systems and Applications*,
Studies in Computational Intelligence 751,
https://doi.org/10.1007/978-3-319-69266-1_8

2 Main Dynamical Systems

The MSSC [1–5] represents the multi-body mechanical system with conjugated pairs of rotors placed on the inertia principle axes of the main body (Fig. 1).

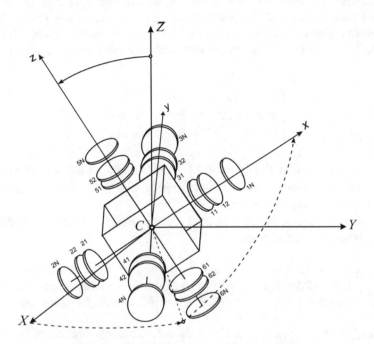

Fig. 1. The MSSC as the multirotor system

The equations of angular motion of MSSC around the "fixed" point O (the center of mass) [1–5] can be presented in the form:

$$\begin{cases} \widehat{A}\dot{p} + \dot{D}_{12} + \left(\widehat{C} - \widehat{B}\right)qr + [qD_{56} - rD_{34}] = M_x^e; \\ \widehat{B}\dot{q} + \dot{D}_{34} + \left(\widehat{A} - \widehat{C}\right)rp + [rD_{12} - pD_{56}] = M_y^e; \\ \widehat{C}\dot{r} + \dot{D}_{34} + \left(\widehat{B} - \widehat{A}\right)pq + [pD_{34} - qD_{12}] = M_z^e \end{cases} \qquad (1)$$

In these equations the following notations are used: $\omega = [p, q, r]^{\mathrm{T}}$—the vector of the absolute angular velocity of the MSSC main body (in projections on the connected frame $Oxyz$); $\widehat{A}, \widehat{B}, \widehat{C}$ are the summarized moments of inertia of the MSSC; M_x^e, M_y^e, M_z^e—the external torques acting on the system. The summarized rotors' angular momentums in the considered case are formed by the control system in the shape:

$$D_{12} = \alpha_p p + \alpha_0; \quad D_{34} = \beta_q q + \beta_0; \quad D_{56} = \gamma_r r + \gamma_0, \qquad (2)$$

The "external" torques also are created (by thrusters) as follows:

$$M_x^e = m_x + \alpha_1 p; \quad M_y^e = m_y + \beta_1 q; \quad M_z^e = m_z + \gamma_1 r, \tag{3}$$

In purposes of the fully description of the attitude dynamics of MSSC the dynamical system (3) should be supplemented by the kinematic equations for Euler (Tait–Bryan) angles defining the rotation of the MSSC connected system $Oxyz$ around the "fixed" point O (the mass center) and relatively the inertial coordinates system $OXYZ$:

$$\begin{cases} \dot{\gamma} = p \sin \varphi + q \cos \varphi; \\ \dot{\psi} = \frac{1}{\cos \gamma}(p \cos \varphi - q \sin \varphi); \\ \dot{\varphi} = r - \frac{\sin \gamma}{\cos \gamma}(p \cos \varphi - q \sin \varphi). \end{cases} \tag{4}$$

So, as we can see from the dynamical Eq. (1) (with the definitions (2) and (3)), the following constant "controlling" terms/coefficients take place: $\{\alpha_0, \ \beta_0, \ \gamma_0, \ m_x, \ m_y, \ m_z, \ \alpha_1, \ \beta_1, \ \gamma_1, \ \alpha_p, \ \beta_q, \ \gamma_r\} \sim$ const.

The system (1) in addition of (2) and (3) can be rewritten in the form of differential equations with quadratic right parts:

$$\begin{cases} \dot{x} = a_0 + a_1 x + a_2 y + a_3 z + a_4 x^2 + a_5 y^2 + a_6 z^2 + a_7 xy + a_8 xz + a_9 yz; \\ \dot{y} = b_0 + b_1 x + b_2 y + b_3 z + b_4 x^2 + b_5 y^2 + b_6 z^2 + b_7 xy + b_8 xz + b_9 yz; \\ \dot{z} = c_0 + c_1 x + c_2 y + c_3 z + c_4 x^2 + c_5 y^2 + c_6 z^2 + c_7 xy + c_8 xz + c_9 yz; \end{cases} \tag{5}$$

where $Coeff = \{a_i, b_i, c_i\}_{0 \leq i \leq 9} \in \mathbb{R}^{30}$ is the set of constant parameters, and where the designation of the variables are used: $p = x; \ q = y; \ r = z$. The following coefficients [1] of the system (5) have exact values:

$$\begin{cases} a_0 = \frac{m_x}{\hat{A} + \alpha_p}; \quad b_0 = \frac{m_y}{\hat{B} + \beta_q}; \quad c_0 = \frac{m_z}{\hat{C} + \gamma_r}; \\ a_1 = \frac{\alpha_1}{\hat{A} + \alpha_p}; \quad b_1 = \frac{\gamma_0}{\hat{B} + \beta_q}; \quad c_1 = \frac{-\beta_0}{\hat{C} + \gamma_r}; \\ a_2 = \frac{-\gamma_0}{\hat{A} + \alpha_p}; \quad b_2 = \frac{\beta_1}{\hat{B} + \beta_q}; \quad c_2 = \frac{\alpha_0}{\hat{C} + \gamma_r}; \\ a_3 = \frac{\beta_0}{\hat{A} + \alpha_p}; \quad b_3 = \frac{-\alpha_0}{\hat{B} + \beta_q}; \quad c_3 = \frac{\gamma_1}{\hat{C} + \gamma_r}; \\ a_4 = a_5 = a_6 = b_4 = b_5 = b_6 = c_4 = c_5 = c_6 \equiv 0; \\ a_7 = 0; \quad b_7 = 0; \quad c_7 = \left(\hat{A} - \hat{B} - \beta_q + \alpha_p\right) \big/ \left(\hat{C} + \gamma_r\right); \\ a_8 = 0; \quad b_8 = \left(\hat{C} - \hat{A} - \alpha_p + \gamma_r\right) \big/ \left(\hat{B} + \beta_q\right); \quad c_8 = 0; \\ a_9 = \left(\hat{B} - \hat{C} - \gamma_r + \beta_q\right) \big/ \left(\hat{A} + \alpha_p\right); \quad b_9 = 0; \quad c_9 = 0. \end{cases} \tag{6}$$

So, taking into our consideration correspondences (6) connecting the dynamical system's coefficients $\{a_i, b_i, c_i\}$ and MSSC parameters $\{\alpha_p, \alpha_0, m_x, \alpha_1, \beta_q, \beta_0,$

$m_y, \beta_1, \gamma_r, \gamma_0, m_z, \gamma_1\}$ with predefined inertia moments $\left\{\widehat{A}, \widehat{B}, \widehat{C}\right\}$, it is possible to find the concrete numerical values of the constants from the set

$$Control = \left\{\alpha_p, \alpha_0, m_x, \alpha_1, \beta_q, \beta_0, m_y, \beta_1, \gamma_r, \gamma_0, m_z, \gamma_1\right\} \in \mathbb{R}^{12} \qquad (7)$$

which can be used to providing the appropriate values of the coefficients $\{a_i, b_i, c_i\}$ close to a "target system" with a strange attractor. Due to the incompatibility of the indicated sets *Coeff* and *Control* we cannot solve this task exactly (as the algebraic equations), therefore, to obtain these values we must use some additional algorithms, e.g. the gradient-search method [3, 5]. Then using the algorithm [1] taking the Newton-Leipnik system as the "target system", it is possible to obtain new concretized dynamical systems with strange attractors inside phase spaces. These new systems and the corresponding properties are presented in the next section of this article.

3 The Numerical Investigation of New Strange Attractors

In this section we focus on the numerical investigation of the dynamical systems for MSSC obtained in the work [1]. These systems contain strange chaotic attractors and/or have the regular but very complex dynamical behavior, that is important not only from the mathematical point of view, but also from the technical point of practical applications. The next subsections of the article present the corresponding blocks of the numerical modeling for itch dynamical system of the Newton-Leipnik type which were found in [1, 5].

Here it is important to remind [6] the structure of the classical Newton-Leipnik system (NL). For the NL equations the following coefficients take place (all other coefficients equal to zero):

$$NL = \left\{ \begin{array}{l} a_1 = -0.4; \ a_2 = 1; \ a_9 = 10; \\ b_1 = -1; \ b_2 = -0.4; \ b_8 = 5; \\ c_3 = 0.175; \ c_7 = -5 \end{array} \right\} \qquad (8)$$

In the NL-system two strange attractors exist (Fig. 2): the upper attractor (black) with initial states (0.349, 0, −0.16), and the lower attractor (blue) with initial states (0.349, 0, −0.18).

So, below in the next subsections six new cases of the dynamical systems of the NL-type will be described, which are called as "SysA", "SysB", "SysC", "SysD", "Complex1" and "Complex2".

3.1 The System SysA Analysis

As it was explored in [1], the dynamical system with new strange attractor can be built at the following concretized set *Coeff* of non-zero numerical coefficients:

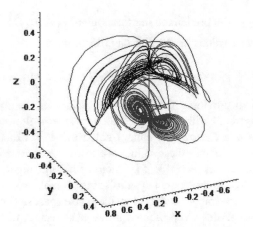

Fig. 2. The classical *Newton-Leipnik* attractors

$$SysA = \left\{ \begin{array}{l} a_1 = -0.4000; \ a_2 = 1.0738; \ a_9 = 10.0403; \\ b_1 = -0.0864; \ b_2 = -0.2471; \ b_8 = 0.1118; \\ c_3 = 0.1752; \ c_7 = -4.7831 \end{array} \right\} \tag{9}$$

These coefficients correspond to the following MSSC parameters [1]:

$$\widehat{A} = 1000; \quad \widehat{B} = 2500; \quad \widehat{C} = 3000 \ [\text{kg} \cdot \text{m}^2];$$
$$ControlSysA = (\alpha_p, \alpha_0, m_x, \ \alpha_1, \beta_q, \beta_0, m_y, \beta_1, \ \gamma_r, \gamma_0, m_z, \ \gamma_1) =$$
$$= (-692.7387, \ 0, \ 0, \ -122.9331, \ 1319.2399, 0, \ 0,$$
$$- 943.7322, \ -2265.7542, \ -329.9222, \ 0, \ 128.6660).$$

Then the new system takes place:

$$SysA = \left\{ \begin{array}{l} \dot{x} = -0.4000x + 1.0738y + 10.0403yz; \\ \dot{y} = -0.0864x - 0.2471y + 0.1118xz; \\ \dot{z} = 0.1752z - 4.7831xy \end{array} \right. \tag{10}$$

The SysA has a new strange chaotic attractor (Fig. 3) at the initial values $x(0) = 0.05; \ y(0) = 0.1; \ z(0) = 1.5$. This strange attractor (red) is depicted (Fig. 3) together with the classical "upper" Newton-Leipnik attractor (black).

The figure (Fig. 3) also contain the co-called hodograph ($\mathbf{e}_z$-hodograph) which represents the space curve corresponding to the trajectory motion of the apex of the Oz-axis of the MSSC (Fig. 1) in the inertial space $OXYZ$, that characterizes the chaotic side of the angular motion of the spacecraft.

In the purposes of the chaotic aspects description of the SysA-system dynamics along its strange attractor, we can present the time-history (Fig. 4) of the spatial angles (4), the Lyapunov's exponents for the SysA strange attractor and the fast Fourier transform (FFT) spectrum (Fig. 5) for the $x(t)$—signal on the attractor.

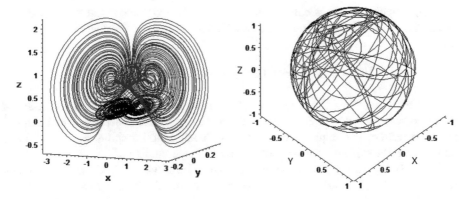

Fig. 3. The *SysA* attractor (red), the *Newton-Leipnik* attractor (black) and the e_z-hodograph

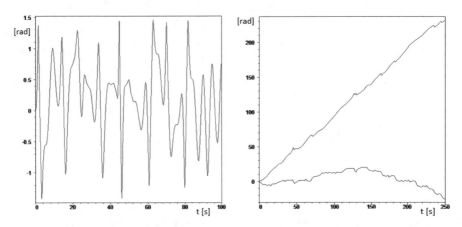

Fig. 4. The time-history of the angle $\gamma(t)$ (red), $\psi(t)$ (black), $\varphi(t)$ (blue) in the *SysA* system

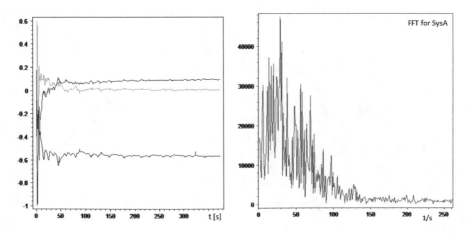

Fig. 5. The Lyapunov exponents and the spectrum of the fast Fourier transform of the $x(t)$-signal

It is important to calculate the Lyapunov's exponents spectrum and the Kaplan-Yorke dimension D_{KY} for the SysA strange attractor (evaluated with the tolerance 10^{-2}):

$$\{\lambda_1 = 0.09;\ \lambda_2 = 0.00;\ \lambda_3 = -0.57\};\quad D_{KY} = 2.17$$

As can we see, the Lyapunov's exponents spectrum has the classical signature for strange attractor in 3D-system $\{+, 0, -\}$; the Kaplan-Yorke dimension is fractional, and the FFT-spectrum is complex (with non-zero "continuous" amplitudes). All of these notations are the typical properties for chaotic strange attractors.

Also it is quite illustrative to show the Poincaré sections for SysA strange attractor (Figs. 6, 7 and 8), that also confirms the fractal nature of the strange attractor.

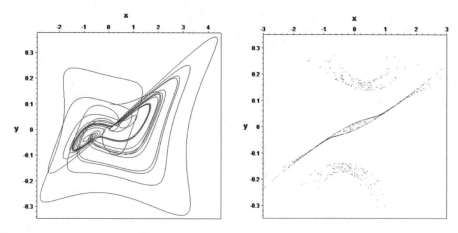

Fig. 6. The xy-projection of *SysA* strange attractor and its Poincaré section (by the plane z = 1)

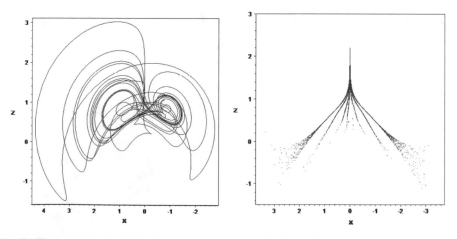

Fig. 7. The xz-projection of *SysA* strange attractor and its Poincaré section (by the plane y = 0)

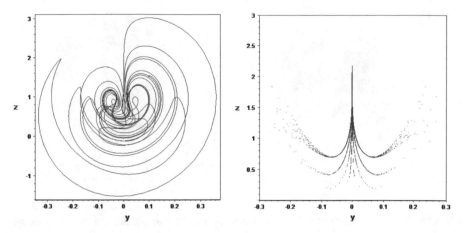

Fig. 8. The yz-projection of *SysA* strange attractor and its Poincaré section (by the plane x = 0)

3.2 The System SysB Analysis

The second case of the MSSC motion with generating the strange attractor realizes at the following parameters [1]:

$$\widehat{A} = 1000; \quad \widehat{B} = 2500; \quad \widehat{C} = 3000 \, [\text{kg} \cdot \text{m}^2];$$

$$ControlSysB = \left(\alpha_p, \alpha_0, m_x, \alpha_1, \beta_q, \beta_0, m_y, \beta_1, \gamma_r, \gamma_0, m_z, \gamma_1\right) =$$

$$= (-695.9057, \ 0, \ 0, \ -121.5977, \ 1281.2392, 0, \ 0,$$

$$- 1467.3693, \ -2272.0667, \ -326.3300, \ 0, \ 199.3635); \tag{11}$$

Then the dynamical system SysB can be indicated with corresponding coefficients:

$$SysB = \begin{cases} a_1 = -0.3999; \ a_2 = 1.0731; \ a_9 = 10.0407; \\ b_1 = -0.0863; \ b_2 = -0.3881; \ b_8 = 0.1121; \\ c_3 = 0.2739; \ c_7 = -4.7767 \end{cases} \tag{12}$$

So we have the new dynamical system

$$SysB = \begin{cases} \dot{x} = -0.3999x + 1.0731y + 10.0407yz; \\ \dot{y} = -0.0863x - 0.3881y + 0.1121xz; \\ \dot{z} = 0.2739z - 4.7767xy \end{cases} \tag{13}$$

with the strange chaotic attractor (Fig. 9) at the initial values $x(0) = 0.05$; $y(0) = 0.1$; $z(0) = 1.5$, depicted together with the "upper" Newton-Leipnik attractor.

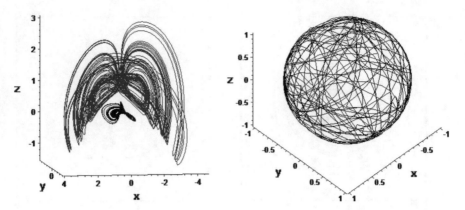

Fig. 9. The *SysB* attractor (red), the *Newton-Leipnik* attractor (black) and the e_z-hodograph

In this case the Lyapunov's exponents spectrum and the Kaplan-Yorke dimension D_{KY} for the SysA strange attractor (evaluated with the tolerance 10^{-2}) are:

$$\{\lambda_1 = 0.14;\ \lambda_2 = 0.00;\ \lambda_3 = -0.61\};\quad D_{KY} = 2.22$$

As in the previous case, we see the typical properties for chaotic strange attractors: the Lyapunov's exponents spectrum has the classical signature for strange attractor in 3D-system $\{+,\ 0,\ -\}$; the the Kaplan-Yorke dimension is fractional, and the FFT-spectrum (Fig. 11) is complex (with non-zero "continuous" amplitudes). The Poincaré sections for SysB strange attractor (Figs. 12, 13 and 14, and the time-history Fig. 10) confirm the fractal nature of the strange attractor.

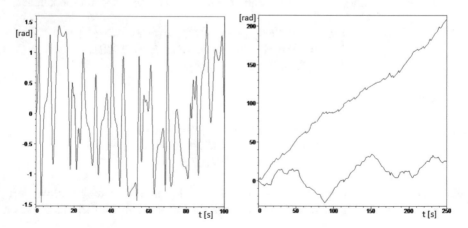

Fig. 10. The time-history of the angle $\gamma(t)$ (red), $\psi(t)$ (black), $\varphi(t)$ (blue) in the *SysB* system

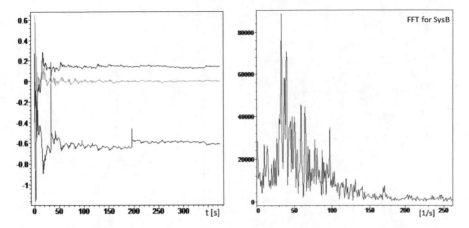

Fig. 11. The Lyapunov exponents and the spectrum of the fast Fourier transform of $x(t)$-signal

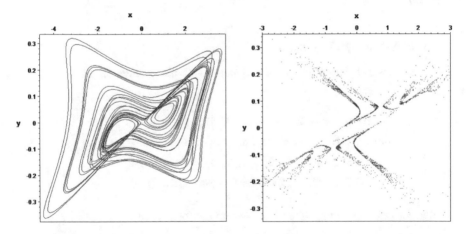

Fig. 12. The xy-projection of *SysB* strange attractor and its Poincaré section (by the plane z = 1)

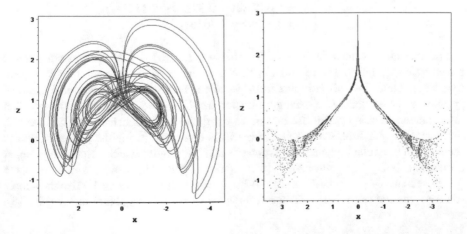

Fig. 13. The xz-projection of *SysB* strange attractor and its Poincaré section (by the plane y = 0)

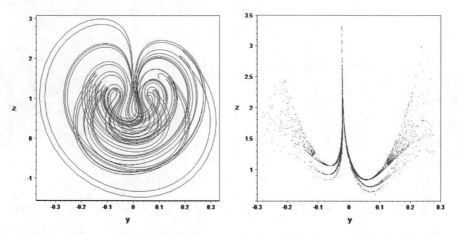

Fig. 14. The yz-projection of *SysB* strange attractor and its Poincaré section (the plane x = −0.25)

3.3   The System SysC Analysis

The MSSC has the complex attitude dynamics at the realization of the following parameters [1]:

$$\widehat{A} = 1000; \quad \widehat{B} = 2500; \quad \widehat{C} = 3000 \ [\text{kg} \cdot \text{m}^2];$$
$$ControlSysC = \left(\alpha_p, \alpha_0, m_x, \ \alpha_1, \beta_q, \beta_0, m_y, \beta_1, \ \gamma_r, \gamma_0, m_z, \ \gamma_1\right) =$$
$$= (-682.4176, \ 0, \ 0, \ -126.8955, \ 1451.8728, 0, \ 0,$$
$$- 1473.7799, \ -2237.0650, \ -340.5281, \ 0, \ 292.1299);$$

Then the corresponding dynamical system can be presented:

$$SysC = \begin{cases} \dot{x} = -0.3996x + 1.0723y + 10.0413yz; \\ \dot{y} = -0.0862x - 0.3729y + 0.1127xz; \\ \dot{z} = 0.3829z - 4.7636xy \end{cases} \tag{14}$$

At the initial values $\{x(0) = 0.05; \ y(0) = 0.1; \ z(0) = 1.5\}$ the system has the phase trajectory (Figs. 15, 16 and 17) similar to the previous strange attractor's form (e.g. Figs. 3 and 9)—this complex form was the main reason to define this dynamical system as the system with strange attractor in [1]. However, at the more detailed investigation of this system the fact of its regularity was confirmed. As can we see, firstly, the regular limit cycle (Fig. 15-black) is contained inside the phase trajectory; secondly, the regular Lyapunov's exponents take place with the corresponding integer dimension of the attractor (the limit cycle in the form of the 3D-closed-curve): $\{\lambda_1 = 0.00; \ \lambda_2 = -0.11; \ \lambda_3 = -0.28\}$; $D_{KY} = 1$. Also the simple (discrete and concentrated) FFT-spectrum is actual for this regular attractor (Fig. 18).

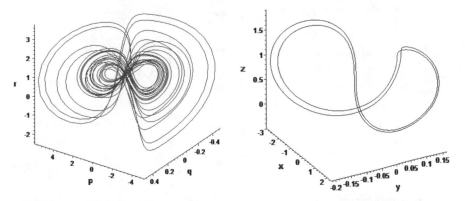

Fig. 15. The phase trajectory (red) close to the limit cycle of the *SysC* system (black)

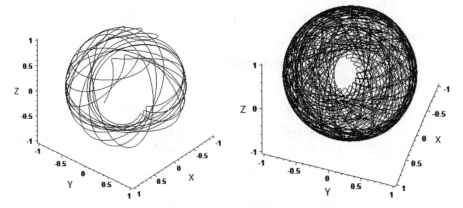

Fig. 16. The e_z-hodograph (and its evolution) of the system *SysC*

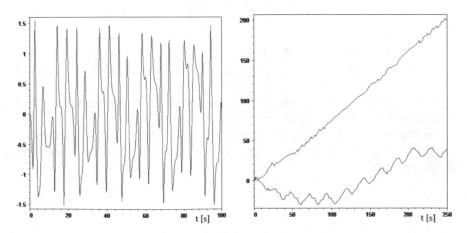

Fig. 17. The time-history of the angle $\gamma(t)$ (red), $\psi(t)$ (black), $\varphi(t)$ (blue) in the *SysC* system

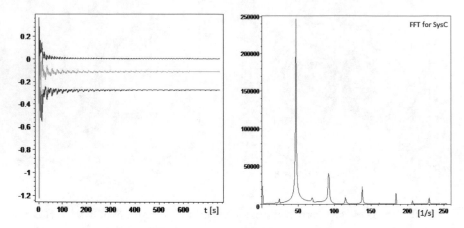

Fig. 18. The Lyapunov exponents and the spectrum of the fast Fourier transform of $x(t)$-signal

3.4   The System SysD

In the work [5] one more dynamical system with strange attractor was found at the following MSSC parameters:

$$\widehat{A} = 90; \quad \widehat{B} = 70; \quad \widehat{C} = 50 \text{ [kg} \cdot \text{m}^2];$$

$$ControlSysD = \left\{ \begin{array}{l} \alpha_p = -80.6893; \ \alpha_0 = 0; \ m_x = 0; \ \alpha_1 = -3.7243; \\ \beta_q = 45.7309; \ \beta_0 = 0; \ m_y = 0; \ \beta_1 = -46.2952; \\ \gamma_r = -27.7522; \ \gamma_0 = -9.9951; \ m_z = 0; \ \gamma_1 = 3.8934 \end{array} \right\} \quad (15)$$

As the result, the dynamical system took the form:

$$SysD = \left\{ \begin{array}{l} \dot{x} = -0.4x + 1.0735y + 10.0403yz; \\ \dot{y} = -0.0864x - 0.4y + 0.1118xz; \\ \dot{z} = 0.1750z - 4.7834xy \end{array} \right. \quad (16)$$

The detailed investigation of the strange attractor (Fig. 19) in this system confirmed the presence of all properties, which are usual for chaotic fractal objects: the Lypunov's spectrum $\{\lambda_1 = 0.10; \ \lambda_2 = 0.00; \ \lambda_3 = -0.59\}$ has the strange attractor's signature $\{+, 0, -\}$; the fractional dimension $D_{KY} = 2.16$; the complex FFT-spectrum (Fig. 21) and fractal pictures at the Poincaré sections (Figs. 20, 22, 23 and 24).

The chaotic angular motion of the MSSC also is illustrated by the complex $\mathbf{e}_z$-hodograph of the Oz-axis apex (Fig. 19), and by the chaotic time-evolutions for the Euler angles (Fig. 20).

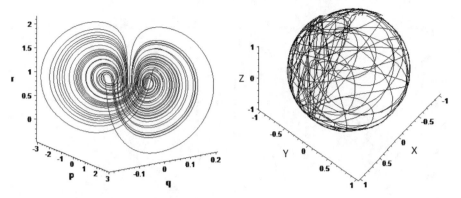

Fig. 19. The phase trajectory (red) and e_z-hodograph (black) for the *SysD* system

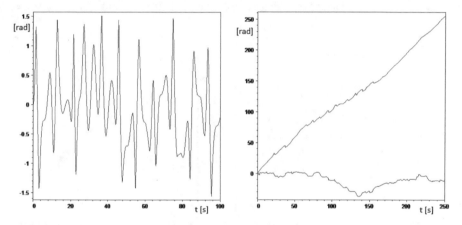

Fig. 20. The time-history of the angle $\gamma(t)$ (red), $\psi(t)$ (black), $\varphi(t)$ (blue) in the *SysD* system

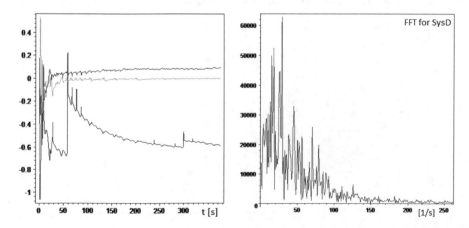

Fig. 21. The Lyapunov exponents and the spectrum of the fast Fourier transform of $x(t)$-signal

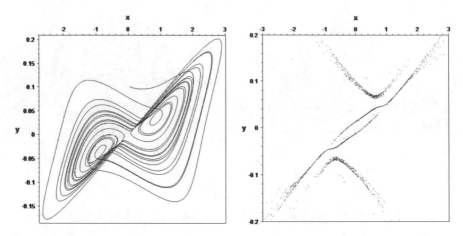

Fig. 22. The xy-projection of *SysD* strange attractor and its Poincaré section (by the plane z = 1)

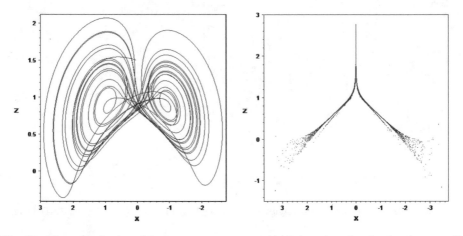

Fig. 23. The xz-projection of *SysD* strange attractor and its Poincaré section (by the plane y = 0)

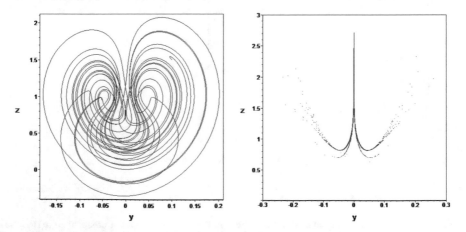

Fig. 24. The yz-projection of *SysD* strange attractor and its Poincaré section (by the plane x = 0)

3.5 The System Complex1

In this section we consider the system which do not include the strange attractors, but it has the complex dynamics of phase trajectories, that can be close to the chaotic dynamics in the sense of a complexity of the MSSC angular motion.

In this case the MSSC has the following parameters [1]:

$$\widehat{A} = 100; \quad \widehat{B} = 250; \quad \widehat{C} = 300 \ [\text{kg} \cdot \text{m}^2];$$
$$ControlComplex1 = (\alpha_p, \alpha_0, m_x, \alpha_1, \beta_q, \beta_0, m_y, \beta_1, \gamma_r, \gamma_0, m_z, \gamma_1) =$$
$$= (-2886.4968, \ 0, \ 0, \ 361.7618, \ 409.0296, 0, \ 0, -263.7884,$$
$$467.3039, -476.9110, \ 0, \ 133.8367); \tag{17}$$

Then the dynamical system can be written:

$$Complex1 = \begin{cases} \dot{x} = -0.1298x - 0.1712y + 0.03886yz; \\ \dot{y} = -0.7237x - 0.4003y + 5.3925xz; \\ \dot{z} = 0.1744z - 4.4904xy \end{cases} \tag{18}$$

In the considered case the complex phase trajectory is generated in the system; and this phase trajectory in the stream of time proceeds to the limit cycle depicted at the figure (Fig. 25).

As it is evaluated for the indicated regular attractor (limit cycle), the Lyapunov exponents spectrum $\{\lambda_1 = 0.00; \ \lambda_2 = -0.09; \ \lambda_3 = -0.28\}$ has the signature $\{0, -, -\}$; the dimensions is integer $D_{KY} = 1$; the FFT-spectrum (Fig. 27) is simple, concentrated. The $\mathbf{e}_z$-hodograph in this case is regularized (Fig. 26).

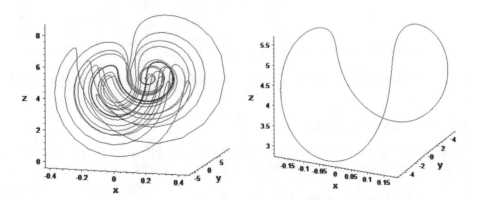

Fig. 25. The phase trajectory of the *Complex1*-system and its limit cycle

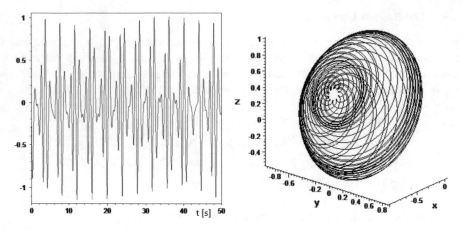

Fig. 26. The time-history of the angle $\gamma(t)$ and e_z-hodograph (black) for the *Complex1* system

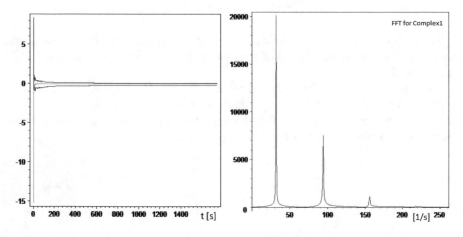

Fig. 27. The Lyapunov exponents and the spectrum of the fast Fourier transform of $x(t)$-signal

3.6   The System Complex2

At the end of the considering set of dynamical systems with complicated dynamics let us present the Complex2 system, which is realized at the MSSC parameters [1]:

$$\widehat{A} = 100; \quad \widehat{B} = 250; \quad \widehat{C} = 300 \ [\text{kg} \cdot \text{m}^2];$$
$$ControlComplex2 = \left(\alpha_p, \alpha_0, m_x, \ \alpha_1, \beta_q, \beta_0, m_y, \beta_1, \ \gamma_r, \gamma_0, m_z, \ \gamma_1\right) =$$
$$= (-2947.8679, \ 0, \ 0, \ 23.5201, \ 430.1965, 0, \ 0,$$
$$- \ 77.3623, \ 500.7775, -105.2338, \ 0, \ 38.9642); \tag{19}$$

$$Complex2 = \begin{cases} \dot{x} = -0.0083x - 0.0369y + 0.0423yz; \\ \dot{y} = -0.1547x - 0.1137y + 5.3641xz; \\ \dot{z} = 0.0487z - 4.4058xy \end{cases} \tag{20}$$

Then in the system's phase space the complex objet is contained; this object can be defined as the complex periodical cycle (Figs. 28 and 31) with two alternate dissipative scrolls. At the figure (Fig. 31) this complex periodical cycle is depicted separately (at the initial values $x(0) = 0.1$, $y(0) = 0.0$, $z(0) = 0.0$). Evaluations give the non-negative Lyapunov exponents $\{\lambda_1 = 3.45; \lambda_2 = 0.36; \lambda_3 = 0.00\}$ for the indicated cycle; and it means the exponential instability of this regime and increasing the phase volume along this complex cycle. Moreover, the dynamical complexity of the regime can be illustrated by the quite complicated FFT-spectrum (Fig. 30), which is rather distributed than concentrated, but decaying. The angular motion of the MSSC is in this case is complex and practically chaotic (Fig. 29), that confirmed by the complex hodograph (Fig. 28).

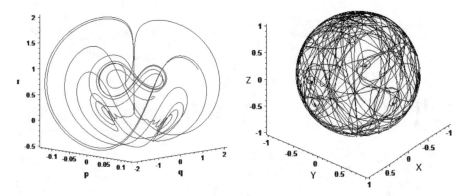

Fig. 28. The *Complex2* phase trajectory (red) and e$_z$-hodograph

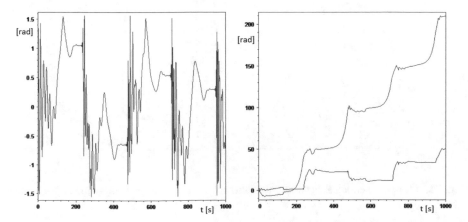

Fig. 29. The time-history of the angle $\gamma(t)$ (red), $\psi(t)$ (black), $\varphi(t)$ (blue) in the *Complex2* system

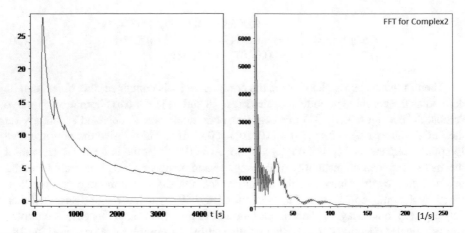

Fig. 30. The Lyapunov exponents and the spectrum of the fast Fourier transform of $x(t)$-signal

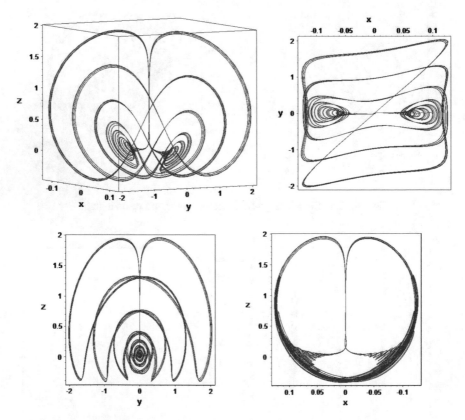

Fig. 31. Complex periodical cycle in the phase space of the *Complex2* system (in projections)

4 Conclusion

In this work the new dynamical systems of the Newton-Leipnik type with strange attractors (or with the complex dynamical behavior) were considered with the detailed numerical investigation of corresponding properties and with the evaluation of theirs characteristics. The phase spaces of the systems and generated strange attractors (and cycles) were explored, including the study of the main characteristics of chaotic/regular dynamics, such as the Lyapunov's exponents, the fractal dimension of strange attractors, the fast Fourier transform of the attractors' signals, the Poincaré sections. All of the considered dynamical systems have the "natural origin" corresponding to the mechanical and mathematical models of the angular motion of spacecraft; and, certainly, the results of the investigation can be applied to solving tasks of the spacecraft chaotic reorientation and chaotic maneuvering.

Acknowledgements. This work is partially supported by the Russian Foundation for Basic Research (RFBR#15-08-05934-A), and by the Ministry of education and science of the Russian Federation in the framework of the State Assignments to higher education institutions and research organizations in the field of scientific activity (the project # 9.1616.2017/ПЧ).

References

1. Doroshin, A.V.: New strange chaotic attractors in dynamical systems of multi-spin spacecraft and gyrostats. In: SAI Intelligent Systems Conference (IntelliSys-2016) (2016)
2. Doroshin, A.V.: Modeling of chaotic motion of gyrostats in resistant environment on the base of dynamical systems with strange attractors. Commun. Nonlinear Sci. Numer. Simul. **16**(8), 3188–3202 (2011)
3. Doroshin, A.V.: Initiations of chaotic regimes of attitude dynamics of multi-spin spacecraft and gyrostat-satellites basing on multiscroll strange chaotic attractors. In: SAI Intelligent Systems Conference (IntelliSys), London, U.K., pp. 698–704 (2015)
4. Doroshin, A.V.: Multi-spin spacecraft and gyrostats as dynamical systems with multiscroll chaotic attractors. In: Science and Information Conference (SAI), London, U.K., pp. 882–887 (2014)
5. Doroshin, A.V.: Initiations of chaotic motions as a method of spacecraft attitude control and reorientations. In: IAENG Transactions on Engineering Sciences, pp. 15–28 (2016)
6. Leipnik, R.B., Newton, T.A.: Double strange attractors in rigid body motion with linear feedback control. Phys. Lett. A **86**, 63–67 (1981)
7. Jafari, S., Sprott, J.C., Nazarimehr, F.: Recent New Examples of Hidden Attractors. The European Physical Journal Special Topics **224**, 1469–1476 (2015)
8. Elhadj, Z., Sprott, J.C.: Simplest 3D continuous-time quadratic systems as candidates for generating multiscroll chaotic attractors. Int. J. Bifurcat. Chaos **23**(7) (2013)
9. Sprott, J.C.: Some simple chaotic flows. Phys. Rev. E **50** (1994)
10. Sprott, J.C.: Simplest dissipative chaotic flow. Phys. Lett. A **228**, 271–274 (1997)
11. Wang, Z., Sun, Y., van Wyk, B.J., Qi, G., van Wyk, M.A.: A 3-D four-wing attractor and its analysis. Braz. J. Phys. **39**, 547–553 (2009)
12. Chen, H.K., Lee, C.I.: Anti-control of chaos in rigid body motion. Chaos, Solitons Fractals **21**, 957–965 (2004)

13. Sheu, L.J., Chen, H.K., Chen, J.H., Tam, L.M., Chen, W.C., Lin, K.T., Kang, Y.: Chaos in the Newton-Leipnik system with fractional order. Chaos, Solitons Fractals **36**(1), 98–103 (2008)
14. Wang, X., Tian, L.: Bifurcation analysis and linear control of the Newton-Leipnik system. Chaos, Solitons Fractals **27**(1), 31–38 (2006)
15. Jia, Q.: Chaos control and synchronization of the Newton-Leipnik chaotic system. Chaos, Solitons Fractals **35**(4), 814–824 (2008)
16. Zhang, K., Wang, H., Fang, H.: Feedback control and hybrid projective synchronization of a fractional-order Newton-Leipnik system. Commun. Nonlinear Sci. Numer. Simul. **17**(1), 317–328 (2012)
17. Richter, H.: Controlling chaotic systems with multiple strange attractors. Phys. Lett. A **300**(2), 182–188 (2002)
18. Yun-Zhong, S., Guang-Zhou, Z., Dong-Lian, Q.: Passive control of chaotic system with multiple strange attractors. Chin. Phys. **15**(10), 2266 (2006)
19. Marlin, B.A.: Periodic orbits in the Newton-Leipnik system. International Journal of Bifurcation and Chaos **12**(03), 511–523 (2002)

Toward Designing an Efficient System for Delivering Contextual Content

Jawaher Al-Yahya, Nouf Al-Rowais, Sara Al-Shathri,
Lamees Alsuhaibani, Amal Alabdulkarim, and Lamya AlBraheem[✉]

Information Technology Department, College of Computer and Information
Sciences, King Saud University, Riyadh, Saudi Arabia
jawaher-khaled@hotmail.com, {nouf.alrowaiss,
lameessuhaibani,amal.alabdulkarim}@gmail.com,
433200155@student.ksu.edu.sa, lamia_ea@yahoo.com

Abstract. It can be seen that a huge attention has been given to the field of location-aware systems. Different technologies such as: RFID, NFC, QR, GPS and iBeacon, have been developed to be used for these systems. However, in order to design an efficient system for location-based content delivery, there is a need to study these technologies with their features and issues. Moreover, this study provides a review and comparison for the available mobile applications that use iBeacon technology for delivering contextual content. According to the result of evaluation and comparison, the idea is developing a location-aware mobile system using the iBeacon technology. This system can help the event's organizers to enhance attendee's experience by providing location based content such as PDFs, image and videos. This is can provide an efficient way for content delivery by reducing the costs and replacing the printed papers.

1 Introduction

Nowadays, it can be seen that the smartphone sales have increased largely all over the world [1]. This increase has contributed mainly to enriching the number of important applications that facilitate the daily life. In addition, there is an interest in developing location-aware applications, which can deliver content to users based on their locations [2]. These applications can be used for different useful purposes in airports, shopping stores and events. In fact, users want to access and find relevant information for a place, on the other hand, content providers want to post their content to the nearby users.

There are different technologies that can be used for delivering content. Fortunately, most smartphones have built-in sensors that can be used as a tool to enhance user experience an provide contextual content. The use of technologies like Quick Response (QR) Code, Radio Frequency Identification (RFID) and Near Field Communications (NFC) allows mobile phones to interact with the surrounding environment and bridge the gap between the physical and digital world [3]. Today, you can find different applications that use these technologies to provide contextual location based content. However, there are some limitations of these technologies which present the need of finding a technology that attempt to address these issues.

© Springer International Publishing AG 2018
Y. Bi et al. (eds.), *Intelligent Systems and Applications*,
Studies in Computational Intelligence 751,
https://doi.org/10.1007/978-3-319-69266-1_9

One of latest location-aware technologies is iBeacon, which is a technology that was developed by Apple to provide services based on location. Since its launch, more and more retailers, event organizers, and museums are using beacons to provide more interactive contextual experience [4]. This growing interest of using iBeacon technology can give attention to the idea of designing an application that use this technology for delivering location based information.

In this study, a location-aware mobile system will be developed using the iBeacon technology. This system can be used in the event sector in order to deliver the contextual content to the event attendees. The system also provides the required functions for event organizers to manage the events and content.

1.1 Problem Statement

In 2012, more than 3.5 million people attended exhibitions and conferences in Saudi Arabia. Also, there are more than 1200 authorized event's organizers in the region. These huge numbers of attendees may not always be able to get contents easily due to the lack of technology. On the other hand, event organizers are usually looking to find efficient ways for content delivery, in order to improve attendee's experience, reduce costs and replace the printed papers. Therefore, there is a need to develop an application to improve event management by allowing the event organizers to offer digital content to attendees based on their locations.

As mentioned previously, there are wide ranges of technologies that used for content delivery. However, they have different limitations and issues. For example, successful communication using NFC technology requires short distance between the communicated devices, which is fewer than 10 cm [3]. Furthermore, QR code operates passively, since it does not send the content automatically to the mobile phones and the users have to scan the code by their phone [5]. This shows the need for finding a better technology for providing location based information.

In addition, it can be seen that there is an obvious deficiency in the number of location-aware applications targeting Arabic users. Therefore, to address all of these issues, there is a need for designing an effective solution that can help Arabic users to get the appropriate content in the events based on their locations.

1.2 Scope

The scope of our project is developing a mobile system that can be used for delivering content to the attendees of event based on their location. The application allows the user to discover, save and share any near-by content. On the other hand, the system can give the authority to multiple organizers to add and manage their content. The application will be developed for IOS platform. In addition, a website that used for managing the system by the organizers. The language that will be used for this application is Arabic. Moreover, the system can be tested for any event.

1.3 Objectives

1. Providing background information that important for building knowledge about the field of study, which covers location awareness, Bluetooth Low Energy (BLE) and Beacon devices.
2. Develop a location-aware system that can deliver content to the event attendees based on their location.
3. Provide an accessible interface that supports the Arabic language for delivering the content in events.
4. Help the organizers to add and manage their events content and give them the ability to measure the success of their events through user's feedback or responses.
5. Presenting the related works that include comparison and discussion.
6. Performing the analysis and design process that required for developing the proposed system.
7. Presenting the required software tools that need to implement the system.

2 Background

In this section, the background information that is required for understanding our project will be presented. First, we will provide a brief definition of the location awareness. Then, we go over the underlying technology behind iBeacon which is the Bluetooth Low Energy. Moreover, description of the iBeacon technology as well as the beacon hardware will be presented.

2.1 Location Awareness

Location awareness is when an application is aware of the current location and acts on that awareness by presenting or retrieving relevant information. As a result, an application that is location aware can adapt better to the user's environment and provide more useful information than a regular application [6]. Location-aware application uses location information that can be obtained by using sensors. An example of such sensor is a Bluetooth sensor, which can sense the surrounding environment for nearby devices and capture location context [7].

2.2 Bluetooth Low Energy

The technology that is used by the iBeacon is the Bluetooth Low Energy. The Bluetooth Low Energy (BLE, often called Bluetooth smart) was introduced as an enhancement of the Bluetooth core specification as Bluetooth 4.0. In fact, BLE consumes little energy as possible comparing with the previous classical Bluetooth [8]. With this enhancement, two modes of communication were introduced, either by broadcasting or by connections. In broadcasting a broadcaster device can send data to any scanning devices in range called observers. Unlike broadcasting, which can send data to more than one peer at time, connections are used so that two devices communicate. Apple's iBeacon uses only the broadcast mode to periodically send a specific advertisement, so that any device that comes in range detects the beacon [8] (Fig. 1).

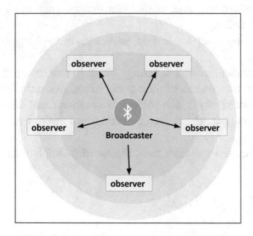

Fig. 1. Broadcast topology

2.3   Beacon

A beacon device is a hardware with BLE that can implement the iBeacon technology to transmit advertising packets. Although devices like phones and laptops that are equipped with a Bluetooth 4.0 or a later version can become a beacon, a dedicated standalone beacon device is cheaper and has a longer battery life [9]. Various manufactures offer dedicated standalone beacon devices in all shapes and sizes however, there are some major factors to consider in a beacon:

1. Battery Life: A continuous steady power source is needed to maintain performance from a beacon. Most beacons use coin cell batteries since they provide the most power in a small size. Some beacons instead use AA batteries but they are larger than coin cells [7, 10].
2. Firmware: A firmware is a programmed code that operates the beacon. Each beacon provides their own configuration method either by using a vendor's app or by an open interface to configure multiple beacons at once [10].
3. Software support (SDK): SDKs can help in the development and the management of an iBeacon application. Therefore, most vendors ship their own SDK with their beacon [7].

   Therefore, it is really important to understand their differences and specifications to choose the most suitable beacon for our application. As shown in the table below, a comparison of beacons from some of the major beacon manufacturers (Table 1).

Table 1. Comparison between beacon devices

Beacon	Description	Range (m)	Battery type	Battery life
Estimote	An enclosed beacon in a colorful silicon case suitable for an outdoor environment. But the case makes the battery unreplaceable [10]	40–50 [11]	Large coin cell [10]	21.4 Months [10]
Gimbal	Although it provides stable signals, it has a large size for its low battery life [10]	Up to 50 [12]	Four AA alkaline batteries [10]	16.4 Months [10]
Kontakt	Provides accurate transmission and power optimization [10]	Up to 70 [13]	Large coin cell [10]	24.3 Months [10]
Bluecats	Enclosed in a casing that makes it water-resistant [14]	30–60 [10]	AA alkaline batteries [10]	23.1 Months [10]
Mpact	Unites BLE and Wi-Fi to provide location-based services [10]	4.57–36.58 [15]	Medium coin cell [10]	2.1 Months [10]

3 Related Works

In this section, the applications that conducted and developed using iBeacon technology to deliver content to users based on their location will be presented. In addition, a comparison and discussion between theses application is provided. This is can be useful to include the best features for our suggested system and find out the limitations that require more improving. It should be mentioned that the details about the review and comparison of the location-based technologies, which are: NFC, RFID, QR code, GPS and iBeacon, are presented and discussed in our previous paper [16].

3.1 BluuBeam

BluuBeam is a mobile application that available for iOS and Android. First, it is used in libraries to notify visitors about related information of the things around their locations. Then, they extend to be use in different categories. BluuBeam provide tow type of users. First type is a regular user who can show, share and save information that get from their around beacons. The other type as an admin user who has beacons in addition with the application. The admin, which called also a (beamer), can create the information content of their things then the application will connect them automatically with his beacons. Latterly, they send it to the users comes around beacon [17].

3.2   TradeShowPro

TradeShowPro is a comprehensive events app developed to improve trade show attendees experience by letting them discover, navigate and save information at trade shows. It allows them to log their notes and save information about booths and exhibitors. On the other hand, It allows exhibitors reach attendees after the show with promotions and special offers [18].

3.3   Webble

Webble is mobile application that available for iOS and Android. The idea for designed is to apply the concept of internet of things using iBeacon technology. It is focus on the entertainment activities like shopping, eating and playing. It is allow users to show, share, save and search for nearby contents beads on their locations or beacons around them. Also, it is allow user to add and share their own contents and link them with their own beacon or current location using Google map [19] (Fig. 2).

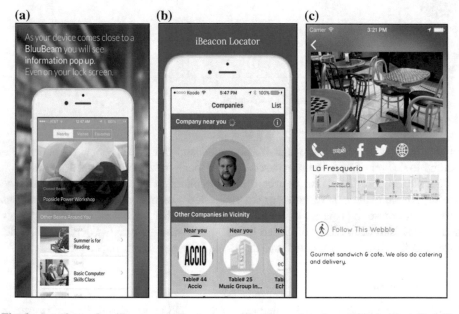

Fig. 2. Interfaces for iBeacon applications: **a** BluuBeam interface [17], **b** TradeShowPro interface [18], **c** Webble interface [19]

3.4   Dubaiculture

Dubai culture is mobile application that available for iOS and Android. It is an interactive app to help people discover Dubai in a different way. By locating attractions and get information about them using iBeacon and interactive maps. It allows the user to show, share and save information. In addition it is allow the user to know about new events in the city and add it to their calendar [20].

3.5 Beaconguides

Beacon Guides is mobile application that available for iOS. It provides a rich and interactive experience in museums, institutions and galleries using beacons. The users are able to determine their locations, and get related information provided by that location. In addition, it allows users to save and search for a specific information. In addition, the users can create their own guides and publish it on the application [21] (Fig. 3).

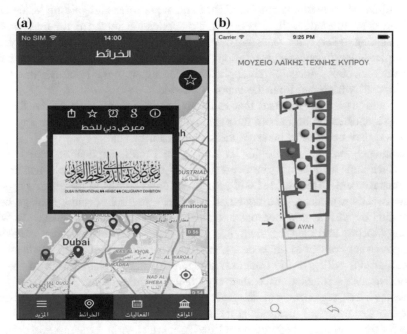

Fig. 3. Interfaces for iBeacon applications: **a** Dubai culture interface [20], **b** Beacon guides interface [21]

3.6 Event Wallet

Event Wallet provides attendees to multiple types of events a simple way to browse, save and share content. It also helps them find out more about booths they visit, save content you can access later, then share it through email or social networks [22].

3.7   Beep

BEEP is a mobile app for enhancing attendees visit by sharing live content and location-based information in the event. BEEP also tracks the attendees' visit and log it to help organizers to identify the major attractions of their event. BEEP uses Beacon technology to improve event management and content sharing [23].

3.8   Comparison Between Mobile Applications that Use iBeacon Technology

In this section, a comparison between the mobile applications that utilize iBeacon technology and our application will be introduced (see Table 2).

Per to this comparison, the only application that supported the Arabic language is Dubai culture. However, it is employed for tourism purposes and guiding inside Dubai. Based on our knowledge, there is no Arabic application intended for location-based content sharing to serve at events and conferences using the iBeacon technology. Also, most applications do not give event organizers the ability to modify their content and create their accounts. Instead, the application developers have to assign most of the event contents which can limit the features provided.

It should also be mentioned that most applications don't provide the feature of receiving a statistical report about the event content to the organizer in the application. This report may be used to measure the success of the event.

Accordingly, we are developing an Arabic iOS mobile application called "Inform Me", which aims to increase the interactivity of events with attendees and improve event management. "Inform Me" will provide several features to the event's organizer and the event's attendees. The organizer can create multiple events, manage beacon information by adding new beacons and linking them to a specific event. As will as updating beacons and deleting them. Moreover, they can manage events' content and connect relevant content to the beacons. Inform Me, unlike most of the applications in the comparison, will give the organizer a report containing statistics about the event and the attendee's interactions with every content. Furthermore, the application will provide the event organizer with the feature of managing multiple events and beacons in different geographical locations at the same time.

As for the Attendees, they will be notified about the nearby beacons. Also, they will automatically receive the content and have different features, such as viewing, sharing and saving the content. Also, they can provide feedback on the content they are viewing. It should be mentioned that some event management applications such as event wallet share many functions with Inform me but the Inform Me system support of the Arabic Language is a remarkable difference for Arabic users [24].

Table 2. Comparison between mobile applications that uses iBeacon technology to deliver location-based content

Features	Trad show pro	Blue beam	Webble	Dubai culture	Beacon guides	Event wallet	BEEP	Inform Me
Platform	iOS	iOS + Android	iOS + Android	iOS + Android	iOS	iOS + Website	iOS + Android	iOS + Website
Category	Event	General	Entertainment	Culture	Culture	Events	Events	Event
Features for organizers								
Arabic language	X	X	X	✓	X	X	X	✓
User account	✓	X	✓	X	✓(admin)	✓	✓	✓
Manage beacon information	X	X	✓	X	X	✓	✓	✓
Manage content	X	X	✓	X	✓	✓	✓	✓
Receive statistical report	X	X	X	X	X	✓	✓	✓
Features for attendees								
View content	✓	✓	✓	✓	✓	✓	✓	✓
Share content	✓	✓	✓	✓	X	✓	X	✓
Save content	✓	✓	✓	✓	✓	✓	✓	✓
Add feedback	X	X	✓	✓	X	X	✓	✓
Receive content automatically	✓	✓	X	✓	X	✓	X	✓
Notify for a nearby beacon	✓	✓	✓	✓	X	✓	✓	✓
Visual map for beacon	X	X	✓	✓	✓	X	X	X

4   System Analysis and Design

There are two main methodologies to system development which are Object-Oriented approach and structured approach. The approach that used for the proposed system in analysis, design and implementation is the Object-Oriented. The system analysis includes two parts which are software requirements and system models. On the other hand, system design process includes the architectural design, data design, component design and finally the interface design. In this section, we will provide the most important parts of the process of system analysis and design.

4.1   User Requirements

See Table 3.

Table 3. User requirements

User	Requirements
Event organizer and attendee	•Register in the system •Log in into the system •Manage profile, and update the personal information •Log out from the system •Recover password
Event organizer	•Manage event information •Manage beacon information •Manage content information •View a report with attendees' feedback and statistics about the event. •Assign beacon to content
Attendee	•Receive content automatically •View content •Save content •Share content •Add a comment to certain content •Delete a comment on certain content •Like certain content •Dislike certain content •Notified about any nearby beacons

4.2   Use Case Diagram

See Figure 4.

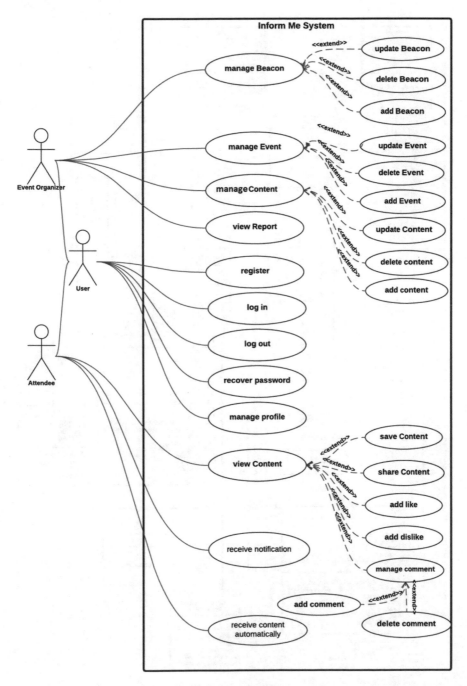

Fig. 4. Use case diagram

4.3   Class Diagram

See Figure 5.

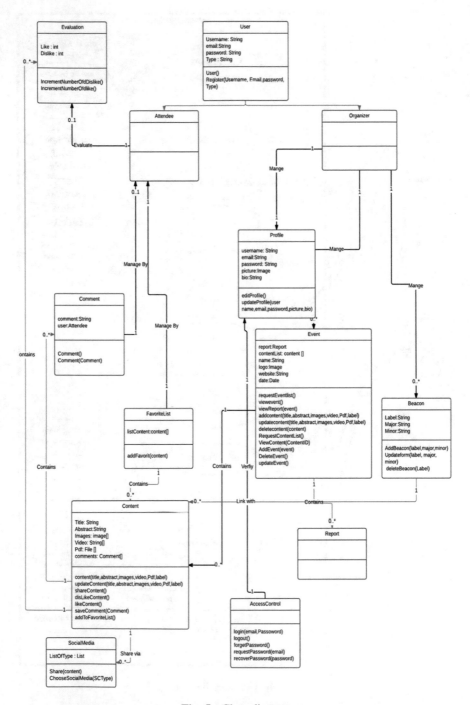

Fig. 5. Class diagram

4.4 Architectural Design

There are different types of the architecture design such as layered architecture, repository architecture, client-server architecture. The proposed system use the client-server architecture which give the ability to store a huge amount of information that can be accessed by the user through a network, so it is the most suitable architecture for our system [25] (Fig. 6).

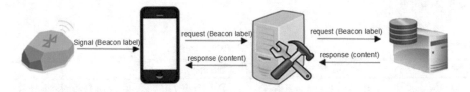

Fig. 6. System architecture

4.5 ER Diagram

See Figure 7.

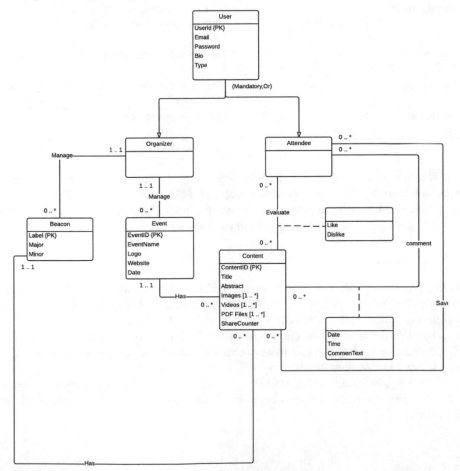

Fig. 7. ER diagram

5   Implementation

The implementation section provides the software tools that used to develop the system. In addition, we provide the code segments for the some components of the system.

5.1   Software Tools

See Table 4.

Table 4. Software tools

Software	Developer	Version	Description
Xcode	Apple Inc.	7.2.1	IDE to develop the InformMe application for iOS
GitHub	GitHub, Inc.	–	A Git repository hosting server, used for source code management and integration
MySQL	Oracle corporation	5.5.42	A RDBMS used to create InformMe database
phpMyAdmin	The phpMyAdmin project	3.5.8.2	A tool to manage MySQL database by creating, modifying or deleting tables or rows
Estimote SDK	Estimote	3.8.2	SDK used to interact with Estimote beacons
Lucidchart	Lucid software	–	To draw use cases, sequence, class, and ER diagrams
MS office word	Microsoft	2013	Write project document

The programming language used for developing the application is Swift. In addition, we used MySQL as a database tool and PHP as the API to allow the iOS application and website to communicate with the project server. JSON parser is used for data interchanging between application and server [25].

5.2   Code

In this project, we used Estimote beacons together with the Estimote iOS SDK to build our location aware application. As for beacon management, Estimote provides an online dashboard called Estimote Cloud. Additionally, Estimote also offers beacon management for nearby beacons by using the Estimote App. From either Estimote Cloud or Estimote App you can simply modify beacons' settings, names, or their major and minor identification numbers.

In the AppDelegate our beacon manager will start scanning for any ibeacon that is related to our application by checking wither it has the same shared UUID for all of our ibeacons.

```
//Set a beacon manager for monitoring
let beaconManager = ESTBeaconManager()
let beacons = Beacon()
func application(application: UIApplication, didFinishLaunchingWithOp-
tions launchOptions: [NSObject: AnyObject]?) ->Bool {
//Set the beacon manager's delegate
self.beaconManager.delegate = self
//Request authorization
self.beaconManager.requestAlwaysAuthorization()
//start region monitoring for the region specified in beacons class
beaconManager.startMonitoringForRegion(beacons.MonitorBeacon())
//To show notifications
UIApplica-
tion.sharedApplication().registerUserNotificationSettings(UIUserNotificatio
nSettings(forTypes: .Alert, categories: nil))}

//When a beacon is nearby show the defined notification in beacons class
func beaconManager(manager: AnyObject, didEnterRegion region: CLBea-
conRegion) {
beacons.BeaconNotification()
  }
```

For our event attendee user, in addition to notifications the list of contents order by the closest will appear. From this list the attendee can choose any content to view or interact with the content or favor it for later.

```
//This manager is for ranging
let beaconManager = ESTBeaconManager()
let beaconRegion = CLBeaconRegion(proximityUUID:
NSUUID(UUIDString: "B9407F30-F5F8-466E-AFF9-24456B57FE6D")!,
identifier: "MyBeacon")
```

The beacon manager delegate is set inside the viewDidLoad function. Whenever a beacon, specific to our application and its content was never request before, is close to the attendee the method below will be called to request its content.

```
//This method will be called everytime we are in the range of a beacon
func beaconManager(manager: AnyObject, didRangeBeacons beacons:
[CLBeacon],
     inRegion region: CLBeaconRegion) {

//Get the array of beacons in range
//For each beacon in array
for beacon in beacons {
//Check if the content was requested

if (!Requested.contains("\(beacon.major):\(beacon.minor)") && ((bea-
con.proximity == .Immediate) || (beacon.proximity == .Near)))
          {//If not request content then add to requested array
             loadContent(beacon.major, minor: beacon.minor)
       Requested.append("\(beacon.major):\(beacon.minor)")
             }
         }
self.tableView.reloadData()
   }
```

This code is added to stop beacon monitoring whenever the attendee leaves the content list page.

```
//To start/stop ranging as the view controller appears/disappears
overridefunc viewWillAppear(animated: Bool) {
super.viewWillAppear(animated)
self.beaconManager.startRangingBeaconsInRegion(self.beaconRegion)
   }

overridefunc viewDidDisappear(animated: Bool) {
super.viewDidDisappear(animated)
self.beaconManager.stopRangingBeaconsInRegion(self.beaconRegion)
   }
```

As for the other user, the application allows the event organizer to manage all of the events by adding, editing, or deleting a selected event. These management functionalities are also available for the beacons and contents for an existing event. Below is the code for the adding an event after the event organizer submits an event.

```
@IBAction func save(sender: AnyObject) {
let name = EventName.text!
let website = EventWebsite.text!
let  date = EventDate.text!
//Check event name field
if (EventName.text == "" || EventDate.text == "") {
displayAlert("", message: "Please complete all fields")
        }
// Check the entered event date
elseif(EventDate.text != ""&& !checkDate(date)){
displayAlert("", message: "Please enter a correct date for the event")}
else { // If there is no problems add the event for this user in the database
print(EventName.text)
let e : Event = Event()
        e.AddEvent (UserID, name: name, web: website, date: date, logo: Event-
Logoo.backgroundImageForState(.Normal)!){
        (flag:Bool) in
//we should perform all segues in the main thread
dispatch_async(dispatch_get_main_queue()) {
self.performSegueWithIdentifier("addEvent", sender:sender)}}
    }
  }
```

6 Conclusion

In the beginning of this study, we referred to some background information and introduced the concept of location awareness, followed by a brief overview of Bluetooth Low Energy and beacon hardware. We presented a comparison between beacons from different manufacturers and choose estimate beacons for the presented study. In addition, we presented some applications that use iBeacon technology for location awareness. In order to highlight the different limitations and issues that require improvement; we compared all of the presented applications as well as our suggested system. We then discussed in detail how our solution advances other solutions and the features included in our applications.

After analyzing and designing the system using object-oriented approach, we ended up with a set of requirements that our system provides. As soon as the application was developed and implemented on iOS, we started testing the system against different test cases and test strategies to insure that the system satisfies its goals and objectives. From the testing results, we concluded that our system fulfils its usability and performance requirements. We performed an experiment that resulted in several findings. For more information about the results of testing and experiment, they presented in details in this paper [25].

We hope that our system will achieve its goal in developing a system that will help the event organizers to deliver digital contents to the attendees based on their locations. This system will use the iBeacon technology to provide efficient ways for content delivery, in order to improve attendee's experience, reduce costs and replace the printed papers.

It should be also mentioned that this project has contributed to scientific field by providing different outcomes as following:

- Publishing a paper in the SAI Intelligent Systems Conference 2016, that held in London, UK [16] that contains a literature review with detailed comparison and evaluation between the location-based technologies in addition to presenting the available mobile applications that can be used for content delivery. This paper was extended to be published as a book chapter in springer.
- Publishing a paper in the 18th International Conference on Information Integration and Web-based Applications & Services (iiWAS2016) [25], that provides the details about the presented system, in addition to the results of the performance and usability testing and the experiments that conducted on real events with the related findings.

References

1. Gartner: Gartner says emerging markets drove worldwide smartphone sales to 19 percent growth in first quarter of 2015, 20 Sept 2015. https://www.gartner.com/newsroom/id/3061917
2. Cheng, N.: A guide to iBeacon hardware and contextually aware platforms, 20 Sept 2014. http://chaione.com/a-guide-to-ibeacon-hardware-and-contextually-aware-platforms/
3. Vazquez-Brseno, M.: Using RFID/NFC and QR-code in mobile phones to link the physical and the digital world. In: Deliyannis, I. (ed.) Interactive Multimedia. InTech, pp. 219–242 (2012)
4. Doyle, K.: The growing reach of beacons: are users ready? 27 Sept 2015. http://newsroom.cisco.com/feature-content?articleId=1590301
5. Sayers, C.: Placing experience within reach: QRC, NFC and the future of contextual activation
6. Levijoki, S.: Privacy vs location awareness. Unpublished manuscript, Helsinki University of Technology, 2001
7. Shahriar, S.: Location based content delivery solution using iBeacon (2015)
8. Kevin Townsend, C.C., Akiba, R.D.: Getting Started with Bluetooth Low Energy. O'Reilly Media, United States of America, Inc., 1005 Gravenstein Highway North, Sebastopol, CA 95472 (2014)
9. Gast, M.S.: Building-Applications-with-iBeacon. O'Reilly Media, United States of America (2015)
10. Aislelabs.: The Hitchhikers guide to iBeacon hardware: a comprehensive report by Aislelabs, 17 Oct 2015. http://www.aislelabs.com/reports/beacon-guide/
11. Beacon Tech Overview, 19 Oct 2015. http://developer.estimote.com/
12. The Gimbal Store, 19 Oct 2015. https://store.gimbal.com
13. Technical Specifications, 19 Oct 2015. http://kontakt.io/our-beacon-ibeacon-technology/technical-specification/

14. Chen, Y., Fard, H.K., Son, K.K.: Indoor positioning of mobile devices with agile ibeacon deployment (2015)
15. MPact-Specification-Sheet Zebra Technologies (2015)
16. AlBraheem, L., Al-Yahya, J., Al-Rowais, N., Al-Shathri, S., Alsuhaibani, L., Alabdulkarim, A.: Location-based content delivery using iBeacon technology. In: SAI Intelligent Systems Conference, London, 2016, pp. 523–531
17. BluuBeam, 1.3.1 ed: BluuBeam LLC (2015)
18. TradeShowPro, 1.0 ed. AirSenze Solutions Inc. (2017)
19. Webble, 2.2 ed: Ength Degree, LLC. (2013)
20. Dubai Culture, 5.1 ed. United Arab Emirates: Dubai Culture & Arts Authority, 2015
21. Beacon Guides, 1.2 ed: Blupath Ltd (2015)
22. Event Wallet, 1.0 ed. London: Earnest Ltd (2015)
23. BEEP, 1.7 ed. Digital Wavefront LLC (2016)
24. AlBraheem, L., Al-Yahya, J., Al-Rowais, N., Al-Shathri, S., Alsuhaibani, L., Alabdulkarim, A.: Location-aware system for content delivery using bluetooth low energy technology. In: The 18th International Conference on Information Integration and Web-based Applications & Services (iiWAS 2016), pp. 396–402, Singapore (2016). https://doi.org/10.1145/3011141.3011187
25. Sommerville, I.: Software Engineering, 9th ed. Library of Congress Cataloging-in-Publication Data, United States of America (2009)

The CaMeLi Framework—A Multimodal Virtual Companion for Older Adults

Christiana Tsiourti[1]([envelope]), João Quintas[2], Maher Ben-Moussa[1], Sten Hanke[3],
Niels Alexander Nijdam[1], and Dimitri Konstantas[1]

[1] University of Geneva, Geneva, Switzerland
{christiana.tsiourti,maher.benmoussa,niels.nijdam,dimitri.konstantas}@unige.ch
[2] Instituto Pedro Nunes, Coimbra, Portugal
jquintas@ipn.pt
[3] Austrian Institue of Technology, Vienna, Austria
sten.hanke@ait.at

Abstract. Artificial Social Companions are a promising solution for the increasing challenges in elderly care. This chapter describes the CaMeLi autonomous conversational agent system which simulates human-like affective behaviour and acts as a companion for older adults living alone at home. The agent employs synthetic speech, gaze, facial expressions, and gestures to support multimodal natural interaction with its users and assists them in a number of daily life scenarios. We present the agent's overall architecture, with a focus on the perception, decision making and synthesis components which give rise to the agent's intelligent affective behavior. The agent was evaluated in an exploratory study where it was introduced in 20 homes of older adults (aged 65+) in three European countries (Switzerland, the Netherlands, Portugal) for a total duration of 12 weeks. We present the results of the evaluation study with regards to acceptance, perceived usability, and usefulness of the agent, and discuss future opportunities for fellow researchers who are striving to bring virtual agents out of the laboratories into successful real world applications.

1 Introduction

To understand the relevance of Artificial Social Companions, first consider the current societal challenge related to the demographic changes happening in Europe. In Europe, the number of people aged 65 and older will almost double over the next 50 years, from 85 million in 2008 to 151 million in 2060. Demographic change and ageing are a common societal challenge for Europe and the socio-economic implications are enormous. For healthcare costs in particular it is estimated that this will lead to an increase of 15–40% on top of current expenditure to maintain existing health services in the EU. The most problematic

© Springer International Publishing AG 2018
Y. Bi et al. (eds.), *Intelligent Systems and Applications*,
Studies in Computational Intelligence 751,
https://doi.org/10.1007/978-3-319-69266-1_10

expressions of population ageing is the concomitant increase in co-morbidities, functional decline and the evoked clinical condition of frailty.

Allowing people to grow old at their homes is a way to promote dignity and wellbeing for the older persons but also a premise for the health and welfare system to subsist. Nursing homes and other residential units are expensive, unable to respond to all the requests and often older persons who have no family support or in social disadvantage use hospitals and long-term care facilities as a solution, increasing health-related costs and not receiving the most suited solution to their needs (i.e. >80% of the older persons do want to receive care in their homes and do not want to become hospitalized). Public and private structures and services are so far not accessible nor inclusive of older people with varying needs and capacities and there is yet a need to encourage social inclusion and participation in order to achieve a sense of a useful meaning of life, also for older people, by empowering them to actively participate in the creation of age-friendly environments.

Technology may offer powerful resources to develop personalised strategies to support active and healthy ageing. These resources may include Artificial Social Companions, as a promising solution to the increasing challenges of elder care. Lets consider the scenario of promoting autonomy at home, which represents a set of relevant interaction contexts satisfying the basic needs of the elderly. In this cases, the main goals are to support the older adult in daily life routines and to promote psycho-social and physical wellbeing. Hence, in this scenario it is safe to assume that the users' basic expectations are avoiding the feeling of loneliness, promoting safety and guidance through activities of the daily life. These expectations may be addressed by the features expected from Embodied Conversational Agents.

Embodied Conversational Agents (ECAs) are computer-animated characters exhibiting a certain degree of intelligence and autonomy, which include social skills that simulate human-like face-to-face conversation and empathize with the user [11]. Based on recent research findings [5,8,14,25], it is anticipated that ECAs, in the future, may come to play an important role on complementing human care in maintaining the health, wellness, and independence of older adults. More specifically in [20,36,39] experiments using the Wizard of Oz methods revealed a positive attitude and high acceptance towards ECAs as companions for older adults. Additionally, we can find a diversity of ECAs [30,36,39] aiming to provide social support to isolated older adults and to address daily needs for an autonomous living. Moreover, these agents have also been used as virtual coaches to motivate the user following healthier behaviours change [4,6].

Despite these promising findings, to date the majority of ECA systems were designed for specific contexts of operation and have been tested mostly on laboratory conditions. These are perfectly valid approaches to test and assess certain aspects of the agents, but does not compare to the uncertainty found in a real-world environment. Consequently, we can find limited information referring to how ECAs perform in relevant environments and which are the major constrains influencing their acceptance by older adults as a new technological aid on private households.

This book chapter describes a fully autonomous agent system developed to simulate a daily life companion for older adults and its assessment and evaluation in relevant environments (i.e. home environment of older adults). The agent operates on a all-in-one device, which provides the user interfaces and interaction modalities needed to realize face-to-face interactions with older adults, in a number of daily life scenarios. For example, it simplifies visualization and query of information related to the user's daily schedule, reminders and social events and helps older adults to connect with others by exchanging messages. We start from commonly agreed requirements, which resulted from following co-design methods. These requirements include some relevant functional aspects for designing an agent that must maintain a long-lasting interaction with the user. For example, the agent should be as natural and believable as possible, it should simulate a real-life human companion; it must support natural speech interaction and it must be capable of recognizing the user's emotional state and delivering appropriately tailored empathetic feedback, including emotional facial expressions. However, the reporting of these types of needs and wishes may result in users getting frustrated and disbelief technology if the agent cannot fully met their initial expectations. Therefore, it becomes very important to have a close and careful follow up by a human actor (e.g. caregiver) that can oversee the introduction of this type of technology for the particular target group of older adults.

In the subsequent sections, after a brief review of related work in the area of ECAs for older adults, we discuss the architectural and conceptual realization of an ECA for older people based on results of three user-centred design studies. We describe how interaction capabilities are achieved and detail the underlying psychology-inspired motivational and affective system, and the multimodal interface. Consequently, we describe an evaluation study showing how deploying the agent into private households was perceived by the users. Finally, we discuss results on acceptance, usability and usefulness, and draw some insightful conclusions that contribute to the knowledge towards successful and acceptable ECA systems for real world applications.

2   Related Work

ECAs are screen-based entities designed to simulate human face-to face conversation skills and thus allow for natural interaction between humans and computers. ECAs are typically represented as humans and are specifically lifelike and believable in the way they behave. Cassell [11] defines ECAs as those virtual characters that have the same properties as humans in face-to-face conversation. An increasing number of researchers are examining the potential of ECAs to enhance human-computer interaction (HCI) by either accompanying or replacing traditional computer interfaces [18].

2.1 Embodied Conversational Agents for Older Adults

ECAs are commonly design to incorporate features that provide non-threatening and familiar user interfaces. These characteristics make ECAs useful for developing systems that are intuitive, engaging and trust worthy. As studied in [23,39] this can happen even with users suffering from age-related or other cognitive impairments. Additionally, ECAs can improve the natural interaction of elderly users with ambient intelligent environments, as shown by Ortiz et al. in [27]. Their study concluded that older adults were capable of associating the facial expressions of an agent to an emotional response, which resulted in following instructions provided by the agent much often when compared to conventional graphical user interface.

Next, we refer to some examples of ECAs that are commonly known in the Active and Assisted Living community, and overview the common set of features that have been integrated into these systems (Table 1).

Rea is a virtual reality embodied conversational agent whose verbal and non-verbal behaviors are generated from underlying conversational functions and representations of the world and information. Rea implements social, linguistic and psychological concepts that is part of any conversation. Rea has a human-like body appearance, which it is used to add body language during conversation. Conversational skills and comprehension is speech based, thus, each user expression is interpreted and the responses are generated according to which function needs to be achieved.

Greta is a embodied conversational agent that uses a virtual three-dimensional model of a female character. Greta uses verbal and nonverbal communication to interact with the user. Greta can be used with different external Text-to-Speech software. Currently available languages include English, Italian, French, German, Swedish and Polish. Facial Animation Parameters are required to animate faces of different sizes and proportions, the FAP values are defined in face animation parameter units (FAPU). Greta is used in various European projects: CALLAS, SEMAINE, HUMAINE, and national French projects: ISCC Apogeste, ANR MyBlog3D, ANR IMMEMO.

DALIA is an assistant for daily life activities at home. It was designed to run on Android based consumer devices commonly available in households (TV, phone, tablet). Older adults can use speech to interact with an avatar, which will answer based on data collected through a set of sensors deployed in the household. This ambient intelligence promote a more independent living for older adults. Moreover, informal carers have access to the same avatar, which can tell them what they should do in different situations or just to talk with the person cared for.

V2me Virtual coach reaches out to me combines virtual and real life social networks to prevent and overcome loneliness of older adults. V2me supports active ageing by improving integration into society through the provision of advanced social connectedness and social network services and activities. V2me was inspired in the **A2E2** project, which is also a financed AAL project. Initially, this system was designed for 7 in. tablets. The key of this choice is portability and low costs; thus, a good graphic processor is needed and the system was extended to All-in-one computers. Professional caregivers and elderly family can monitor users activity with the system by a designed web interface. By this way is also possible to adjust and configure the way the assistant interacts with the old users. One-on-one test sessions had shown that a lot of elderly are not familiar with new technologies, this includes touch interface of the device, thus preferring human support.

Table 1. Comparison between different ECAs implementations. * - (Without avatar on portable devices)

	CaMeLi	DALIA	Rea	V2Me *	Greta
LipSync	Yes, BML	Yes	Yes	Yes	Yes (BML and FML)
Platform	Windows	Android, Windows	Android	Android	Windows
Speech recognition	Windows SDK	Built-in	N/A	Third-party services	Built-in
Emotional state	Read and expression	Read	Read and expression	Read and expression	Expressions
Multi language	Yes	Yes	No	No	Yes
Multi avatar	No	No	No	No	No
Social networks	Built-in	No	No	Yes	No
Camera	Yes, Kinect	Yes	Yes	Yes, Kinect	No
Video conferencing	Yes	Yes	N/A	Yes	No
Portable	Yes	Yes	No	Yes	No
Emergency mechanisms	Yes	Yes	N/A	N/A	N/A
Internet connection	Recommended	Required	Optional	Recommended	N/A

3   The CaMeLi Daily Life Companion for Older Adults

In the scope of the European project CaMeLi (*www.cameli.eu*) we investigate the use of ECAs as long-term assistive companions for older adults (aged 65 and above). Specifically, we developed an autonomous agent to motivate older adults self-manage their care and remain socially and physically active [34]. The agent is seamlessly integrated into the living environment and at any point in time, a user can request assistance. The agent delivers appropriately tailored empathetic feedback, in a similar way a real-life human companion would do.

3.1   Involvement of End-Users

To understand how to design the agent, we conducted three user-centered design studies in collaboration with three care organizations which provide daily assistance to elderly in Switzerland, the Netherlands and Portugal. The studies were based on focus groups, individual interviews and paper based surveys with older adults with age above 65 (N = 20), professional caregivers (N = 12) and psychologists specialized in the aging process (N = 2). The study was performed on three different user environments: (1) Independent elderly living at home in an urban setting, (2) elderly home and (3) care apartments. The detailed methodology and results on how we carried out these initial part of the studies are described in previous work [33].

These studies resulted in a list of needs and requirements as expressed by older adults. The complete list of the expected features to include in an ECA is provided in Table 2. For example, reminder functions, brain training or other games, daily agenda, fall detection and call for help were identified as needs in all three user settings, whereas other services appear only in one or two user settings.

User: Avatar, show me the agenda for today.

Agent: On [activityDate], from [activityStartTime] to [activityEndTime], the activity [activityName] will take place in the location [activityLocation].

User: Ok, please invite my friend [friendName] for [activityName]

Agent: I sent an invitation message of [activityName] for your friend [friendName]

User: Thank you!

Agent (Facial expression [Joy]): You are welcome. I will notify you when you receive a message from your friend.

3.2   The Interaction Process

In CaMeLi, users interact with the ECA using a multi-modal interface that includes a graphical touch-based user interface (GUI) and automatic speech recognition. In spite of both modalities can be used to complement each other, we mapped all graphical artefacts associated with actions to a voice command. In this way, we simplified the learning process and allow a broader group of users to use the system. This means that, by providing a redundant interaction

Table 2. User surveys and list of services

Services	Setting1	Setting2	Setting3
Agenda for daily living activities (wake up)	−	+	+
Breakfast reminder	−	+	−
Medication reminder	+	+	−
Daily schedule reminder	+	⊢	+
Reminder prior to appointments or activities	+	+	+
Add agenda items themselves	+	+	+
Program of activities nearby or in the city	+	−	+
Notify carer or family member when running late or cancelling a planned event	−	+	+
Messages from the system	−	+	−
Do not forget things when going out	+	−	−
Fall detection and call for help	+	+	+
Call for help in case of an emergency	+	+	+
Noise detection (for safety purposes) and call for help	+	+	+
Safety alarm follow-up	+	+	+
Behaviour analysis and motivation functionality	+	+	+
Emotion recognition by means of facial expression and speech analysis	+	+	+
Brain training or other games	+	+	+
Playing recorded relaxations, meditations	+	−	−
Stimulation for physical exercise (Yoga or going out)	+	+	+
Finding things detected by the camera	−	+	+
Object storage memory	−	+	−
Dinner menu of the restaurant (read out)	−	+	−
Shopping list	+	−	−
Skype functionality	+	+	+
Communication with friends via CaMeLi system messages	+	+	+
Communication with friends and family via text messages (sms)	+	+	−
Retrieve information online	+	−	−
Bus/public transport time tables and routes	+	−	−

mechanism we allow people with different capabilities to still use the system. For example, visually impaired people or illiterate can use speech to activate the same functionalities as another user that might not be able to express himself in word can, by touching virtual buttons. On the other hand, although our agent does not perform at a human level conversational skills, the natural speech interaction interface was based on human-to-human dialogues from scenarios of daily life interaction. While the dialogues are highly situation and task dependent, we engaged in the co-design process involving both older adults and caregivers to define dialogue scripts that would cover the assistive care services mentioned above. In most scenarios, interaction is started by the user, who initiates the conversation with the agent, following a question-answer sequence.

4   Implementation

The CaMeLi agent is controlled by a cognitively motivated agent architecture that resembles what has been proposed as a reference architecture for ECAs [10]. As outlined in Fig. 1, the architecture consists of three key components responsible for perception, decision making, and synthesis. The well-orchestrated interplay of these components gives rise to the human-like cognitive capabilities of the agent. Figure 1 outlines the components in more detail.

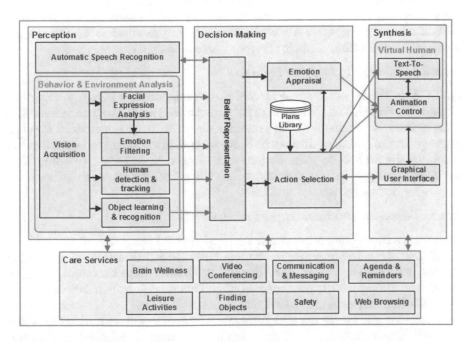

Fig. 1. The system architecture of the CaMeLi system

4.1   Hardware Setup

The hardware requirements for the agent to operate properly include a all-in-one stationary computer (e.g. we selected Lenovo ThinkCentre Edge 93z All-in-One), that can be mounted on some support at a height ranging from 80 to 120 cm, allowing the user to interact from a standing or sitting position. Moreover, the preference for an all-in-one system was also given by built-in integration of a set of hardware components that are required to provide multi-modal interaction. In particular, a touch-screen and graphical display, a built-in microphone, a built-in video camera and speakers. Additionally, we used a Microsoft Kinect device to acquire depth (RGB-D) visual data. On the particular case of using this device for Human Behaviour and Environment Analysis, we realized that the appropriate mounting height should be around 130 cm. Given that this device has a viewing angle approximately of 60 vertical by 70 horizontal, we ensured a minimal field of view in which we could detect standing and laying down persons and detecting objects at heights of up to 90 cm (e.g. small room table).

4.2   Perception Components

The Perception components collect sensory information from different sources e.g., speech recognizer (Automatic Speech Recognition), RGB cameras tracking the user and depth cameras that recognize the environment (Vision Acquisition). The perceived information, such as the users speech commands and perceived emotion, directly triggers reactive responses from the agent or is send to the Decision Making components for further processing.

Automatic Speech Recognition (ASR)

The ECA developed in the scope of CaMeLi can support four languages: English, Portuguese, French and Dutch. In order to activate the different functionalities of the system, we created custom grammar files (i.e., language models) for each language. In the grammar files, we indicate what utterances are recognized. Also, it includes the probability distributions associated with sequences of words. This was done to constrain (i.e. defining rules) the search among alternative word hypotheses during speech recognition.

Human Behaviour and Environment Analysis

The agent includes human-behaviour and environment analysis features, built on top of vision-based approaches for human body detection and facial expression analysis, and objects recognition. Regarding Human Detection And Tracking we followed the approach presented in [21]. This uses a probabilistic approach to classify human behaviour based on movement features like the height variations, velocity, and acceleration of limbs. We acquired visual data (RGB-D) using Microsoft Kinect and the classifiers were implemented as a Bayesian Network.

The main interest of using this component was to recognize abrupt body motions that could relate to a sudden fall. The Facial Expression Analysis module extends the FaceReader [35]. FaceReader is a commercial product from Noldus. It was developed with the objective of emotion recognition in laboratory settings; to be used in behaviour analysis studies. Using this component, our ECA can detect the six basic emotions described by Ekman [15] (happy, sad, angry, surprised, scared, disgusted) and a neutral state. The approach used for environment analysis focus mostly on Object Recognition. We implemented our approach based on the Global Hypothesis Verification, proposed in [1], using shape descriptors obtained from different views of the object, represented in point cloud data (PCD) [29]. This approach is especially useful on a household, which we considered to be an unstructured and uncontrolled environment; given it has the inherent ability to detect significantly occluded objects without increasing the number of false positives.

4.3   Dialogue Management Components

The dialogue management components (Fig. 2) manage the flow of the interaction with the user, by initiating and managing the agent's interaction strategies in order to simulate human-like behaviour. Our implementation follows a Belief-Desire-Intention (BDI) approach and includes the agent's beliefs, desires (goals/motives) and plans for creating intentions. The approach, as proposed in [2], depends on several high-level goals and motives that constrain decisions at every moment. The set of goals/motives includes features associated with the users well-being and user satisfaction with the system (e.g. "User is feeling well", "User is socially active", "User is physically active"). This set is used as a descriptor vector for the emotional and adaptive behaviour of the companion, which will maximize the user physical, social and cognitive status. User status is described in terms of Beliefs, which reflect perceived user actions as well as situation events and can emerge from internal processing but also through interactions with the environment or the user.

Action Selection

The action selection component of the architecture is responsible for selecting and executing AI plans on the belief representation system. Based on the current state of the interaction, the action selection component selects an agent behaviour (intention) as a situated response to the environment and the user.

Behaviours instantiated at the action selection component are activated dynamically when certain preconditions have been satisfied and then they are communicated to the behaviour synthesis components to control the agent's verbal and non-verbal expressions (i.e., facial expressions, speech).

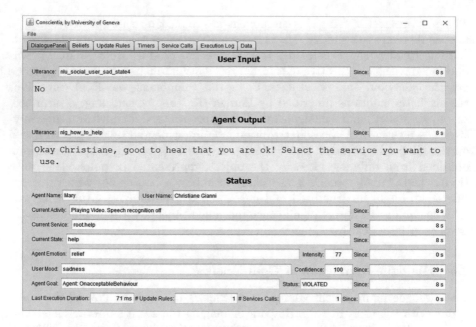

Fig. 2. Dialogue management component GUI

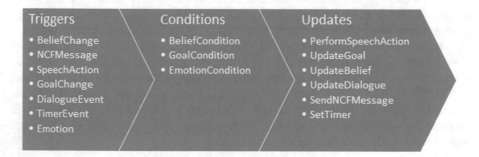

Fig. 3. AI Plans structure

As illustrated in Fig. 3 an AI plan in our implementation has three main sections: triggers, conditions and updates. triggers and conditions are used to test the plan against the beliefs and other criteria, while effects are used to perform actions or update beliefs if the plan is executed. A condition indicates that an plan is eligible for triggering, while a trigger indicates when a plan should be triggered (e.g. plan WelcomeNewParticipant would only be triggered directly after a new user appears). The dialogue manager implements several types of triggers, conditions and updates as illustrated in Fig. 3. These triggers, conditions and updates represent components in the dialogue manager such as the goal management, emotion simulation, time management, speech input and generation as well as components for communication network protocols.

Belief Representation System

The agent's belief system uses a structured knowledge representation, which drives decision-making and action selection. As illustrated in Fig. 4, this includes a representation of the agent's perceived events and the conversation status as well as the user's cognitive model. This implementation was inspired in David Traums information state approach and other structured knowledge base approaches [32].

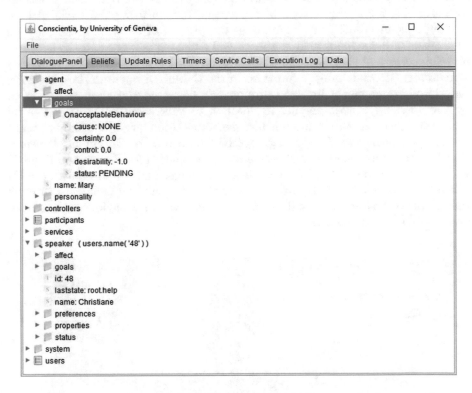

Fig. 4. The agent's belief representation system, including perceived events, the conversation status, and the user's cognitive model

Emotion Behaviour System

The agents emotional reactions at each moment of the interaction is controlled in the Emotional Behaviour System. It appraises and simulates this reaction based on an evaluation of its internal goals and motives [3]. This evaluation takes into account several appraisal variables (e.g., desirability, expectedness, agency, and control potential) to assess an event and determine the appropriate emotion [16].

4.4   Behaviour Synthesis Components

The behaviour synthesis components are in charge of generating intelligible verbal, gestural and facial expressions and combining them into a continuous flow of human-like multi-modal behaviours. These components are built on top of SmartBody [31]. SmartBody is responsible for generating the gestures and facial expressions of the virtual companion and is controlled by the CaMeLi's framework through the Behavioural Markup Language (BML). Consequently, it enables easy control of each action unit (AU) to achieve particular emotions by describing facial expressions of the virtual human using the Facial Action Coding System (FACS). This simplifies the process of describing the physical realization of behaviours such as speech and facial expressions and the synchronization constraints between these behaviours. A set of pre-set emotional expression configurations is pre-programmed in the CaMeLi framework. Furthermore, SmartBody implements lip resynchronisation based on the visemes generated by the speech synthesis engine. In addition to generating gestures and facial expressions, the animation module also generates secondary behaviours to make the agent appear more lifelike, e.g., eye blink and breathe. In our implementation Microsofts native text-to-speech engine is employed for French and English, while CereProc was utilised for the Dutch language. Figure 5 illustrates the 3D virtual human, which closely simulates human conversational behaviour through the use of synthesized voice and synchronized non-verbal behaviour such as head nods and facial expressions.

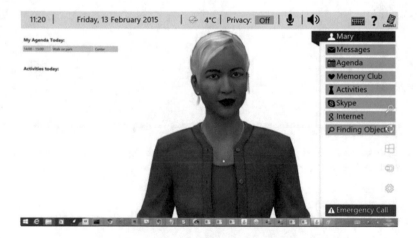

Fig. 5. The agent and the GUI

4.5   Graphical User Interface (GUI)

In addition to the natural speech interaction, the users can issue commands to the agent using buttons and elements on the GUI. The design of the GUI (Fig. 5) took into consideration relevant standards of User Interface Design (e.g. such as defined by W3C) and the main input was results of the iterative process of the co-design with the end-users. The later, allowed us to identify clear requirements and to collect relevant feedback from older adults. The result was a GUI with three main areas: (1) on the right-hand side there is a menu that provides access to the care services (note the big and clearly spaced buttons to facilitate their activation); (2) on the top, we put a status bar that provides information about current time and date, the weather forecast, and notifications. Also, on the right-hand side of the status bar, the visible icons inform the user about the agents status (e.g. enabled/disabled, currently talking/listening for commands). Additional buttons are provided to enable the on-screen keyboard and the help menu and to exit the system; (3) the central area of the screen is reserved to display content. In the home screen, the avatar will appear maximized. When accessing to a service, the graphical interface for that service appears in the center and the avatar appears in a smaller version at the bottom of the menu column. This design ensures that the users can interact with the touch-based interface intuitively after a short introduction.

5   Agent Evaluation Study

One important matter in designing and implementing ECAs, or Artificial Social Companions in general, is to understand the impact they have in the daily life of the users for who they are designed for. In this sense, system evaluation and assessment is vital to understand if the system (i.e. agent) answers the user needs and if its behaviour met user expectations. Consequently, we conducted a longitudinal evaluation study to assess the effects of the agent in the daily life of older adults. Our main goal was to verify empirically how our target user group would interact with the ECA-based agent, when installed in their private space. Also, we wanted to understand the subjective opinions users had in terms of acceptance, perceived usability and usefulness while operating a relevant environment. This means, operating in a real scenario where various types of interaction and tasks could occur during daily life activities.

5.1   Setting

The study took place in three distinct test-beds in three European countries. In Switzerland, where elderly were living alone in apartments of an assisted living complex; in the Netherlands where participants were living alone in independent apartments; and in Portugal, where elderly living independently stay in a home care during the day period (e.g. Fig. 6).

Fig. 6. User evaluation is one of the pilot settings

5.2 Participants

We considered the target group to be healthy older adults that might have light physical or light mental health problems, who live alone at home and want to be helped or stimulated to carry out their daily activities. The inclusion criteria for older adults to be included in the study were:

- be able to put glasses off when necessary for the system;
- should not use a wheelchair inside;
- living alone.

We estimate the approximate number of population with this characteristics of the inclusion criteria is between 5 and 7 Million people, for the sum of the three settings. The total number of users involved of 36 participants, which correspond for a confidence level of 95% and confidence interval of 16.33%. The distribution per respective sites is: 13 (Netherlands), 11 (Switzerland), 12 (Portugal).

The characteristics of the users involved in the evaluation and assessment of the system is shown in Table 3.

5.3 Evaluation Phases

The total duration of the this evaluation study was 12 weeks. We divided evaluation into three parts (see Fig. 7). First, in the Introduction phase (T0), we carried out the baseline measurements, system installation and users' training. Regarding the baseline measurements, we adopted the World Health Organization Quality of Life (WHOQOL) questionnaire [17] and conducted semi-structured interviews. Second, after 4 weeks, in the intermediate evaluation (T1), we used the System Usability Score (SUS) questionnaire to gather user's impressions about the system [9]. This phase was important to collect data during the period were

Table 3. User baseline measurements

Topic	Netherlands	Switzerland	Portugal
Participants	N=13	N=7 (4 drop-outs)	N=12 (2 drop-outs)
Average age	79,4	76,4	79,9
Experience computers (1–5)	3,1	4	1
Experience tablets (1–5)	3,8	2,6	1
Experience avatars (1–5)	1,4	1,7	1
Satisfaction daily life	100%	87,5%	NA
Memory (1–10)	6,9	Qualitative	NA
Quality of life (1–10)	7,5	Qualitative	NA
Expectation of becoming more autonomous with the system (1–5)	2,6	2,9	NA
Expectation of becoming more organized with the system (1–5)	2,3	3,4	NA
Expectation of becoming more active with the system (1–5)	1,8	2,5	NA
Expectation of improving memory with the system (1–5)	3,0	4,1	NA

the users might have experience the novelty effect, which would be useful to compare with an equivalent assessment after a longer period of use. Third, in the final evaluation (T2), we applied both of the previous questionnaires and conducted semi-structured interviews. This last phase was performed after at the end of the 12 weeks period of daily basis interaction with the system.

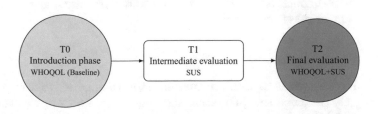

Fig. 7. The three phases of the evaluation study

5.4   User Acceptance

Overall, users were positive regards accepting the agent in their households. However, after sometime interacting with the system, passed the novelty effect, we observed, in the final evaluation (T2), a decrease in acceptance. The most reported reason was that the agent did not fully reach users' original expectations. Among the main reasons for this we note the mismatched expectations related with verbal communication capabilities. In fact, the majority of the users in the target group, on facing a human-like character expected a more natural interaction in terms of speech dialogues. Although, we could not find an ASR system that could fulfil such expectation. Participants easily got frustrated after a few unexpected verbal behaviour by the agent. On the other hand, we noticed that users from the target groups face some challenges regarding training the interaction with the agent. This led to a higher number of repetitions that desirable. From these consideration, we draw two main conclusions.

1. An ideal solution would be to have more flexibility and variety in the speech commands.
2. it is of utmost importance that all the interaction components run as robustly as possible, are fault-tolerant, and support repair mechanisms.

5.5   Assessing Usability

Usability was assessed using the System Usability Scale (SUS). The average SUS scores of each setting were: 52 for the Netherlands, 62.2 for Switzerland and 67 for Portugal. The combined average score was 60.4 (range from 37.5–80). Thus, overall we conclude that the interaction with the system was perceived as "average" positive. However, when considering these results individually we observe a high variation between settings. While the system was perceived as positive in all settings, the users in Portugal were more positive towards the system than the users from Switzerland and Netherlands (i.e. the less positive towards the system). At this point, we must argue that cultural differences may have influenced this result. Recall from the user characteristics, in Table 3 that users in the Netherlands were more experienced with technology than the rest. This fact may be enough reason for them to be more demanding and have higher expectations from novel technology.

5.6   Perceived Usefulness

We asked all participants to assess the usefulness of the digital services provided with the agent. Based on the qualitative and quantitative results, we observed a significant variability, between the three settings and between users. In Switzerland, the users favoured more services for memory training, agenda, simplified

Internet browsing and Skype integration. In the Netherlands, the most popular services were agenda, social activities, and message exchange between friends. Finally in Portugal, the more positive services were the agenda, skype connection and internet browsing. Summarizing, these findings revealed that usefulness is tightly coupled with the specific context of each older adults. This means, it depend highly on each personal care needs and lifestyle choices. Therefore, to remain its usefulness in the long-run, the system should be capable to learn and adapt to its users context (i.e. personal needs, and social and physical situation).

5.7 Ethical Considerations

During the development of CaMeLi, we looked into several ethical issues that cannot be disregarded when using socially intelligent ECAs. Given the specificities associated with our target group, first and foremost, any artificial agent should not be obtrusive or stigmatizing for the users, nor restrict their privacy. In fact, during the study, we registered some of the worries that cross older adults minds, and any other person, regarding the installation of cameras and the associated uncertainty about whether or not their interactions with the agent would be recorded. Additionally, it was uncomfortable for the user when, during the installation of the agent, we found any constrain in the household layout that required to rearrange the space for the system to work properly. For example, insufficient space or inappropriate location for mounting the computer and additional devices at the optimal positions (functional for the system vs. comfortable for the user). From this study, we learned that introducing hardware in the household is inevitable, but ideally, no recorded data should be stored, and all the devices should be integrated into the existing furnishings. Additionally, following the good practices of interaction design, it is essential that older adults are not deceived into thinking they are interacting with a agent that is capable of doing things that only humans can do in an interaction. This means, it must be clearly explained to the user that the agent is a computer-driven character with limited capabilities.

6 Conclusion and Future Work

In this paper, we start by presenting the motivation that support the common belief that ECAs are a promising solution to the challenges faced by the ageing population. The purpose of this study was to conduct a 12 weeks evaluation where we could observe how older adults interact with an artificial social companion; an ECA in the particular case of this study. The study was conducted in three distinct settings in three European countries (Netherlands, Switzerland and Portugal) and involved 36 participants. This sample size was calculated based on the approximate population that met the inclusion criteria for the study and had a confidence level of 95% and a confidence interval of 16.6%. We followed a qualitative and quantitative methodology, using standardized assessment scales WHOQOL and SUS and looked at the users interacting in daily-life scenarios with the agent. Results confirmed that overall users are positive regarding accepting novel technology, in particular regarding interactive systems. Usability and perceived usefulness was also positive. However, unmet expectations in terms of level and maturity of interaction resulted in borderline average results for usability, as shown in the results from SUS score. Nevertheless, we noticed some differences on how users from different settings responded to the same technology. Hence, cultural, socio-economical and personal preferences (i.e. context) can influence the opinions towards this type of technological approaches. The results suggest that older adults with lower digital literacy are more optimistic regarding ECAs, when compared with older adults more familiar with Information and Communication Technologies. Hence, the results, conclusions and lessons learned in this study represent a contribution to the designing and implementation of ECAs for real-world applications. Moreover, this work have been serving as a framework that supports some works in progress that aim to overcome some of the current limitations in terms of decision making mechanisms for interaction planning [28] and the validations of assessment instruments, like usability scales [13].

Acknowledgements. This work was supported by the European research projects CaMeLi (Grant No. 010000-2012-16), Miraculous Life (Grant No. 616421), Vizier (Grant No. AAL-2015-2-145), GrowMeUp (Grant No. 643647) and GEO-SAFE (Grant No. 691161). We express our gratitude to all the study participants and to all the project partners, including SIEMENS AG and Noldus IT who led the design of CaMeLi's GUI and the end-users VIVA and Zuyderland who performed tremendous work in supporting the user studies and trials.

References

1. Aldoma, A., Tombari, F., Di Stefano, L., Vincze, M.: A Global Hypotheses Verification Method for 3D Object Recognition, In: ECCV, pp. 511–524 (2012)
2. Austin, J.T., Vancouver, J.B.: Goal constructs in psychology: structure, process, and content. Psychol. Bull. **120**(3), 338–375 (1996)
3. Ben Moussa, M., Magnenat-Thalmann, N.: Toward socially responsible agents: integrating attachment and learning in emotional decision-making. Comput. Animat. Virtual Worlds. vol. 24, no. 34, pp. 327–334 (2013)
4. Bickmore, T.W., Caruso, L., Clough-Gorr, K., Heeren, T.: Its just like you talk to a friend relational agents for older adults. Interact. Comput. **17**(6), 711–735 (2005)
5. Bickmore, T.W., Gruber, A., Picard, R.W.: Establishing the computer-patient working alliance in automated health behavior change interventions. Patient Educ. Couns. **59**(1), 2130 (2005)

6. Bickmore, T.W., Silliman, R.A., Nelson, K., Cheng, M., Winter, M., Henault, L., Paasche-Orlow, M.K.: A randomized controlled trial of an automated exercise coach for older adults. J. Am. Geriatr. Soc. **61**(10), 1676–1683 (2013)
7. Bickmore, T.W.: Relational agents: effecting change through human-computer relationships. Massachusetts Institute of Technology (2003)
8. Bickmore, T.W., Schulman, D., Sidner, C.L.: A reusable framework for health counseling dialogue systems based on a behavioral medicine ontology. J. Biomed. Inform. **44**(2), 183197 (2011)
9. Brooke, J.: SUS —a quick and dirty usability scale. Usability Eval. Ind. **189**(194), 47 (1996)
10. Cassell, J., Bickmore, T., Campbell, L., Vilhjlmsson, H., Yan, H.: Human conversation as a system framework: designing embodied conversational agents. 29–63 (2001)
11. Cassell, J.: More than just another pretty face: embodied conversational interface agents. Commun. **43**, 70–78. ACM (2000)
12. Colby, B.N., Ortony, A., Clore, G.L., Collins, A.: The Cognitive Structure of Emotions. Contemp. Sociol. **18**(6), 957 (1989)
13. Dantas, C., Jegundo, A.L., Quintas, J., Martins, A.I., Queirós, A., Rocha, N.P.: European portuguese validation of usefulness, satisfaction and ease of use questionnaire (USE). In: World Conference on Information Systems and Technologies, pp. 561–570 (2017)
14. de Rosis, F., Novielli, N., Carofiglio, V., Cavalluzzi, A., De Carolis, B.: User modeling and adaptation in health promotion dialogs with an animated character. J. Biomed. Inform. **39**(5), 51431 (2006)
15. Ekman, P., Keltner, D.: Universal facial expressions of emotion. Calif. Ment. Heal. Res. Dig. **8**(4), 151–158 (1970)
16. Fagel, S., Ben-Moussa, M., Cereghetti, D.: How avatars in care context should show affect. In: Pervasive Health 2016 Workshop on Affective Interaction with Virtual Assistants within the Healthcare Context (2016)
17. Group, T.W.: Development of the world health organization WHOQOL-BREF quality of life assessment. The WHOQOL Group. Psychol. Med. **28**(3), 551–558 (1998)
18. Hanke, S., Tsiourti, C., Sili, M., Christodoulou, E.: Embodied ambient intelligent systems. in ambient intelligence and smart environments: recent advances in ambient assisted living, bridging assistive technologies, e-health and personalized health care, vol. 20, pp. 65–85. IOS Press (2015)
19. Kasap, Z., Magnenat-Thalmann, N.: Building long-term relationships with virtual and robotic characters: the role of remembering. Vis. Comput. **28**(1), 87–97 (2012)
20. Kelley, J.F.: An iterative design methodology for user-friendly natural language office information applications. ACM Trans. Inf. Syst. **2**(1), 26–41 (1984)
21. Khoshhal, K., Aliakbarpour, H., Quintas, J., Drews, P., Dias, J.: Probabilistic LMA-based classification of human behaviour understanding using power spectrum technique. In: 2010 13th International Conference of Information Fusion, pp. 17 (2010)
22. Klein, J., Moon, Y., Picard, R.W.: This computer responds to user frustration—theory, design, results and implications. In: Proceedings of CHI 99 Extended Abstracts on Human Factors in Computing Systems, pp. 242 (1999)

23. Kramer, M., Yaghoubzadeh, R., Kopp, S., Pitsch, K.: A conversational virtual human as autonomous assistant for elderly and cognitively impaired users?. Social acceptability and design considerations. Lect. Notes Inf. (2013)
24. Lim, M.Y.: Memory models for intelligent social companions. Human-Computer Interaction: The Agency Perspective, pp. 241–262. Springer (2012)
25. Lisetti, C., Yasavur, U., de Leon, C., Amini, R., Rishe, N., Visserv, U.: Building an on-demand avatar-based health intervention for behavior change. In: Proceedings of Twenty-Fifth International Florida Intelligence Research Soceity Conference, no. Mi, pp. 175–180 (2012)
26. Nijholt, A.: Disappearing computers, social actors and embodied agents. In: Proceedings of the International Conference on Cyberworlds, pp. 128–134 (2003)
27. Ortiz, A., Carretero, P., Oyarzun, D., Yanguas, J.J., Buiza, C., Gonzalez, M.F., Etxeberria, I.: Elderly users in ambient intelligence : does an avatar improve the interaction? Intelligence **4397**, 99–114 (2002)
28. Quintas, J., Paulo, M., Jorge, D.: Information model and architecture specification for context awareness interaction decision support in cyber-physical human-machine systems. IEEE Trans. Hum. Mach. Syst. **47**(3), 323–331 (2016)
29. Rusu, R.B., Cousins, S.: 3D is here: point cloud library. In: IEEE International Conference on Robotics and Automation, pp. 1–4 (2011)
30. Sakai, Y., Nonaka, Y., Yasuda, K., Nakano, Y.I.: Listener agent for elderly people with dementia. In: 2012 7th ACM/IEEE International Conference on Human-Robot Interaction (HRI), pp. 199–200 (2012)
31. Thiebaux, M., Marsella, S., Marshall, A.N., Kallmann, M.: SmartBody: behavior realization for embodied conversational agents. In: Proceedings of the 7th International Joint Conference on Autonomous Agents and Multiagent Systems (AAMAS '08), pp. 151–158 (2008)
32. Traum, D., Larsson, S.: The information state approach to dialogue management. Current and New Directions in Discourse and Dialogue, pp. 325–353 (2003)
33. Tsiourti, C., Ben Moussa, M., Joly, E., Wings, C., Wac, K.: Virtual assistive companions for older adults: qualitative field study and design implications. In: 8th International Conference on Pervasive Computing Technologies for Healthcare (PervasiveHealth) (2014)
34. Tsiourti, C., Ben-Moussa, M., Quintas, J., Loke, B., Jochem, I., Lopes, J.A., Konstantas, D.: A virtual assistive companion for older adults: design and evaluation of a real-world application. In: Proceedings of SAI Intelligent Systems Conference 2016, London, UK (2016)
35. Van Kuilenburg, H., Wiering, M., Den Uyl, M.: A model based method for automatic facial expression recognition. In: Proceeding of Machine Learning: ECML 2005, pp. 194–205. Springer (2005)
36. Vardoulakis, L.P., Ring, L., Barry, B., Sidner, C.L., Bickmore, T.: Designing relational agents as long term social companions for older adults. In: Proceedings of the 12th International Conference on Intelligent Virtual Agents, vol. 7502, pp. 289–302 (2012)
37. Verberne, F.M.F., Ham, J., Ponnada, A., Midden, C.J.H.: Trusting digital chameleons: the effect of mimicry by a virtual social agent on user trust. In: Berkovsky, S., Freyne, J. (eds.) Persuasive Technology, vol. 7822, pp. 234–245, Springer, Berlin, Heidelberg (2013)

38. Woelfel, M., McDonough, J.: Distant Speech Recognition. Wiley (2009)
39. Yaghoubzadeh, R., Kramer, M., Pitsch, K., Kopp, S.: Virtual Agents as daily assistants for elderly or cognitively impaired people. In: Proceedings 13th International Conference of Intelligence Virtual Agents, vol. 8108, pp. 91 (2013)

Emotional Domotics: Inhabitable Home Automation System for Emotion Modulation Through Facial Analysis

Sergio A. Navarro-Tuch$^{(\boxtimes)}$, M. Rogelio Bustamante-Bello, Javier Izquierdo-Reyes, Roberto Avila-Vazquez, Ricardo Ramirez-Mendoza, Pablos-Hach Jose Luis, and Yadira Gutierrez-Martinez

Escuela de Ingenieria y Ciencias, Tecnológico de Monterrey, Calle del Puente 222, Tlalpan, Ejidos de Huipulco, 14380 Mexico City, Mexico
senatuch@hotmail.com, rbustama@itesm.mx

Abstract. This research proposed working with an influence on the subject mood, presenting an approach to state the subjects analysis when the light hue is varied. The experimental results led to the finding of the emotional response time dynamics. Such dynamics are important for future design and implementation of the control loops in-house automation systems for emotion modulation. Throughout this document, the details and progress of the research in emotional domotics, with the aim of developing a controlled algorithm for living space based on the user's emotional state, will be illustrated and detailed. This project is centered on domotics (home automation) systems, which is, a set of elements installed, interconnected and controlled by a computer system. After introducing the investigation's core, general preview, and the experiment's description conducted with light hue variation, the description is followed by a presentative approach to state the subjects analysis when light hue is varied. The experimental results led to the time dynamics of emotional response findings. Such dynamics are important for future design and implementation of the control loops in house automation systems for emotion modulation.

Keywords: Emotional domotics · Intelligent ambient · Facial analysis
Facial action coding system

1 Introduction

From the automatization perspective, domotics or control of inhabitable spaces is an interdisciplinary idea that allude to the joining of various advances inside the home range, with the sole point of enhancing the human way of life. Home Automation is understood as a set of elements installed, interconnected and controlled by a centralized system. These systems provide energy, intercoms, welfare and security management services. Home Automation seeks the integration of sensing and control in the least intrusive manner possible systems [1]. With this in mind, this research proposed the idea of working with an influence on the mood of the subject. An individual's mood evaluation can be achieved through various strategies; some are of intrusive, such as the variable physiological monitoring of the subject from sensors carried by the user. Others, are of the non-intrusive kind, for example, voice analysis [2] and the facial analysis [3].

From the point of view of the home automation, domotics is an interdisciplinary concept that refers to the integration of different technologies in the home by the use of simultaneous telecommunications, electronics, the computer and electricity, to improve the quality of life of human beings [4]. Home Automation provides us with the tools necessary to allow for the flexibility of the use of various technologies for the control of a space and therefore the altered state of subjects in the controlled space. However, it is due to its interdisciplinary nature, requiring the implementation of its management and knowledge of different areas for its optimal deployment. To meet the objectives of deployment, the research was previously divided into three central pillars for this proposal based on the emotional analysis of the facial analysis. The three central areas of research work are analyzed in the following sections.

2 Investigation Areas

2.1 Facial Recognition

The area of facial recognition refers to the identification of faces within an image or sequence of digital images. For the case of the emotion home automation, it is required that the processing is performed in real time, by what is known as the starting point and the focus on one of the first algorithm of facial recognition in real time should also be considered with the algorithm of Viola, Jones [5].

For the development of its analysis algorithm, Viola and Jones developed what they called "integral image representation" [6], in the integral image representation, rectangular subdivisions are generated, where the comparison of two adjacent rectangles from the contrast of their areas, are evaluated. This classification of features by contrast of rectangles allows the identification of characteristics scalable to the different sizes of the located face. Once the integral image is obtained, as well as the features detached of the same, the AdaBoost learning algorithm is implemented. The use of this learning algorithm helps to obtain a set of weak classifiers. They are called weak classifiers since none gives answers with precision high by itself, i.e. the percentage of success rare time exceeds the 50%, however, its action combined, allows a higher success index each time. The percentage of error decreases exponentially with each execution of the Adaboost as proved by Feund and Schapire in Schapire et al. 1997 [7]. Each feature analyzed represents a node in the tree of decisions. The main advantage in the use of the AdaBoost as a selector of features against other similar algorithms, lies in their speed of learning, this obeys to the made of that in each round, the dependency of the feature previously evaluated is encoded using the weight of the previous examples in a compact way [6]. According to reports, this algorithm showed a percentage of detection of the 81.1% against the 79.9% of the test implemented by the MIT. According to their article *Robust Real-Time Face Detection* [6], in the development of a real time face recognition algorithm, it is necessary to consider the following elements:

1. Number of features: Referring to the features to search and evaluate in order to determine the hand if the feature is a face or environment and by the other, relate it with a name or label differentiating.
2. Type of processing: image processing can follow different strategies, among which is the decision tree [5] (cascade), total processing andor multiple neural networks [6].
3. Tolerance and parameters: In order to get an optimal training, it is necessary to define the index of detection and the index of false positive. These parameters delimit the parameters of error acceptable to define an acceptable algorithm.
4. The Processing time: optimization of the cost vs benefit (the greater the number of features the greater the time of processing). A high processing time limits the capacity of response in real time, as well as the number of pictures per second to the rating. These requirements limit and allow defining the requirements of software and hardware for the implementation.

For a better understanding of this area of research Fig. 1 can act as an aid. In this diagram, the area of facial recognition and identification is resumed. In the first section, reference articles were discovered of some authors whose article had impacted directly in the area of investigation for the facial recognition. For example, Sun and Poggio [8] present an example-based learning approach for locating vertical frontal views of human faces in complex scenes. The technique models the distribution of human face patterns by means of a few view-based face and nonface model clusters. Rowley et al. [9] present a neural network-based upright frontal face detection system. Schneiderman and Kanade [10] describe an algorithm for object recognition that explicitly models and estimates the posterior probability function, they used this method for detecting human faces from frontal and profile views. Schneiderman and Kanade [11] describe a statistical method for 3D object detection, their approach is to use many such histograms representing a wide variety of visual attributes. Using this method, they developed an algorithm that can reliably detect human faces with out-of-plane rotation. Roth et. al [12] A novel learning approach using linear units is presented and tested against methods that use neural networks or Bayesian methods among others, all focused on facial detection. Viola and Jones 2001 [13] describes a machine learning approach for visual object detection which is capable of processing images extremely rapidly and achieving high detection rates. Also in a paper of 2004 Viola and Jones [6], describes a face detection framework that is capable of processing images extremely rapidly while achieving high detection rates. This system yields face detection performance comparable to the best previous systems [8, 9, 11, 12]. In a more recent reference, Kwang and Myung [14] introduce a novel architecture of the future interactive TV and propose a real time face analysis system that can detect and recognize human faces and even their expressions. Face recognition and facial expression recognition are employed to identify specific TV viewers and recognize their internal emotional states, which is necessary in order to enable personalized user interfaces and services. Jiang et al. [15] investigate the merits of the family of local binary pattern descriptors for Facial Action Coding System (FACS). They compare Local Binary Patterns (LBP) and Local Phase Quantization (LPQ) for static Action Units (AU) analysis. The algorithms section of such diagrams refers to the types of analysis strategies that may be implemented, for example, a by multiple neural networks training or by a decision tree training in order to clean and reduce the number of characteristics being analyzed each level, in order to reduce the computational cost and obtaining a multilayer classification. The boosters aid the algorithms by providing either simplification strategy or by lowering the complexity of the analyzed system.

In the Database section of the Fig. 1 the reader can find some of the existing databases used for training and testing of facial recognition and facial detection algorithms. For example, the MIT-CMU database has been used by authors like [6, 13, 16, 17]. The GEMEP-FERA database use can be found in the publications of various authors like [18–23], the need for various databases has led some authors to the complement of working databases for facial recognition and facial expression recognition like the case of Cohn Kanade [24]. The implementation requirements mentioned in the figure are discussed previously and are based on the article *"Robust Real-Time Face Detection"* [6].

2.2 Emotions Recognition

The traditional methods of extraction of features in the facial analysis are divided into two streams. Methods based on geometric features and methods based on the appearance of the features [15]. The methods based on geometric features, measure the speeds and distances or separation between points of interest in the face. As regards the methods based on changes of appearance, look for changes in the texture of the image, as are caused by wrinkles or skin tension to the collapse in different groups of muscles. Several authors have demonstrated that both

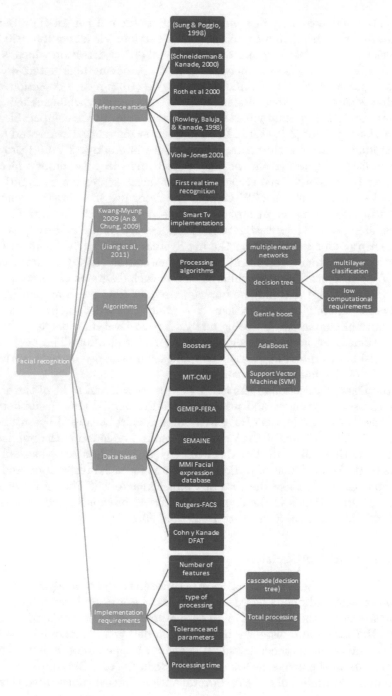

Fig. 1. Facial recognition area reference diagram [6,8,9,11,14,15]

methods can have a very efficient performance at the time of finding the AU [25] necessary to determine the mood based on the FACS [3]; is by this that an algorithm that combines the benefits of both methods is highly efficient [15]. among the various algorithms that exist for the facial and emotional analysis, there are those that work on static images and those which work with video sequences. In the first case there are those with descriptors of local static appearance [15]; this type of algorithms work on the analysis of small static neighborhoods. Examples of these algorithms are the Local Binary Patterns (LBP) [26] and the Volume Local Binary Patterns (VLBP) [15]. This type of descriptors define a central pixel, evaluating it against its outline. The first algorithms created neighborhoods of 3×3, with the central pixel as a reference, giving each a value of zero to 255 based on a histogram. Versions later of the algorithm are based on circular regions with Center in the central pixel of analysis. These new versions allow covering multiple pixel RADIUS in a single neighborhood. After the comparison, the value of each pixel is transformed into ones and zeros, depending on the comparison of values with respect to the central bit. If the pattern is shown at most two transitions of the value of the bit in its analysis, it is a uniform pattern said. In a neighborhood of 8 pixels (i.e. a box of 3 for 3 with a pixel central), there is a total of 256 patterns possible. The analysis by LBP generate histograms by blocks that allow the analysis of the region, depending on the type of AU that is being analyzed, the area of interest in the face is determined [15]. For algorithms that work with movement, sequences, and images, they allow the recognition of facial activities, without the need of having a neutral model as a reference. The local phase quantisation (LPQ)is presented as an example of this type of algorithms [15]. These algorithms were proposed originally by Ojansivo and Heikkil, and were used for the processing of textures blurred in the picture. This algorithm work was transformed from two-dimensional to short-term Fourier transform (STFT), implemented in a rectangular neighborhood. For the analysis of variations in the surface, the values recovered from Fourier Transforms and the values of the neutral value are compared. The process of object identification correlates the information obtained from an object with the objects contained in a database, in order to find if there is any correlation. This is to say that the process of object recognition involves two phases. In the first phase, it extracts the characteristics of objects known and is stored in order to generate the database of reference. In a second phase, it compares the characteristics obtained from the unknown objects, to those features that make up the base of the known objects, in order to achieve identification in the case of exit. This comparison can be applied to the analysis of videos for the identification of those individuals in a video. To understand the basic operation of an algorithm of extraction and facial recognition the following diagram help. Figure 2 (based on the diagram by [27]):

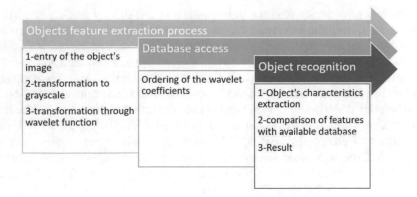

Fig. 2. Outline of operation of an algorithm of objects recognition [27]

For a better understanding of this area of research, the Fig. 3. can be of aid. In this figure we can appreciate first of all the different types of emotional recognition strategies that can be implemented, although the voice analysis and biometric signal analysis were not mentioned, they are not deepened because they are considered secondary to the focus of this investigation which is centered on the facial expression analysis. In the facial expression analysis section there is an area defined as temporal dynamics, the terms of onset, apex and offset [28] refer to the cycle of the expression, being the onset, the beginning of the expression, the apex of the peak or climax of the facial expression and the offset the ending of the facial expression. The duration of each of the three terms determines the kind of expression being dealt with. These expressions may be spontaneous, posed or micro expressions. Another section on the same level as the temporal dynamics, the primary emotion can be found being analyzed by the proposed system, based on the FACET [29] module developed with the aid of Paul Ekman's [3,25,30] group even though different systems and works propose variations in the emotions analyzed for example the inclusion of stress analysis [31], or the omission of the contempt [32] or even the inclusion of an expression defined as neutral [33].

The system requirements refer to the desired characteristics of the system being developed and with the kind of working material available for emotion recognition through facial analysis. An ideal home automation system based on facial response needs to provide a real timely response to the subjects emotions. This will lead to the kind of entrance signal to be analyzed, because different kinds of entrance alters the way the image is analyzed, for example, while a video entrance allows the system to detect changes in the subjects expression through the comparison of previous frames with current ones. A system analysis based on images depends on the analysis of the area based on contextual analysis as previously mentioned. An ideal system for facial expression recognition should be capable of surpassing some obstacles such as illumination variation or diversity, for this, the system can use different strategies like varying their

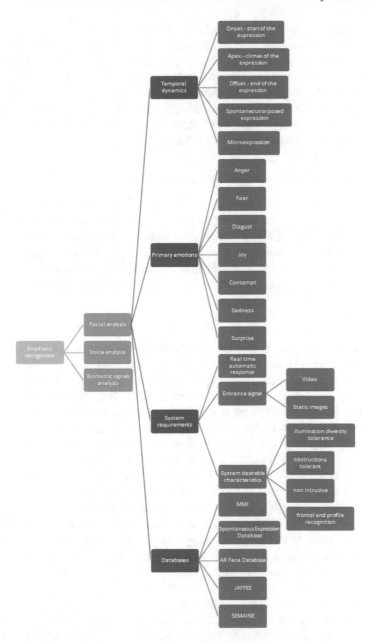

Fig. 3. Emotions recognition investigation diagram

luminosity sensitivity or using the alternative radiation sensors. Obstructions tolerance refers to the capability of the system to analyze the face by the use of the non-obstructed areas, obstructions may refer without limiting to hands, glasses or caps. An ideal system may be able to analyze the expression from a frontal or profile view, although the profile view limits the capability of asymmetric analysis referred, for example, the disgust and contempt expressions, because it may assume a symmetry in the facial expression. Finally, the databases section refers to some of the available databases for the training and testing of facial expression analysis algorithms. For example the MMI [21,34,35], the AR Face Database [36,37] can be found, the Japanese Female Facial Expression (JAF-FEE) [36] and the Semaine [15,38,39] among others.

2.3   Facial Action Coding System (FACS)

The FACS, allows the recognition, the categorization and the qualification of the changes in the appearance of the facial result activation and facial muscles relaxation [40]. Some of the features that tend to affect the results of a coding manual of the face are the structure bone of the subject, deposits of fats in the face, as well as differences in the muscle development of the face of the individual. Understanding of each action unit, in particular, entails a process.

2.4   Domotics

As previously seen, from the point of view of automation, the home automation or control of spaces, is an interdisciplinary concept that refers to the integration of the different technologies in the home and its end improves the human quality of life [4]. The control of spaces allows an optimized management of energy, an improved life quality of the individuals, as well as ensure its security. The domotics seeks the integration of those systems of sensing and control of it ways less intrusive possible [1]. For the development of the emotional induction algorithm, you can use various strategies of sensory stimulation. For the case of the ear, it can stimulate the production of music previously studied, or with tones that are pure as in the case of the binaural beats [41], or work with noise suppression. In the case of the olfactory stimulus, it can work with suppression of smells or use the principles of the aromatherapy. However, along the consultation with experts in the area of the aromatherapy, the use of this strategy in automation was downcast because it can easily trigger very acute emotional response and potentially negative, given the physiological nature of the said sense.

The senses of touch and sight can be stimulated by controlling the environment variables such as light intensity, the color of the light, the temperature and the humidity in the air. All of these factors together will enable not only a change in the mood but can also help in the control of the energy consumption in the area.

For a better understanding of this area of research, the Fig. 4 can act as an aid. The home automation covers the control starting from the technologies of digitization, i.e. the conversion of data and measurement analog, for the conversion of data and variable digital, allowing thus its processing and treatment by means of a driver or processor. To achieve such control and communication, there are different communication strategies, which can be wired or wireless via local area as the LAN connections. This leads to the concept of the Internet of Things (IoT). Some of the tasks that can be covered through the installation of automated control systems are the following [42]:

1. The interconnectivity of equipment: by means of local area networks, it is possible to monitor the status of the devices in the living area, as well as their operational state, allowing the control and the interconnection of data provided by each device. This structure among other advantages allows the monitoring of the consumption of energy resources and therefore the costs and implications of each in the work area. This task acquires greater importance when it reflects on the possibility of their scaling to areas of greater coverage, as is in the case of building automation inmotic and the domotic (interconnection and automation of populated areas).
2. The security systems: Besides the use of integrated systems for the monitoring of resources, domotic systems allow the integration of monitoring and security systems for the detection of the presence and intrusions in the controlled space, as well as the detection of risks or accidents as it can also be caused by power failure, floods and fires. These applications can allow either the automatic response of the system to stop or fight the emergency, as well as to implement early warning systems to the user.
3. Systems of automation and control for the home: the systems of control automated in the home have the objective of providing comfort to the user. Thus, by means of control system programming reagents to the physical condition of the environment or with the programming of schedules and scenarios that request the user. These systems can be centralized control and decentralized control as in the case of KNX systems that is addressed later. Some of the elements that may be included in the automated home systems are the control lights, the temperature, and the blinds, as well as the timing for the transmission of alerts by personal sensors such as medical alert devices.

The home automation systems can be divided into the different components base. The components here mentioned refers to the more generalized structures of implementation, no desire to exclude or limit to these components:

1. The processors: as all computational systems, processors are the components responsible for the reception, the interpretation and the management of instructions according to a previously designed program. The processor relies on the instructions that you provide to them various actuators integrated within the system.
2. The channels of communication: The communication links can be by wire or wireless. The intention of such components is only to collect and transmit the data received by the sensors, towards the processors, and towards the actuators.
3. The sensors: The sensors of the system are responsible for its interaction between the physical space and the the system program, the physical reaction of the sensor part with their components to the environment, for the signal analog generation that the processor uses to determine the environmental state and react according to your program established, some of the

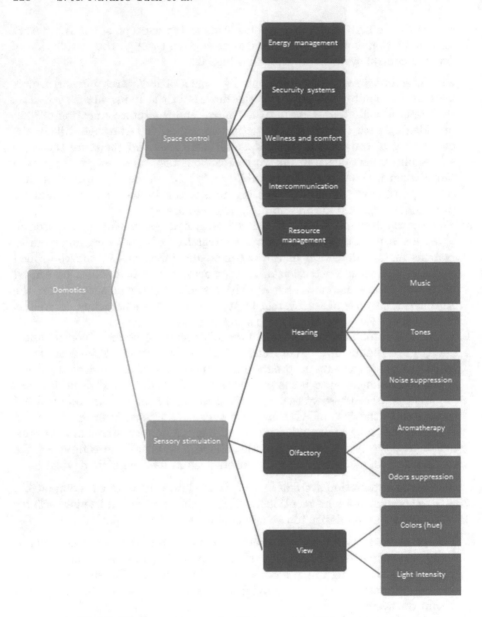

Fig. 4. Diagram of research of home and building automation

variables that a sensor can detect are the humidity, temperature, light intensity, movement, voltage, among others.
4. The actuators: actuators transform the instructions transmitted by the processor acting on the environment, i.e. they are responsible for the activation of the systems of the physical effect such as lights, air conditioning, ventilation, or motors for roller shutters among others.

5. The communication interfaces: the communication interfaces, as well as sensors, allow the processor to know the state of the environment at any time to take decisions, interfaces become deployed information processor for human consumption. The basic function of these elements is that it allows the user to know the state of the variables that the system works with and sometimes also can allow you to modify the program.

The implementation of the project was initially evaluated as a global home automation application protocol, which is the standard KNX. This standard allows the implementations at various home automation levels, building automation and future domotic which is the implementation and automation of projects. This standard allows both the implementation and the installation with a minimum space intrusion to modify. This works from the unification of a programmable communication using a common software, the ETS5 system. The knowledge of the system allowed users to measure in home automation implementation current capacities and to become familiar with the most common implementations and needs that are often from both ends customers and users.

2.5   Domotics and KNX

Previously discussed, the themes of what is home automation, as well as its implications and objectives, however, a problem that had the first domotic designs was the lack of interaction between systems of different manufacturers and suppliers. These had consequence limiting the potential domotic installations, as well as the need to install several different systems for the control of each elements or subsystems that they want to control. As a result of this disadvantage, the facilities as well as having a high cost, had a high level of complexity for the installation. Added to these disadvantages at the time of installation the limited useful life of the system, derived from the rapid evolution of products and its lacks of compatibility with previous versions.

As a response to these problems different quality standards to the home automation system was proposed, this frame gives the creation of the first standard Open KNX of the Association Konnex, [43]. This standard complies with the requirements of other international standards especially the European standards such as the ISO/ICE 14543, el CENELEC EN 50090,CEN EN 13321-1 y GB/Z 20965 [44]. This standard proceeds as a basis for its creation, a combination of the European Bus installation (EIB), BatiBus and the European Home Systems (EHS) [45].

The KNX standard allows different interconnection methods for communication, with the twisted pair wire or the Powerline. The first of the two is carried by means of four wires, the power in one of the pairs and the data or telegrams by a second pair. The Powerline allows transmission of telegrams through the cable power supply. As media, a wireless communication has as options, the communication via infrared or the communication via wifi and ethernet. The advantages offered by wireless means and the ethernet for communication is the ability to minimize the installation requirements for cabling and system connection, however, has a weakness need for various receiver modules for these signals.

As a main feature of the KNX systems it was bumped into that they are decentralized control systems. Meaning that, each components either sensors or actuators possesses its own driver that can be scheduled to act according to a routine preset and to keep to the telegram pending of the sensors in order to adapt its routine according to the user's instructions or to the environmental physical condition. The KNX communication system has a low-speed data transmission, with a speed of 9600 bits per second [46], even though this allows the system a high tolerance to noise and response to physical changes in the environment, it is true that it limits their ability to implementation at the level of industrial machine control, that is why their approach is usually housing.

The topology of a KNX system preset has the following features. As spine or Backline, a maximum of fifteen areas can be interconnected, each area can contain up to 15 lines, each line integrated within an area can have a maximum of sixty-four devices connected to it, that is for every Backline there can be a maximum of fourteen thousand four hundred elements. These limitations of the number of items come as a result of the the telegram's structure communication system. The said telegram is to indicate the address of the device to which it makes reference, has with two bytes assigned of which four bits identify the area referred, four bits identify the line and a byte sets the number of the device [46]. The programming of the KNX system is made through the ETS (Engineering Tool Software) software program which allows not only the individual response of each device programming but also the generation of groups in response to scenarios, i.e. allows you to program multiple drivers to expect a telegram and act together. This advantage is the possibility to minimize the number of telegrams used to control and maximize the speed and response of the integration.

A minimal KNX system has sensors, a source, an actuator and a communication bus. As the sensor counts anything that allows the system to communicate with the physical environment, this includes keypads, touchscreens, or physical changes in the environment sensing components. Source is in charge of providing the necessary power to components within the KNX network, is worth mentioning that, while this source remains at operating the actuators and sensors, as well as provide the high voltage to the communication bus, that source is not the energy provider for the operation of the actuators, this power is taken except in special cases of public wiring power supply of the building. The bus of communication is the route that was mentioned previously, that allows the transmission of the telegrams in the system, in addition to carrying the power to the drivers. However, for a bus of twisted pair, a device maximum distance to the source without a significant operation capacity loss is of three hundred and fifty meters, therefore, the maximum distance between two devices is seven hundred meters. One of the benefits that the system KNX offers is the characteristic of free topology.

Between existing implementations of KNX systems, a variety of examples can bump up, from control offices and business centers [47], up to applications in the control of spaces habitable as it is in trains and means of transport [48]. In addition to the comfort offered by these systems for homes and buildings [46], the implementation for the development of areas of support to older people or with self-sufficiency problems derived from his age or health as sought [49]. Some of the advantages offered by these implementations are the possibility to develop areas with high degree of efficiency making it an ideal system for certifications such as, LEED certification.

3 Experimentation in Gilbreth and Taylor Booths

Taylor and Gilbreth experimentation, are two cabins that are prepared and designed for implementation in various tasks of industrial productivity test subjects, subject to environmental conditions that simulate various working conditions, where you can modify and control part of the positioning and distribution of parts and tools, the physical variables of the environment. These cabins allow the experimenter set conditions of work where the temperature, the relative humidity, intensity luminous in the area of work intensity sound are preset and maintained along the test. These cabins should give their names to investigators in movements and productivity Frederick Winslow Taylor and the couple Frank and Lillian Gilbreth. Both Taylor and the couple developed research on productivity and temporary cost of industrial production. In the case of the couple Gilbreth, they developed the theory of micro-actions that is called therbligs [50,51] to describe the basic movements and tasks which submits an operator in their space of work. With regard to Frederick Taylor, the development accompaniments work in relation to the productivity and the scientific analysis of its productivity measuring of the way analytical times of execution and working [51], considering Frederick Taylor as the father of the scientific administration.

3.1 Planning and Preparation of the Tests

For the development of tests experimentation in cockpits, opted for taking as a basis of work and employee experimentation protocol for the formation of the students of industrial engineering. In this test, the subject prompted the assembling of a Lego vehicle not more than thirteen assembly steps. Each subject is subjected to a total of three sessions in the assembly whose specifications are explained later in this section. For the selection of the conditions of the test, they are defined based on previous experiments realized in some Gesell booths, from the experience prior to the experimentation of PhD. Aguayo for the development of this thesis [52] and of the experience of eng. Yadira based on standards [53,54]. Each of these experiences provided the following information to be considered for the design of the tests. Experience in testing with system cameras of Gesell Emotions grasped the importance of maintaining processing equipment and computing in an environment set correctly in order to avoid the impact of their noise and heat emissions in the subject's experimentation environment, of the same tests learned the distracting elements elimination importance in the subject's environment as well as the elimination advantages of non-active subjects in the experimentation area, to eliminate the emotional alterations striking the subject on which the tests are underway. To resolve both problems, a decision was taken to keep all the computation that are not vital out of the experiment booth, which is met by leaving only the camera or video captures inside the cabin, connected to a USB port extended towards it's computer outside the cabin. Thus, the same emissions had no noticeable effect on the experimental environment. In terms of external distractions and interactions elimination with researchers outside the zone of the experiment, the cabin was filled as a whole with a translucent membrane. Such membrane allows the lamps to diffuse the light emitted from the color and white lamps eliminating the thermal effect of the lamps.

For the experimental design, the current official norms in Mexico was referred to, the Official Mexican Norm (NOM). These standards that were permitted set the temperature range [53], the lighting [54] and the humidity used for experimentation (the choice and their values are specified in the descriptions of each session later). Finally, the maximum duration of the experiments, as well as the key being used was the result of experience provided by PhD. Aguayo in the development of his thesis. Based on his results, he established that the maximum duration that a test could have, taking the acute hue color effect was about 20 min [52]. Since the assembly of Lego vehicles tests was designed for a duration of 5 min, it was chosen by that time, as the time for the development of subjects activity. Experience with the development and implementation of therapies acoustic research protocols, the measuring importance and the acquisition of an initial baseline subject state was used in order to have a starting point for the comparison of the relative changes in the tests. It is for this reason that the current implementation protocol established two periods of a minute each in which the subject remained without making any manual activity, in order to be able to determine its initial and final state. Also, the experimentation in the therapies acoustic culture, the importance of the standardization in the instructions that you give to the subject, so for these tests, it is opted by leaving the instructions in a voice generated by a computer. This decision helped to ensure that all the subject received exactly the same instructions, in addition, to allow that all had the same time of execution, captured by a program integrator of audiovisual projects, Camtasia^{TM}.

3.2   Development of Tests

Prior to the development of the tests and based on the clinical protocol developed for the research in previous acoustic therapies, prepared an informed consent for the subject that would participate in the tests, where it least advises about not only the procedures to perform, but also the possible risks and reasons by which they could be excluded off the study, including pre-existing physiological* and psychological conditions*, or hidden effects that are displayed during the testing. Pain, anxiety, and dizziness as not severe effects are among the possible adverse effects that might occur during the test. In any of those cases, the team of research is committed to being prepared to respond and provide its medical assistance of emergency required. The same document also contains details of the commitment from the team to ensure the privacy and anonymity of the subjects with respect to the results and data collected from each. Then, it describes the procedures made during each and of the sessions of experimentation to which were the subjects participating.

3.2.1   First Session of Experimentation

This session starts with the subject on the list of attendance, on arrival to the place of experimentation, which is the Industrial Engineering laboratory. As personal data requested the subjects to provide their age, sex, race, courses matter, and contact email, as well as indicate if they had already signed the corresponding consent. If any subject complied with all the requirements of health to proceed with this session, i.e., if the subject came without a headache or under the influence of any drug or physical illness, he then proceeds to put the lab coat that was delivered to him and entered the cockpit of experimentation, which had been prepared and aligned with a preset height Chair and a shelf with parts sorted according to the color of a distance and position preset for testing. Once the subject is placed in the corresponding site, the test site begins to play and the software video captures Camtasia^{TM} which explains the subject activity and provides instructions to follow during the (armed with a collapsible cart toy based on the instructions in the printed sheets in front of him). During this session, the environmental conditions of experimentation are as follows: optimum lighting (300 lux white light), temperature (25 °C) and humidity (40± 5% relative humidity). Also, find a sheet that indicates the subject and how he should assemble the toy. This first session was conducted without any problems in their implementation, given the simplicity of the ideal conditions according to the relevant standards and instructions [53,54]. the reason by which the subjects were asked to wear a laboratory coat during the session of experimentation was by recommendation of a professor, specialist in ergonomics who us said that such clothing allows both neutrality and eliminating possible distractions for the subject, as a refraction uniform of the light tonality. In order to maintain consistency throughout the study, such clothing will be provided to subjects during each of the sessions.

3.2.2   Second Session of Experimentation

During the execution of the first tests with an intense blue shade given, the disposition of lights, was captured on video camera, the facial features were eliminated so there was an advance to lower the intensity of the lampshade that impacted the face of the subjects making it closest to being white, but kept the shade of blue in the area of work and instructions. Although carried out modifications in hue, and brightness of about 300 lx were preserved.

3.2.3   Third Session of Experimentation

The third session proceeded without technical difficulties. However, multiple subjects reported light headaches at the end of the tests, the complaint is a valuable feedback to remember, to limit the intensity if red hues in the environmental control algorithms are used, once the testing booth is assembled.

3.3   Preliminary Analysis of the Results

The procedures to follow for the full analysis of the captured videos during testing in the cabins of Taylor and Gilbreth, as well as observations during the initial processing of the same using iMotions™ software are mentioned in this section. Prior to the correlated analysis of environmental variables and emotional States, imported videos captured during the sessions in the cabins of Gilbreth and Taylor to the analysis iMotions™ software, where videos are processed to give an evidence of emotional states, along with the facial activity, even though the program already gives a logarithmic scale of the probability of occurrence of each emotion analysis based on action units detected in the face decided to include in the report of data exported by the same evidence of pure action units, akin to make a more detailed analysis of each and to determine those influenced by environmental conditions more later if required. With regard to the emotional evidence referred to, the software delivers a table with prints of time relating each evidence of emotion together with it instantly in the video, given its high speed of response, those measurements that appear multiple times in a picture repeated the value prior, however, in those instants where the subject came out from the area of video captures, precluding him from the program, the analysis of the emotions is reported as spaces in white, causing the export to excel that these boxes did not report any value. The boxes in white presented the challenge that can be exported and analyzed in the software's analysis of Matlab which marks errors of captures to the bumping with these features. Initially, it was proposed that to avoid the said errors, the boxes in white would be substituted with the value zero, thinking that its impact would be negligible. However, given that the emotional evidence of the scale, present values that have covered less than seven to three values, zero as an intermediate point, and affects normal distributions of the evidence. For this reason, they chose to give a value at the end of the distributions and trim the extreme values; the value that these boxes will be filled with will be less than ten, which represents values so small that they are negligible, slash in this analysis by at least nine values. The first observations of the videos captured allowed observing why the first period of basal state was designed to obtain the basal state without alterations of the subject receiving instructions, resulting in the response of the subject having as a consequence an altered state. By this, the initial basal state, possibly would not allow making the filtered comparative of variations that was initially planned. The final basal state must be analyzed as a possible alternate filter, in order to be able to detect the variations that occurred during the test. Is discussed also is the possibility of using the basal states of the second and third test to find the subject's state basal pair, of these sessions in which the subject already knew the general dynamic of its activity, by what their initial instructions could not alter so it marked its basal state. Other proposals of analysis include the comparison between the initial basal states and the end in order to detect alterations in the mood as a consequence of the type of tonality to which they subjected the experimental subjects. The first analysis with Matlab processing software showed that, by omitting the impact of the empty boxes, subjects show a tendency to normal distributions of emotional states through the stage of measuring basal, and it may be considered that the impact of emotions is null or non-existent, the smaller value of the logarithmic scale reported. With the values reported, in order to know the behavior of the emotions and with the help of each individuals histograms, it seeks to locate the predominant emotions in each one, make a distribution by frequency of all subjects, in order to verify if there are general patterns of emotions occurrence as indicated in the previous theories or if the performed task had predominance on the emotional state of the subject. If you have a predominance type of activity on in the emotional state it would indicate the need for testing and further research with passive tasks in order to be able to continue placing values of importance and emotional impact of the environmental variables. Where activity is independent of the emotional state of patients, it would allow us to continue investigations with the tasks of a more active nature, allowing the approach to control in an active work environment, or in classrooms for subsequent phases of research.

4   Experimental Model

The experimental testing in the booths of Gilbreth and Taylor in which the three treatments were applied: in the first white light, in the second blue light and on the third red light, each one applied to forty subjects. The students of an industrial engineering course in ages between 18 and 19 years in the first semester of their career. Those who participated voluntarily filled a consent form in the experiment. Each subject received three treatments. It is worth mentioning that there were subjects who received only one or two treatments. Each subject will apply at least one of the treatments and some received three treatments. By what is considered and that corresponds to the design known as completely randomized design (CDR), in accordance with [55]. Forty tests correspond to the number of treatments with a separation of at least two weeks between applications in order to eliminate the residual effects of the treatments. Originally, there were sixty subjects, however, there was a defection of a third of the population, leaving in the end with only forty subjects by the test. Given the above subjects participated in all three lights while others only participated in two or just in a treatment.

$$Y_{ij} = \mu + \tau_i + \epsilon_{ij} \qquad 1 \le i \le 3, 1 \le j \le 40 \qquad (1)$$

where:

μ = general median
τ_i = effect of the i-light
ϵ_{ij} experimental error of the i-light over the j-individual

$$i = \begin{cases} 1 = \text{white light} \\ 2 = \text{blue light} \\ 3 = \text{red light} \end{cases}$$

$$j = individual_n umber \rightarrow 1 \leq j \leq 40$$

The error $_{ij}$ is a random variable with

$$E(\epsilon_{ij}) = 0 \tag{2}$$

$$Var(\epsilon_{ij}) = \sigma^2 \tag{3}$$

And they are not correlated.

Where: y_{ij} corresponds to the value of the variable depending on the study of the j-th repetition of the i-th

$$treatment \begin{Bmatrix} 1 \leq i \leq 3 \\ 1 \leq j \leq 40 \end{Bmatrix} \tag{4}$$

Which is a random variable with

$$E(y_{ij}) = \mu + \tau_i \\ Var(y_{ij}) = \sigma^2 \tag{5}$$

and are non correlated observations.

4.1   Statistical Analysis of the Results

For the analysis of the results of the tests it was decided to use, the SPSS^{TM} tool with the tool of analysis of non-parametric tests, in this way, the differences in the medians of the final state against the initial were discussed. The final state corresponds to the emotional state in 3 s after finishing the activity, while the initial state includes the 15 s before the activity. This analysis was detached following the resumed table.

From this table, it is appreciated that according to the test of medium there is a difference present to the emotions of sadness, surprise and fear with a level of significance of 10. Further analysis allowed the realization of the significant difference in the sad emotion when contrasted with the effects of light blue against the white, while the emotions of surprise and fear were presented with the contrast of the white and red lights. The videos of the subject's face during the activity in the three sessions is processed using the module FACET of the tool iMotions^{TM} for its emotional interpretation of each subject. From the processed videos, are extracted two periods of time, the first period was in the second fifteen of the second thirty to establish the basal states of the emotions of each subject. The second period of time initially was the 15 s after the Lego assembly. However, ANOVA analysis showed some significance, are not there where there was a graphical analysis as shown in the Fig. 5.

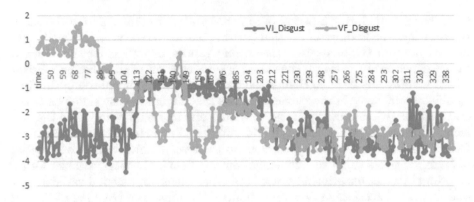

Fig. 5. Graph of behavior of the emotion disgust in periods of time initial and final

Table 1. Test of medians results with null hypothesis rejection table

Null Hypothesis		Test	Significance	Decision
Median of FV are the same for the three treatments	Surprise	Test of the median of independent samples	$P < 0.046$	Null hypothesis rejected, red light has a major impact over surprise versus other lights
	Fear		$P < 0.018$	Null hypothesis rejected, red light has a major impact over fear versus other lights
	Sadness		$P < 0.007$	Null hypothesis rejected, blue light has a major impact over sadness versus other lights

From this graph we can find that the impact of the activity on emotion tends to decline in a time period not longer than 3 s, so for the following analysis took the 3 s period after the assembly activity analysis of the medians of the final values of the subjects are as shown in Table 1.

5   Conclusions and future work

The tests carried out in the testing booths of fixed environment (cabins of Gilbreth and Taylor), led to the development of the strategy of processing of the raw data. As well as the generation of diffused model for the interpretation of the emotions and implementation in the control loop of the control of emotions. In addition to this, the analysis of the results obtained with the tool iMotions^{TM} of the experiments in the cabins has allowed the appreciation of a phenomenon particularly in the conduct of the emotions, which tend to return to its basal state during a period of time not greater to than 3 s from the end of the activity. The findings of the said pattern, has led to the need to consider the period, both for the analysis of the current results, as well as for the design of future experiments, since the duration period is so short of the emotion reflected which involves the need to change the paradigm of experimentation. Finally, a significant difference in the impact that has the red light on the emotions of fear and surprise, while blue light tend to induce an emotion of sadness on the subjects, which could be checked with iMotion^{TM} tool and an analyzed by SPSS^{TM}. Given the above results, the next steps in the research involve the construction of an experiments cabin with dynamic control of environmental variables for the emotional impact experimentation of the other the environmental variables and search their exacerbation or attenuation by the impact of the kind of light to use. Starting from the experiences acquired and reported in previous periods of the research, defining the technical requirements and variable to manipulate the experimentation and implementation of the control algorithm spaces are based on emotions. A time defined as the variable to control the desired temperature, relative humidity relative [5], light intensity [6], As well as its hue [7], seeking a thermal and acoustic insulation of the medium. In additional to the technical requirements of the environment, is a less raise to the teams of design following the requirements. The space to design would be a space focused on its experimentation complying to have a proper reflection of the light within, isolated from the external lighting effects and temperature, trying to minimize or eliminate also the exchange of heat in the measure of its possibility between the environment and the inside. Space should be optimized for the distribution inside the simulation hearing, as well as having the possibility of controlling from the outside the gaseous exchange starting from the actuators for regularization of the temperature and humidity. Besides, the space can be detachable, it should count inside with the furniture necessary to allow the subject test the realization of the tasks and tests both written as interaction with a screen or computer. The systems of computation and control should locate the outside of the area of experimentation and while the furniture should be ergonomically suitable for the type of task, this should of the result be emotionally neutral (i.e. avoid induces relaxation of other alterations in the basal state).

References

1. Coral Lozada, M.A.: In: 2014 IEEE Colombian Conference on Communications and Computing (COLCOM), pp. 1–6 (2014). https://doi.org/10.1109/ColComCon.2014.6860413. http://ieeexplore.ieee.org/lpdocs/epic03/wrapper.htm?arnumber=6860413

2. Lutfi, S., Lucas, J.M., Montero, J.M.: pp. 1–5. http://www-gth.die.upm.es/research/documentation/AN-090Des-09.pdf

3. Ekman, P., Rosenberg, E.: What the Face Reveals, 2nd edn. Oxford University Press, United States of America (1997)

4. Herrera Quintero, L.F.: Revista Ingeniería e investigación **25**(2), 45 (2005). http://www.itescam.edu.mx/principal/sylabus/fpdb/recursos/r101024.PDF

5. Wang, Y.Q.: Image Processing On Line **4**, 128 (2014). http://www.ipol.im/pub/art/2014/104/article.pdf
6. Viola, P., Jones, M.J.: Int. J. Comput. Vis. **57**(2), 137 (2004). http://www.vision.caltech.edu/html-files/EE148-2005-Spring/pprs/viola04ijcv.pdf
7. Freund, Y., Schapire, R.: Comput. Learn. Theory **55**, 119 (1997). https://doi.org/10.1006/jcss.1997.1504. http://link.springer.com/chapter/10.1007/3-540-59119-2_166
8. Sung, K.K., Poggio, T.: IEEE Trans. Pattern Anal. Mach. Intell. **20**(1), 39 (1998)
9. Rowley, H.A., Baluja, S., Kanade, T.: IEEE Trans. Pattern Anal. Mach. Intell. **20**(1), 23 (1998). https://doi.org/10.1109/34.655647. http://ieeexplore.ieee.org/lpdocs/epic03/wrapper.htm?arnumber=655647
10. Schneiderman, H., Kanade, T.: Proceedings. 1998 IEEE Computer Society Conference on Computer Vision and Pattern Recognition (Cat. No.98CB36231), pp. 45–51 (1998). https://doi.org/10.1109/CVPR.1998.698586. http://ieeexplore.ieee.org/lpdocs/epic03/wrapper.htm?arnumber=698586
11. Schneiderman, H., Kanade, T.: Proceedings. IEEE Conference on Computer Vision and Pattern Recognition. CVPR 2000 (Cat. No.PR00662), vol. 1, p. 2 (2000). https://doi.org/10.1109/CVPR.2000.855895
12. Roth, D., Yang, M.-H., Ahuja, N.: A Snow-Based Face Detectos. Neural Inf. Process. Syst. 12 (n.d.)
13. Viola, P., Jones, M.: Comput. Vis. Pattern Recogn. (CVPR) **1**, I (2001). https://doi.org/10.1109/CVPR.2001.990517
14. An, K.H., Chung, M.J.: pp. 2271–2279 (2009). http://ieeexplore.ieee.org/stamp/stamp.jsp?tp=&arnumber=5373798
15. Jiang, B., Valstar, M.F., Pantic, M.: In: Face and Gesture 2011, pp. 314–321 (2011). https://doi.org/10.1109/FG.2011.5771416. http://ieeexplore.ieee.org/lpdocs/epic03/wrapper.htm?arnumber=5771416. http://www.doc.ic.ac.uk/mvalstar/Documents/laud_submission.pdf
16. Littlewort, G.C., Bartlett, M.S., Fasel, I.R., Chenu, J., Kanda, T., Ishiguro, H., Movellan, J.R.: Adv. Neural Inf. Process. Syst. **16**, 1563–1570 (2004). http://papers.nips.cc/paper/2402-towards-social-robots-automatic-evaluation-of-human-robot-interaction-by-facial-expression-classification.pdf
17. Bartlett, M., Littlewort, G.: In: 7th International Conference on Automatic Face and Gesture Recognition, 2006. FGR 2006, pp. 223–230 (2006). https://doi.org/10.1109/FGR.2006.55. http://ieeexplore.ieee.org/lpdocs/epic03/wrapper.htm?arnumber=1613024. http://ieeexplore.ieee.org/xpls/abs_all.jsp?arnumber=1613024
18. Chew, S.W., Lucey, P., Lucey, S., Saragih, J., Cohn, J.F., Matthews, I., Sridharan, S.: IEEE Trans. Syst. Man Cybern. Part B, Cybern.: Publ. IEEE Syst. Man Cybern. Soc. **42**(4), 1006 (2012). https://doi.org/10.1109/TSMCB.2012.2194485. http://www.ncbi.nlm.nih.gov/pubmed/22581139
19. Senechal, T., Rapp, V., Salam, H., Seguier, R., Bailly, K., Prevost, L.: IEEE Trans. Syst. Man Cybern. **42**(4), 993 (2012). http://ieeexplore.ieee.org/stamp/stamp.jsp?tp=&arnumber=6202713
20. Wu, T., Butko, N.J., Ruvolo, P., Whitehill, J., Bartlett, M.S., Movellan, J.R.: In: Face and Gesture 2011, pp. 889–896 (2011). https://doi.org/10.1109/FG.2011.5771369. http://ieeexplore.ieee.org/lpdocs/epic03/wrapper.htm?arnumber=5771369
21. Wu, T., Butko, N.J., Ruvolo, P., Whitehill, J., Bartlett, M.S., Movellan, J.R.: IEEE Trans. Syst. Man Cybern. Part B: Cybern. **42**(4), 1027 (2012). https://doi.org/10.1109/TSMCB.2012.2195170

22. Gehrig, T., Ekenel, H.K.: In: Cvpr 2011 Workshops, pp. 1–6 (2011). https://doi.org/10.1109/CVPRW.2011.5981817. http://ieeexplore.ieee.org/lpdocs/epic03/wrapper.htm?arnumber=5981817
23. Littlewort, G., Whitehill, J., Wu, T.F., Butko, N., Ruvolo, P., Movellan, J., Bartlett, M.: In: Face and Gesture 2011, pp. 897–902 (2011). https://doi.org/10.1109/FG.2011.5771370. http://ieeexplore.ieee.org/lpdocs/epic03/wrapper.htm?arnumber=5771370
24. Kanade, T., Tian, Y., Cohn, J.F.: In: Proceedings of the Fourth IEEE International Conference on Automatic Face and Gesture Recognition 2000, FG '00, pp. 46–53. IEEE Computer Society, Washington, DC, USA (2000). http://dl.acm.org/citation.cfm?id=795661.796155
25. Ekman, P.: Emotions Revealed, 2nd edn. Henry Holt and Company, LLC, New York (2007)
26. Hadid, A.: In: 2008 First Workshops on Image Processing Theory, Tools and Applications, pp. 1–9 (2008). https://doi.org/10.1109/IPTA.2008.4743795. http://ieeexplore.ieee.org/lpdocs/epic03/wrapper.htm?arnumber=4743795
27. Bui, T. T. T., Phan, N. H., Spitsyn, V. G.: Face and hand gesture recognition algorithm based on wavelet transforms and principal component analysis. Strategic Technology (IFOST), 2012 7th International Forum on, pp. 1–4 (2012). https://doi.org/10.1109/IFOST.2012.6357626
28. Pantic, M., Patras, I.: IEEE Trans. Syst. Man Cybern. 36(2), 433 (2006). http://ieeexplore.ieee.org/stamp/stamp.jsp?tp=&arnumber=1605389
29. FACET Module Facial Expression Analysis.: 1(617) (2014)
30. Donato, G., Bartlett, M.S., Hager, J.C., Ekman, P., Sejnowski, T.J.: IEEE Trans. Pattern Anal. Mach. Intell. 21(10), 974 (1999). https://doi.org/10.1109/34.799905
31. Jang, E.H., Park, B.J., Kim, S.H., Eum, Y., Sohn, J.: In: 2011 International Conference on Engineering and Industries (ICEI), pp. 1–6 (2011)
32. Mano, L.Y., Faiçal, B.S., Nakamura, L.H.V., Gomes, P.H., Libralon, G.L., Meneguete, R.I., Filho, G.P.R., Giancristofaro, G.T., Pessin, G., Krishnamachari, B., Ueyama, J.: Comput. Commun. 89–90, 178 (2015). http://dx.doi.org/10.1016/j.comcom.2016.03.010
33. Barmaki, R.: Gesture Assessment of Teachers in an Immersive Rehearsal Environment. Ph.D. thesis, University of Central Florida Electronic (2016). http://stars.library.ucf.edu/cgi/viewcontent.cgi?article=6067&context=etd
34. Cruz, A.C., Bhanu, B., Thakoor, N.S.: IEEE Trans. Affect. Comput. 5(4), 418 (2014). https://doi.org/10.1109/TAFFC.2014.2316151
35. Khowaja, S.A., Dahri, K.: pp. 4–9 (2015). https://doi.org/10.1109/ICET.2015.7389223
36. Bettadapura, V.: arXiv preprint arXiv:1203.6722, pp. 1–27 (2012)
37. Xie, X., Lam, K.M.: Pattern Recogn. 42(5), 1003 (2009). https://doi.org/10.1016/j.patcog.2008.08.034. http://linkinghub.elsevier.com/retrieve/pii/S0031320308003397
38. Picard, R.: Crowd sourcing facial responses to online videos. J. LATEX Class Files 6(1), 456–468 (2007)
39. Fernández-Caballero, A., Martínez-Rodrigo, A., Pastor, J.M., Castillo, J.C., Lozano-Monasor, E., López, M.T., Zangróniz, R., Latorre, J.M., Fernández-Sotos, A.: J. Biomed. Inf. 64, 55 (2016). https://doi.org/10.1016/j.jbi.2016.09.015. http://linkinghub.elsevier.com/retrieve/pii/S1532046416301289
40. Ekman, P., Friesen, W.V., Hager, J.C.: Facial Action Coding System, vol. 1. Research Nexus division of Network Information Research Coorporation, Salt Lake City (2002)

41. Atwater, F. H.: Inducing altered states of consciousness with binaural beat technology binaural beats and the physiology of the brain. In: Proceedings of the 8th International Symposium on New Science, pp. 11–15 (2003)
42. Quentin, W.: Guide to Digital Home Technology Integration, 1st edn. Cengage Learning, New York (2009)
43. Lee, W.S., Hong, S.H.: In: 4th IEEE Conference on Automation Science and Engineering, CASE 2008, pp. 750–755 (2008). https://doi.org/10.1109/COASE.2008. 4626433
44. Woo, S.L., Seung, H.H.: In: Digest of Technical Papers—IEEE International Conference on Consumer Electronics (2009), pp. 545–549. https://doi.org/10.1109/ ISCE.2009.5156866
45. Ning, H. N. H., Ya-Hu, W. Y.-H. W., Yi, T. Y. T.: Research of KNX device node and development based on the bus interface module. In: Control Conference (CCC), 2010 29th Chinese, pp. 4346–4350 (2010)
46. Martirano, L., Marrocco, R., Liberati, F., Di Giorgio, A.: KNX protocol compliant load shifting and storage control in residential buildings. In: Industry Applications Society Annual Meeting, IEEE, pp. 1–6 (2015)
47. Bujdei, C., Moraru, S.A.: In: Proceedings—2011 7th International Conference on Intelligent Environments, IE 2011, pp. 222–229 (2011). https://doi.org/10.1109/ IE.2011.29
48. Sita, I.V.: In: Electrical Systems for Aircraft, Railway and Ship Propulsion. ESARS (2012). https://doi.org/10.1109/ESARS.2012.6387411
49. Vanus, J., Cerny, M., Koziorek, J.: The proposal of the smart home care solution with KNX components. In: 2015 38th International Conference on Telecommunications and Signal Processing, TSP 2015 (2015). https://doi.org/10.1109/TSP.2015. 7296410
50. Gilbreth, F.B., Gilbreth, L.M.: Applied Motion Study: A Collection of Papers on the Efficient Method to Industrial Preparedness. Sturgis & Walton Company, New York (1917). https://doi.org/10.1017/CBO9781107415324.004
51. Taylor, F.W.: The Principles of Scientific Management, vol. 5 (1911). https:// doi.org/10.1016/0016-0032(50)90610-7. http://elibrary.kiu.ac.ug:8080/jspui/ bitstream/1/516/1/ThePrinciplesOfScientificManagement.pdf
52. Aguayo González, R.: El revestimiento del espacio energético. Ph.D. thesis, Universidad Politécnica de Cataluña (1999)
53. Abascal Carranza, C. M.: NORMA Oficial Mexicana, Condiciones de seguridad e higiene en los centros de trabajo donde segenere ruido. NOM-011-STPS-2001 (2002). Estados Unidos Mexicanos
54. Lozano Alarcon, J.: NORMA Oficial Mexicana, Condiciones de iluminación en los centros de trabajo, NOM-025-STPS- 2008, 1 §(2008). Estados Unidos Mexicanos
55. Kuehl, R.O.: Diseño de experimentos, 2nd edn. Thomson Learning, Mexico city (2001)

Measuring Behavioural Change of Players in Public Goods Game

Polla Fattah[1(✉)], Uwe Aickelin[2], and Christian Wagner[1]

[1] School of Computer Science, The University of Nottingham, Nottingham, UK
{psxpf1,pszcw}@nottingham.ac.uk
[2] School of Computer Science, The University of Nottingham Ningbo, Ningbo, China
uwe.aickelin@nottingham.edu.cn

Abstract. In the public goods game, players can be classified into different types according to their participation in the game. It is an important issue for economists to be able to measure players' strategy changes over time which can be considered as concept drift. In this study, we present a method for measuring changes in items' cluster membership in temporal data. The method consists of three steps in the first step, the temporal data will be transformed into a discrete series of time points then each time point will be clustered separately. In the last step, the items' membership in the clusters is compared with a reference of behaviour to determine the amount of behavioural change in each time point. Different external cluster validity indices and area under the curve are used to measure these changes. Instead of different cluster label comparison, we use these indices a new way to compare between clusters and reference points. In this study, three categories of reference of behaviours are used 1- first time point, 2- previous time pint and 3- the general overall behaviour of the items. For the public goods game, our results indicate that the players are changing over time but the change is smooth and relatively constant between any two time points.

1 Introduction

There are multiple studies in the economics regarding the behaviour and strategy change of players in the public goods games [9]. The players are classified into four types conditional cooperators, free riders, triangular contributors and others. The change of players behaviour means jumping from one class to another during the gameplay rounds. This jump can be considered as concept drift. As defined by Gama et al. [13], concept drift learning under is an unexpected change from the targeted future estimation due to uncalculated hidden contexts in the system. Tsymbal [31] identified two types of concept drift: sudden and gradual. This work presents a method to measure the quantity of the change occurring within populations in any two different time points.

Different methods are developed to detect drifts in the data streams such as [3,17] and special classification methods are introduced to precisely predict

© Springer International Publishing AG 2018
Y. Bi et al. (eds.), *Intelligent Systems and Applications*,
Studies in Computational Intelligence 751,
https://doi.org/10.1007/978-3-319-69266-1_12

items with the existence of concept drift [6,14,33]. Moreover measuring changes in clusters for different time points are well studied in data analysis, especially for data streams [26,29,34]. However, these methods aim to find overall patterns of change in the clusters rather than presenting a measure of how much change has occurred in each cluster.

External cluster validity is primarily used to check the performance of clustering algorithms by measuring the difference between ground truth labels given to the items by experts and the group in which they have been placed by a clustering algorithm [28]. This study uses external cluster validity measures like Variation of Information [24] VI to show the amount of items that jumped from one cluster to another between two consequent time points. Moreover, for comparison with these criteria we use area under the curve of Receiver Operating Characteristic ROC [5] which is originally a measure for classification algorithms to quantify change of behaviours as well. To accomplish this measurement, the items should be clustered separately in every time point. As the clustering is performed at a single time point, which eliminates the time dimension for the collected data about items, theoretically any traditional (non-temporal) clustering algorithm should be sufficient. After clustering, an external clustering validity measure can quantify the amount of changes between clusters at any two time points.

To compare our results, public goods games data were also tested using MONIC, which is a method of detecting changes among clusters in the data stream. The results show that there is a periodic change in clusters as they disappear and other clusters are emerging, but this is inconclusive as there is no indication of whether the change originated from players' strategies or from the nature of algorithm, as it reduces the effect of the old items in the cluster and removes them after two time points.

2   Background and Related Work

In the following subsections, a brief description of the public goods games and the used data sets for the experiments will be explained as well as describing the available techniques and methods to measure changes over time.

2.1   Public Goods Games Data

Public good is any service or resource that cannot be withheld from any individuals due to inalienable characteristics relating to citizens' rights [21]. Examples of public good resources include city parks, street lighting and roads, which are funded by the state but which are available to all. The public goods game is an experimental game that simulates real situations of public good in a lab with controlled conditions and focused purposes of conducting experiments. There are many slightly different variations of this game, but the data been used in this paper as a case study is based on the model of Fischbacher et al. [10].

There are many variations and set-ups for the public goods game experiment, However, the data which has been used in this study is collected through experiments conducted by Fischbacher et al. [10]. Their experiment for public goods game consists of two sub-experiments; P-experiment and C-experiment, both of which every participant (player) has to accomplish. In the following sections, we will explain how these two sub-experiments are conducted, and then describe the collected data which will be used in later chapters.

2.1.1   Game Set-Up

Prior to each sub-experiment of P-experiment and C-Experiment, experimenters explain the rules of the game for the participants so that they understand the rules, and how their decision will affect their result and the number of points available. Participants should answer a number of control questions correctly to demonstrate their comprehension of the game. Experimenters make every effort to ensure that the players are paying attention and playing thoughtfully by rewarding them extra points for correct guesses and well-thought out decisions during the game.

In P-experiment, four players start the game with 20 points each in their private account and they can contribute any amount they deem necessary to a project which represents public good. The amount which they do not contribute in the project will be kept only for the players themselves. The collected amount from the project will be distributed among all players regardless of their contribution to the project. The amount of points each player can accrue from the project is determined by this equation:

$$PlayerShareFromProject = TotalAmountOfAllPlayersContribution * 0.4$$

So that each players total point after the game will be:

$$Player'sPoints = (20 - ContributionInTheProject) + PlayerShareFromProject$$

For each player, only one of the two contributions will be selected by the computer as their final contribution to the project. One of the four players' conditional contribution will be randomly chosen to be used as their final contribution. while for the other three players their unconditional contribution will be used. This random selection of players' contributions is one of the mechanisms that experimenters have used to make sure that players are thinking thoroughly about their decision for the contribution to the project.

When the P-experiment is completed, players start C-experiment. C-experiment is similar to a repeated sequence of unconditional contribution except this time the player, in addition to their own contribution, will be asked to guess other players' rounded average of contribution. After each round of the game, players will be notified of their total points in that particular game. The sequence length of the games can vary from one experiment to another. In this study, we will use data sets with 10 and 27 series of rounds of the game. In each round, four different random players will play the game so that players can not predict others' contributions in advance. Players will gain extra points if they make correct guesses about other players' rounded contributions. They will, therefore, not fill in the boxes randomly.

2.1.2 Data Set Attributes

To measure and classify the behaviour of players in public goods games, this study used two different data sets. These experiments are conducted on different samples of players, so the first data set has 140 players and the second data set 128 players. These data sets have the same attributes and follow exactly the same experiment procedures, except for the P-experiment length, as the first one consists of 10 rounds while the other has 27 rounds.

The attributes of the data sets can be divided into two types the temporal and non-temporal attributes. The temporal attributes are generated in the P-experiment as it contains multiple rounds and non-temporal attributes are generated in C-experiment. The following is the list of all the attributes of the data sets. Please notice that the temporal attributes are underlined:

- **Idtyp**: labels for players categories assigned by experts. The categories are: conditional contributors = 1, free riders = 2, triangle contributors = 3, and others = 4. These categories are generated depending entirely on the static data of the contribution table.
- **Idsubj**: a unique identifier for each player during both C and P experiments.
- **b0–b20**: twenty one attributes representing the contribution table for each player as their response in C-experiment to every possible rounded average of other players' contribution.
- **u**: the unconditional contribution of the player for C-experiment during the actual game.
- **Predictedcontribution**: Players' prediction about other co-players rounded average of contribution to the project.
- **Period**: the session number for P-experiment. As P-experiment for each player consists of multiple rounds, each players' playing times are recorded to keep track of the number of games played.
- **Contribution**: players' actual contribution to the project in each round of the P-experiment.
- **Belief**: players' beliefs about other players average contribution in each session.
- **Otherscontrib**: Other co-players' rounded average contribution.

2.2 External Cluster Validity

External criteria validate the results of clustering based on some predefined structures of the data which is provided from an external source. The most well-known example of structural information is labels for the data provided by experts (called true classes). The main task of this approach is to determine a statistical measure for the similarity or dissimilarity between obtained clusters and labels [16,27]. According to the methods incorporated in the external criteria, they can be divided into three types: pairwise measures, entropy-based measures and matching based measures [35].

As mentioned previously, the four types of classification guesses evaluation are true positive, true negative, false positive and false negative. These terms are used in the terminology of external cluster validity, especially when using pairwise measures, but with slightly different meanings to enable the evaluation of clusters in the same manner as classification [35]:

- True Positives **TP**: Any two instances with the same label that are in the same cluster.
- False Negatives **FN**: Any two instances with the same label that are not in the same cluster.
- False Positives **FP**: Any two instances with different labels that are not in the same cluster.
- True Negatives **TN**: Any two instances with different labels that are not in the same cluster.

In this study we use various external cluster validity indices to determine differences between a reference of behaviour for items in a temporal data and clusters of items in each time point. The method is discussed in more detail in chapter three, and implemented in chapter four for public goods games and chapter six for stock market data. The used criteria in the thesis are listed below:

2.2.1   Jaccard Coefficient

This coefficient is a pairwise measure representing the degree of similarity between clusters. With this coefficient each cluster is treated as a mathematical set and the coefficient value is calculated by dividing the cardinality of the intersection of the resultant cluster with the prior cluster to the cardinality of the union between them [32]:

$$Jaccard = \frac{TP}{TP + FP + FN}$$

With a perfect clustering, when false positives and false negative equal to zero, the Jaccard coefficient value equals 1. This measure ignores the true negatives and only focuses on the true positives to evaluate the quality of the clusters [35].

2.2.2   Rand Statistic

The Rand statistic measures the fraction of true positives and true negatives over all point pairs; it is defined as

$$Rand = \frac{TP + TN}{N}$$

where N is the total number of instances in the data set. This measure is similar to Jaccard Coefficient, so its value equals 1 in perfect clustering [35].

2.2.3 Fowlkes-Mallows (FM) Measure

FM define precision and recall values for produced clusters [12]

$$FM = \sqrt{prec.recall} = \frac{TP}{\sqrt{(TP + FN)(TP + FP)}}$$

where $prec = \frac{TP}{TP+FP}$ and $recall = \frac{TP}{TP+FN}$. For the perfect clustering this measure equals 1 too [35].

2.2.4 Variation of Information VI

This index measure is based on contingency table which is a matrix with $r \times k$, where r is number of produced clusters and k is the number of externally provided clusters. Each element of this matrix contains a number of agreed instances between any two clusters of the externally provided and produced clusters. As introduced by Meila [23], this index calculates mutual information and entropy between previously provided and produced clusters derived from the contingency table:

$$VI(C, T) = 2H(T, C) - H(T) - H(C)$$

where C is produced clusters, T is ground truth clusters, $H(C)$ is entropy of C and $H(T)$ is entropy of T [35].

2.3 Measuring Cluster Changes in Data Streams

Spiliopoulou et al. [30] introduced the MONIC model, which finds cluster transition over accumulating data sets, providing an ageing function for clustering data that prioritizes new records over old ones and eliminates records older than two time points. Matching for clusters in one time point to the next one is carried out by passing a threshold that determines normalized maximum number of records that exist in both matched clusters in the two time points. This model defines two kinds of transitions, external and internal. In external transition clusters may survive, split, be absorbed, disappear or emerge, while in internal transition clusters can change in size, compactness or location.

According to MONIC, each cluster has a lifetime, which is the number of time points throughout which it can survive. Longer cluster lifetimes enable more predictable clustering while short lifetimes lead to volatile and unpredictable clustering.

It can be observed that this model relies on accumulated data over time to detect cluster matches, therefore it cannot be used with non-accumulated data. Moreover, it emphases the measurement of cluster changes and cannot detect changes in cluster membership for individual items clustered over time points.

Gunnemann et al. [15] introduced a method which traces cluster evolution as change in behaviour types indicated by the value of objects (e.g. persons) in high-dimensional data sets. Different types of mapping function were introduced to map clusters according to their values in different dimensions and subspaces instead of object identifier. Using this method cluster evolutions were detected and counted in the forms of emerge, disappear, converge and diverge. Moreover, the loss and gain of dimensions of subspace clusters were calculated. This method counts the number of various changes that occur to clusters of any high dimensional data set, but it lacks to any mean by which to quantify the changes themselves; in other words, there is no indication of the quantity of change that happens to any cluster in two consecutive time points.

Hawwash et al. [18] proposed a framework for mining, tracking and validating clusters in a data stream using statistical cluster measures like cardinality, scale and density of clusters to detect milestones of clusters change and monitor the behaviour of cluster. This framework targets accumulative clustering on data streams, but instead of using fixed-time window for clustering it uses milestones to detect the next-best clustering time. They used a linear model in their metrics, which cannot represent real-life situations. They made this concession due to time limitations and the memory complexity of higher degree models. With some enhanced models this method could be profitably used to determine critical time points in the data stream clustering and to track clusters behaviour in general using statistical measures for representative numbers pertaining to the situation of clusterings.

Kalnis et al. [20] introduced a method to discover moving objects in the snapshots of spatio-temporal data using cluster mapping function, treating clusters as sets and calculating the cardinality ratio of intersection for each two time constitutive clusters over their union; if the ratio passes a certain threshold the cluster is considered to be a moving cluster. This method detects move in overall clusters and provides visual aids enabling human experts to grasp changes in the underling data [4,26]. This method is excellent for tracking moving cluster change [25], but it still lacks a method to quantify the magnitude of change for overall clustering objects.

Aggarwal [1] introduced a new method to detect changes for single clusters in the data streams that also works for snapshots of data as special cases. This method uses forward and reverse time slice density estimates based on fixed length time window to calculate velocity density at time and space dimensions. By calculating velocity density three types of change can appear on the clusters in evolving data streams: (1) they may coagulate if the value passed a user specified threshold; (2) they may decay if the value does not pass the threshold; or (3) they may shift their location to another. This method is particularly germane to visually understanding the characteristics of underlying data.

The mentioned methods: are mostly designed to work with data streams or snapshots of spatio-temporal data sets. They detect changes inside data by monitoring cluster change in terms of split, absorbed, disappear and emerged etc., which is a good indication for detecting existence of change, but which does not specify the magnitude of change. Our aim is to create a simple factor (scalar) to express the magnitude of change among members of clusterings in temporal data sets.

3   Methodology

Measuring behaviour differences of items between time points requires three steps: The first step is to address time points, the second step is grouping similar behaviour and the last step is to find and measure the amount of differences between these time points.

The temporal data has to be split into separate time points. If the temporal data has discrete records of time then each timestamp can represent a single time point. If the data set has continues timestamps, then it might be converted to a discrete set of time points using fixed intervals of time windows as used by many studies like [30]. It might be preferable that the time points have similar intervals between them so that the behavioural change measure M can represent the difference between any two time points in the same data set uniformly. For illustration consider $t \in T$ and the time intervals between [t-1, t] and [t, t+1] are equal which makes $m1 = \delta(t-1, t)$ and $m2 = \delta(t, t+1)$ so that m1 and m2 can represent the two defined time intervals uniformly.

The second step is grouping similar behaviours of the items in the data so that we can identify each items' category of behaviour at every particular time points. As defined by Jain [19] clustering is the task of finding groups of similar members in an item set. So that each time point is clustered using one of the clustering methods to find similarly behaved groups at each time point. The clustering algorithms which are used in the process of measuring the difference between time points in this study are Kmeans, PAM, and hierarchical clustering.

Clustering items in each time point eliminate two issues regarding items' drift between time points. These issues are minor changes of items' behaviour and shifting all of the items in the same group. These problems will be solved by clustering each time point separately without any influence of other time points. Because the clustering will ensure that the minor changes of items in the values of attributes will not affect their membership in the group and by clustering each time point's data independently, we ensure that the entire movement of a group will not affect the measures of items' behaviour change.

The last step is to find the number of items which are changed their behaviour significantly so that they can be counted as they are in other groups or using the percentage of items' behaviour change. This means finding δ function. It is also possible to use AUC of ROC to find the difference between items' clusters in any two consequent time points by using cluster labels of t and t+1 instead of true class labels and predicted classes by a classification model as inputs to the AUC function so that it finds the difference between these two time points.

3.1   Choosing Clustering Algorithms

As each of the produced subsets of data represents one time point of the temporal dataset so that each subset alone do not carry any information about time dimension. This means it is possible to use non-temporal clustering algorithms to cluster items in each subset of the temporal datasets.

Clustering algorithms can be categorised according to their method of finding similarities between items in the data. These categories are partitional, hierarchical, density-based, grid-based and fuzzy clustering [35]. However, the main clustering categories which we used are partition based clustering, hierarchical clustering and fuzzy algorithms. For the tests in this experiment, we used k-means and PAM as methods of partitioning clustering, hierarchical clustering with Euclidean distance and c-means as fuzzy clustering. As we aim to find similarities between items according to their distance from each other, so that we did not use density-based and grid-based clustering methods.

To find similarities between items, clustering methods can use linear distance measures like Euclidean distance or use non-linear kernels to cluster complicated patterns in the data items. In the tests, we only used linear distance based clustering methods because the aim is to find the similarity of behaviour based on the overall proximity of the attribute values of items. For the same reason, we do not use density-based clustering like DBSCAN and grid-based clustering like STING as these methods do not depend on the mutual proximity of cluster items to a centroid which represents a behaviour category.

3.2   Choosing External Cluster Validity Indices

As explained before we propose using external cluster validity indices and area under the curve AUC to measure changes which might occur in the behaviour of the items between multiple time points in a temporal data. Many external cluster validity indices are available [2] to measure the validity of clusters produced by clustering methods compared with the natural partitions exist. In Chap. 17 of their book Zaki et al. [35] categorised the external clustering validities into three types matching based measures, entropy-based measures and pairwise measures.

As for matching based measures external cluster validity indices are calculating the match of the clusters to the partitions. This means that the measure do not concern about individual elements' differences between clusters and partition, so that this category might not be beneficial to calculate the changes over time.

The second category of external cluster validity indices, entropy-based measures, calculates the difference of entropy between clusters and ground truth partitions. This method also does not concern about individual items in the clusters and partitions. However, we used one measure of this category which is Variation of Information VI because the entropy of the clusters might be affected by the change of items within the clusters. We also used VI for comparison with other indices.

The last category, pairwise measures, measures cluster validity by comparing the produced clusters and original labels of items' classes. As this category calculates the validity using all elements of the dataset, so that it might be the most appropriate category to be used to calculate the items' changes over time points. Three instances of pairwise measures are used in this study which they are Jaccard Coefficient, Rand Statistic and Fowlkes-Mallows Measure.

To maintain standard criteria for different external cluster validity indices so that the final result which quantifies the amount of change in each time point reflects the actual change which is happened to the groups' items regardless of the measure which is used. To make sure that the measures are standard they should follow two rules (1) the scale of the measure should be between zero and one (2) with being zero as the total change and one is the perfect match between any time point and reference of behaviour. However, not all measures are following these rules. For example, in the selected measures the VI is not bounded to any scale and zero is considered as a perfect match so that the results of this measure should be first scaled to fit in the range of [0–1] and then reversed.

3.3 Using Different Reference of Behaviours

This study considers three different references of behaviours (1) the first time point is used as a reference of behaviour for all other time points (2) The previous time point is used to be the reference of behaviour for the current time point. (3) a temporal classification method will be proposed to classify items in temporal datasets.

Each of these different references of behaviour imposes different meaning and can be used for various reasons. The first time point can be used as a reference of behaviour to quantify the progress of change which happens to the items in any later time points in the dataset. An example of that if we want to quantify the change of behaviour of players in public goods game from the first round of the game to any round of the game. Using previous time point as a reference for the current time point means we aim to measure stepwise changes of items' behaviour between any time point. This can be used to measure the stability of change over time. An example of using this method is when we want to check the stability of changes which can occur in player's behaviour between time points. Items' classes can be used as a reference of behaviour to quantify items' deviation from their generalised behaviour at any time point.

3.4 Classifying Players of Public Goods Game

To be able to use the third type of reference of behaviour, players of the public goods game have to be classified according to their temporal data to reveal their general overall behaviour. The available data sets do not contain labels for players according to their temporal behaviour so that we use the rule-based classification method which is introduced by Polla et al. [7]. This classification method consists of two steps: initial rule generation and rule optimisation.

In the first step, the experts of the area of data set provide classes for items in the data set with initial rules separating these classes from each other. The rules are expressed in the form of aggregated attributes which are derived from the temporal attributes. The aggregation ensures simplicity of the rules so that it can be understandable by human experts. The aggregation for temporal attributes can be done using one of the mathematical or statistical functions like minimum, maximum, average and standard deviation. Moreover, each rule might have a range of values to separate between adjacent classes which can be expressed as [min, max]. This range might exist due to the uncertainty of the experts of the best cutting position between two classes or disagreement among different studies [7].

The second step optimises the initial rules by finding the optimum classifier among all available ranges. After this step all [min, max] ranges of the initial rules will be replaced with a single value. This step uses the temporal attributes directly. The optimum classifier is the classifier minimises the distance (dispersion) among items of each class at all time points.

The players of public goods games are classified into four classes according to their contribution and belief attribute the classes are:

- Free Riders: players who contribute by equal or less than one point on average for all rounds or who are not contributing in most rounds. This class corresponds to the traditional category of Free Riders.
- Weak Contributors: players who contribute between 1 and 5 or they are not contributing in half of the rounds.
- Normal Contributors: players who contribute on average around 5 points. This class is strongly related to conditional contributors as it fits the same criteria.
- Strong Contributors: players who contribute more than 10 points on average. This class relates to conditional comparators and others in the classical categories.

4   Measuring Players' Strategy Change over Time

The main objective of this study is to quantify players' strategy change in the public goods game, which may lead to contribute to the understanding of the players' behaviour and present a tool for economists to measure the amount of change for different setups of their experiments. Another objective is our aim to demonstrate the ability of the proposed method to produce quantifiable measures for items' changes in the temporal data and provide interpretable results. We compare the results and findings of our method with the MONIC method which is originally used to measure cluster changes in data streams [30].

For this experiment both datasets to measure players behaviour and strategy change during the consequent rounds of the game. Players contribution attribute and their expectation for other players contribution at each time point are used by this method to find the magnitude of the change. These two datasets have different groups of players and different lengths as the first dataset is 10 rounds length and the second is 27 so that these two datasets are used separately and treated as different datasets in this experiment. We used four clusters to cluster players in each time point using kmeans, PAM, cmeans and hierarchical clustering methods as it has been selected based on our discussion in Sect. 3.1.

As both of datasets of public goods game share the same experiment settings and setup so that it can be hypothesised that the results of the behaviour change should be consistent with regards to the length of the experiment which might affect the behaviour of players [8]. While we use all previously selected external cluster validity index in Sect. 3.2 and AUC of ROC, however, we will depend on AUC and Rand results to compare the behaviour of players in these two different datasets.

4.1   Using Proposed Method

Different types of reference of behaviours reveal different aspects of players' strategy change. By using first time point as the reference of behaviour, we can

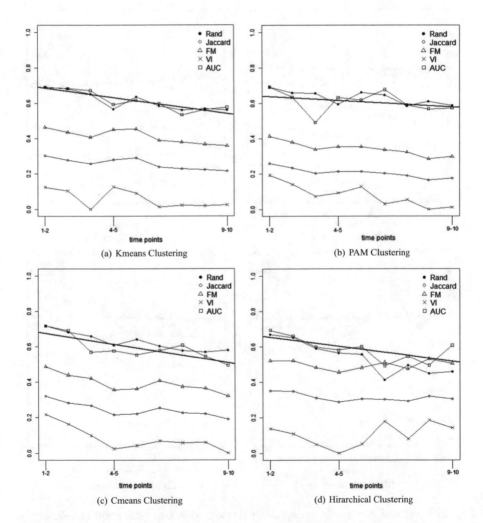

Fig. 1. Results of various clustering methods using first time point as reference of behaviour to calculate the amount of changes which happen to the groups of items in consequent time points in the 10 rounds public goods game dataset

detect drift of players' behaviour from the initial expectation and contribution. As shown in Fig. 1 for 10 rounds dataset and Fig. 3 for 27 rounds, players in both datasets are gradually drifting away from their initial game plan and expectation. This trend can be seen with all four clustering methods with different measuring methods of external cluster validity index and AUC. Results of AUC and Rand are consistent across all clustering methods so that we used AUC to calculate the linear regression of the results. The negative results of linear regression is an indication that players change their behaviour by drifting away from their original gameplay (Fig. 2).

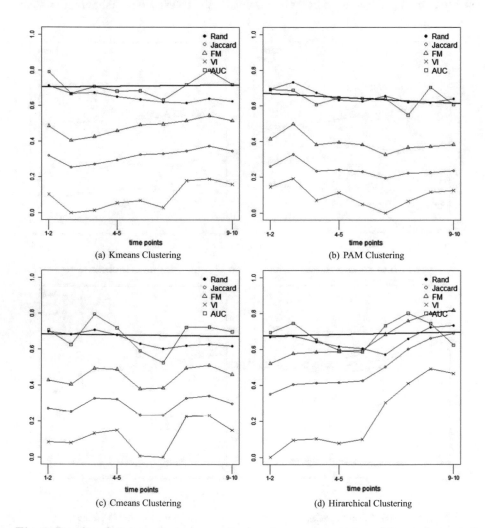

(a) Kmeans Clustering

(b) PAM Clustering

(c) Cmeans Clustering

(d) Hirarchical Clustering

Fig. 2. Results of various clustering methods using previous time point as reference of behaviour to calculate the amount of changes which happen to the groups of items in consequent time points in the 10 rounds public goods game dataset

By using the previous time point as the reference of behaviour we can measure the amount of change between any two consecutive time points. This allows detection of players' behaviour transition from one time point to another. Figure 2 of the 10 rounds dataset shows that players strategy changes from one time point to another is constant which indicated by the linear regression of AUC and Rand measures. In contrast Fig. 4 of 27 rounds shows that the change between time points are decreasing throughout the progress of the game.

At first glance, the results of 10 and 27 rounds datasets are not consistent. However, with a closer look at the results, we can detect that the players' behaviour change in 27 rounds dataset is stable without any decrease until round 10 of the game. This decrease might be due to the fact that most of the players dropped their contribution to zero when they reached a certain round of the game. Which means no further room is left for change in the game rounds, except some players randomly start to rise their contribution again but the rise is not constant so that after round 10 we detect less change than expected.

We hypothesised in the previous section that the players' behaviour has to be consistent in both datasets. The results for measuring changes using first time point as the reference of behaviour are compatible in both cases as it can be noticed that the players' contribution drops gradually. The results of using previous time point as the reference of behaviour show that players strategy change are constant till round 10. In 27 rounds dataset, after round 10 most players' contribution dropped to zero so that there is no room for further change in their strategy. Hence, their strategy decreases and their game pattern starts to be similar between any two consequent time points.

The results of the proposed method for both datasets are compatible with the findings of the economists [9,11]. However, this method provides a tool which enables them to quantify changes in players behaviour. Quantifying behaviour change is important so that they can measure the nuanced differences between various gameplay setups like the length of the rounds, the percentage of the rewards from the public project, and knowing the identity of other players.

After players were classified according to their temporal attributes which reflect their contribution behaviour over all time points, we can use the new players' classes as a reference of behaviour. As shown in the Figs. 5 and 6, there are significant difference between players' classes and their temporal behaviour. This difference can be seen with the low value of the behavioural change measures across all clusterings and all different external cluster validity indices external cluster validity index (less than 0.6). This indicates that the players are not always playing the same strategy. Instead, they try and explore other strategies which contribute in their learning process to identify different strategy results. However, the regression of the behavioural change for all cases are small (near zero), which indicates the difference is stable throughout all time points. This is another indication that despite their temporary strategy change, however, these changes do not affect their general playing behaviour.

Despite the sensitivity difference between external cluster validity indices, all the results of different clusterings and external cluster validity indices are simi-

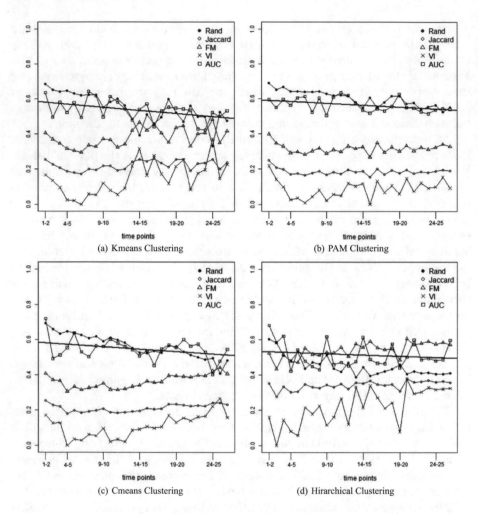

Fig. 3. Results of various clustering methods using first time point as reference of behaviour to calculate the amount of changes which happen to the groups of items in consequent time points in the 27 rounds public goods game dataset

lar with regression slope equal to zero. This might be an indication that using items' overall general behaviour in the temporal attributes can create more stable estimation than other two reference of behaviours on the items' behavioural change. However, each reference of behaviour can be useful for situations. It can be noticed that different reference of behaviours might reveal different aspects of the players' behaviour. Using the first time point as the reference of behaviour will demonstrate how items are deviating from their initial behaviour. Using the previous time point shows the stability of the items during different stages of the temporal data. While using players temporal classes as reference of behav-

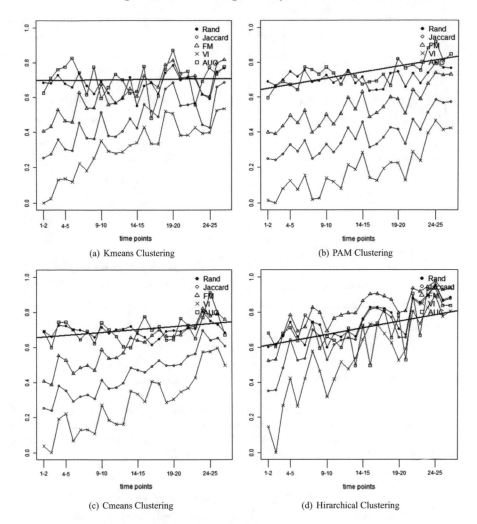

(a) Kmeans Clustering

(b) PAM Clustering

(c) Cmeans Clustering

(d) Hirarchical Clustering

Fig. 4. Results of various clustering methods using previous time point as reference of behaviour to calculate the amount of changes which happen to the groups of items in consequent time points in the 27 rounds public goods game dataset

iour demonstrates items behavioural variability in various stages related to their overall behaviour across all time points.

4.2   Comparing Results with MONIC

MONIC[1] is used to gain more insights about the public goods games data sets and to compare our results with the existing methods of measuring cluster changes in different time points. The data for each time period were clustered

---

[1] Available at http://infolab.cs.unipi.gr/people/ntoutsi/monic.html.

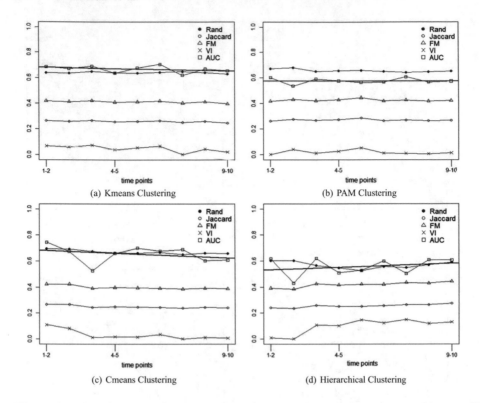

Fig. 5. Results of various clustering methods using players' classes as a reference of behaviour to calculate the amount of changes which happen to the groups of items in consequent time points in the test dataset. The amount of change is measured by using different external cluster validity index and AUC of ROC

separately using k-means with four clusters. The clustering was carried out on the main temporal attributes of the data, namely belief and contribution, then the data and cluster labels of items in each consequent pair of time points are fed to the MONIC algorithm to calculate changes to clusters from one time point to another. The method calculated the number of survived, appeared and disappeared clusters, as shown in Figs. 7 and 8, for the ten rounds of the game.

In the 10 rounds dataset, the number of survived clusters reduced from four clusters between the first and second time points until it reached zero, while new clusters appeared in the middle of the fifth and sixth game rounds, then the number rose again until the end of the game. This might be due to the fact that players are changing their strategies and exploring new options until they ultimately settle on a certain strategic pattern. This change is consistent with our findings, as the measures slightly increase between the fifth and seventh time points, which might be an indication of players changing their strategy back to their original ones. As Keser and Winden [22] suggest, this change might be due

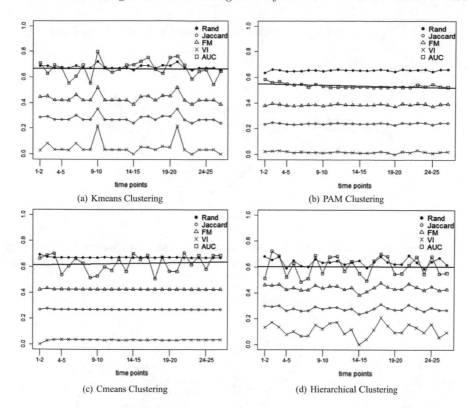

(a) Kmeans Clustering

(b) PAM Clustering

(c) Cmeans Clustering

(d) Hierarchical Clustering

Fig. 6. Results of various clustering methods using players' classes as reference of behaviour to calculate the amount of changes which happen to the groups of items in consequent time points in the test dataset. The amount of change is measured by using different external cluster validity index and AUC of ROC

to the players responding to the average contribution of other players in the previous round.

The results for the 27 rounds dataset is not straightforward, as the numbers of cluster survivals, appearances and disappearances change more frequently. However, the cyclic pattern of increasing and decreasing the number of survived clusters might be an effect of changing players' strategies or due to the underlying algorithm, as it provides an ageing factor to the items.

As the MONIC algorithm was originally introduced to detect cluster changes in a data stream, it uses an ageing factor which reduces the effect of older items in the cluster and removes items older than two time points [30]. This ageing factor is essential for the algorithm to keep up-to-date with the flowing data stream and give the right results for the current status of the clusters. However, this might not be useful for public goods games data, as there is a fixed number of players which might result in the removal of players who stay in the same cluster for long time points. The effect of the ageing might not be obvious in

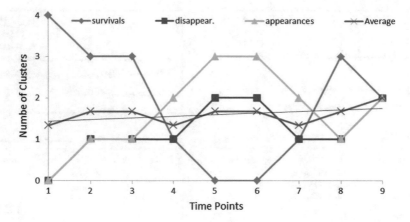

Fig. 7. Number of survival, appearance and disappearance of clusters between every tow consequent time points for ten rounds public goods game as measured by MONIC

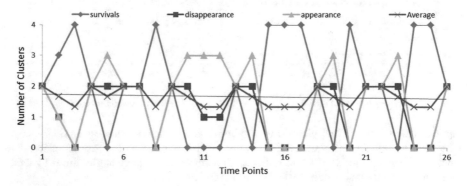

Fig. 8. Number of survival, appearance and disappearance of clusters between every tow consequent time points for 27 rounds public goods game as measured by MONIC

the 10 rounds game due to the limited number of time points, but it might undermine players' strategies.

While the proposed method assumes a fixed number of clusters to calculate items membership change, the MONIC algorithm is a good method to have insights on the available clusters and their stability by measuring the number of survived clusters between two time points. However it does not measure the amount of items drifting from one cluster into another, which can be detected by the proposed method, as it introduces a specific ratio between each consequent time point, indicating the amount of change happening to the items in the clusters by calculating their membership change among clusters.

MONIC can be compared with the proposed method especially with the case of previous time point as a reference of behaviour because both of these methods compare the current clusters with the previous time point. The regression result for the average of cluster moves (appear, disappear and survive) is near zero (-0.00547 for 27 rounds and 0.03889 for 10 rounds of the data set), which is compatible with the proposed method results using previous time point as reference of behaviour except for 27 rounds clustered by PAM and hierarchical clustering. By comparing results from the proposed method and MONIC we can conclude that the players slightly and gradually change their cluster membership. However, the change amount of change is stable from one time point to another. The proposed method gives an exact number for the change while the MONIC presents overall clusters movement and change.

5   Summary and Conclusions

This study presents a method to quantify changes over time for items across multiple readings at different times for the same attributes using external cluster validity measures and area under the curve of receiver operating characteristic. The proposed method consists of three steps: the first step separates individual time points from the temporal data, the second step clusters each time point so that items with the similar behaviour can be grouped together. The third step quantifies changes of items' membership in the clusters using different reference of behaviours. Two data sets are used to determine behavioural changes of players in the public goods games.

Various clustering algorithms are used to cluster items at each time point. Then these clusters are compared with the the reference of behaviours using different external cluster validity indices like VI, FM and Jaccard. Three references of behaviours are used which they are first time point, previous time point and players overall behaviour as classes for their temporal attributes. We used a rule-based temporal classification method to classify players in both data sets.

The results are compared with the MONIC method which is designed to detect changes in the number of clusters in a data stream. The results were compatible with the findings using our proposed method especially with the results of previous time point as the reference of behaviour. All of the results indicate that the players' strategy are changing and the change is subtle and gradual.

Acknowledgements. The authors record their thanks to Simon Gaechter and Felix Kolle in the School of Economics at the University of Nottingham for providing us with data from the public goods experiment and taking time to explain it to us.

References

1. Aggarwal, C.C.: On change diagnosis in evolving data streams. IEEE Trans. Knowl. Data Eng. **17**(5), 587–600 (2005)
2. Arbelaitz, O., Gurrutxaga, I., Muguerza, J., Pé rez, J.M., Perona, I.: An extensive comparative study of cluster validity indices. Pattern Recogn. **46**, 243–256 (2012)

3. Baena-Garcia, M., del Campo-Avila, J., Fidalgo, R., Bifet, A., Gavalda, R., Morales-Bueno, R.: Early drift detection method. In: 4th ECML PKDD International Workshop on Knowledge Discovery from Data Streams, pp. 77–86 (2006)
4. Böttcher, M., Höppner, F., Spiliopoulou, M.: On exploiting the power of time in data mining. ACM SIGKDD Explor. Newsl. **10**(2), 3–11 (2008). Dec
5. Bradley, A.P.: The use of the area under the ROC curve in the evaluation of machine learning algorithms. Pattern Recogn. **30**(7), 1145–1159 (1997)
6. Elwell, R., Polikar, R.: Incremental learning of concept drift in nonstationary environments. IEEE Trans. Neural Netw. **22**(10), 1517–1531 (2011)
7. Fattah, P., Aickelin, U., Wagner, C.: Optimising rule-based classification in temporal data. Zanco J. Pure Appl. Sci. **28**(2), 135–146 (2016)
8. Figuières, C., Masclet, D., Willinger, M.: Weak moral motivation leads to the decline of voluntary contributions. J. Public Econ. Theory **15**(5), 745–772 (2013)
9. Fischbacher, U., Gächter, S.: Social preferences, beliefs, and the dynamics of free riding in public goods experiments. Am. Econ. Rev. **100**(1), 541–556 (2010)
10. Fischbacher, U., Gachter, S., Quercia, S., Gächter, S.: The behavioral validity of the strategy method in public good experiments. J. Econ. Psychol. **33**(4), 897–913 (2012). Aug
11. Fischbacher, U., Gächter, S., Whitehead, K.: Heterogeneous social preferences and the dynamics of free riding in public good experiments about the centre or contact. Am. Econ. Rev. **100**(1), 541–556 (2010)
12. Fowlkes, E.B., Mallows, C.L.: A method for comparing two hierarchical clusterings. J. Am. Stat. Assoc. **78**(383), 553–569 (1983)
13. Gama, J., Zliobaite, I., Bifet, A., Pechenizkiy, M., Bouchachia, Abdelhamid: A survey on concept drift adaptation. ACM Comput. Surv. (CSUR) **46**(4), 44 (2014)
14. Garnett, Roman., Roberts, S.J.: Learning from Data Streams with Concept Drift. Technical Report PARG-08-01, Department of Engineering Science, University of Oxford (2008)
15. Günnemann, S., Kremer, H., Laufkotter, C., Seidl, T.: Tracing evolving clusters by subspace and value similarity. Adv. Knowl. Discov. Data Min. **6635**, 444–456 (2011)
16. Halkidi, M., Batistakis, Y., Vazirgiannis, M.: Cluster validity methods: part I. ACM Sigmod Rec. **31**(2), 40–45 (2002)
17. Harel, M., Mannor, S., El-Yaniv, R., Crammer, K.: Concept drift detection through resampling. In: Proceedings of the 31st International Conference on Machine Learning (ICML-14), pp. 1009–1017 (2014)
18. Hawwash, B., Nasraoui, O.O.: Stream-dashboard: a framework for mining, tracking and validating clusters in a data stream. In: Proceedings of the 1st International Workshop on Big Data, Streams and Heterogeneous Source Mining: Algorithms, Systems, Programming Models and Applications, pp. 109–117 (2012)
19. Jain, A.K.: Data clustering: 50 years beyond K-means. Pattern Recogn. Lett. **31**, 651–666 (2010)
20. Kalnis, P., Mamoulis, N., Bakiras, S.: On discovering moving clusters in spatio-temporal data. Adv. Spat. Temporal Databases **3633**, 364–381 (2005)
21. Kaul, I., Grungberg, I., Stern, M.A.: Global public goods. In: Global Public Goods (1999)
22. Keser, C., van Winden, F.: Conditional cooperation and voluntary contributions to public goods. Scand. J. Econ. **102**, 23–39 (2000)
23. Meil, M.: Comparing clusterings an information based distance. J. Multivar. Anal. **98**(5), 873–895 (2007)

24. Meila, M.: Comparing clusterings by the variation of information. In: Learning Theory and Kernel Machines: 16th Annual Conference on Learning Theory and 7th Kernel Workshop, COLT/Kernel 2003, Washington, DC, USA, 24–27 Aug 2003, proceedings, p. 173 (2003)
25. Ntoutsi, I., Spiliopoulou, M., Theodoridis, Y.: Tracing cluster transitions for different cluster types. Control Cybern. **38**(1), 239–259 (2009)
26. Ntoutsi, I., Spiliopoulou, M., Theodoridis, Y.: Summarizing cluster evolution in dynamic environments. In: Computational Science and Its Applications—ICCSA 2011, vol. 6783, pp. 562–577. Springer, Berlin, Heidelberg (2011)
27. Rendón, E., Abundez, I.: Internal versus external cluster validation indexes. Int. J. Comput. Commun. **5**(1), 27–34 (2011)
28. Rezaei, M., Franti, P.: Set matching measures for external cluster validity. IEEE Trans. Knowl. Data Eng. **28**(8), 2173–2186 (2016)
29. Spiliopoulou, M., Ntoutsi, E., Theodoridis, Y., Schult, R.: MONIC and followups on modeling and monitoring cluster transitions. Mach. Learn. Knowl. Discov. Databases **8190**(2013), 622–626 (2013)
30. Spiliopoulou, M., Ntoutsi, I., Theodoridis, Y., Schult, R.: Monic: modeling and monitoring cluster transitions. In: Proceedings of the 12th ACM SIGKDD International Conference on Knowledge Discovery and Data Mining, pp. 706–711 (2006)
31. Tsymbal, A.: The problem of concept drift: definitions and related work. Computer Science Department, Trinity College Dublin (2004)
32. Vendramin, L., Campello, R.J.G.B., Hruschka, E.R.: Relative clustering validity criteria: a comparative overview. Stat. Anal. Data Min. **4**(3), 209–235 (2010)
33. Xiaofeng, L., Weiwei, G.: Study on a classification model of data stream based on concept drift. Int. J. Multimedia Ubiquitous Eng. **9**(5), 363–372 (2014)
34. Yang, D., Guo, Z., Rundensteiner, E.A., Ward, M.O.: CLUES: a unified framework supporting interactive exploration of density-based clusters in streams. In: Proceedings of the 20th ACM International Conference on Information and Knowledge Management, pp. 815–824 (2011)
35. Zaki, M.J., Meira Jr., M.: Data Mining and Analysis: Fundamental Concepts and Algorithms. Cambridge University Press, New York (2014)

Object Segmentation for Vehicle Video and Dental CBCT by Neuromorphic Convolutional Recurrent Neural Network

Woo-Sup Han[1] and Il Song Han[2(✉)]

[1] ODIGA Ltd, London, UK
phil.han@odiga.co.uk
[2] Graduate School for Green Transportation, Korea Advanced Institute of
Science and Technology, Daejon, Korea
i.s.han@kaist.ac.kr

Abstract. The neuromorphic visual processing inspired by the biological vision system of brain offers an alternative process into applying machine vision in various environments. With the emerging interests on transportation safety enhancement of Advanced Driver Assistance System or a driverless car, the neuromorphic convolutional recurrent neural networks was proposed and tested for the night-time vehicle or VRU detection. The effectiveness of proposed convolutional-recurrent neural networks of neuromorphic visual processing was evaluated successfully for the object detection without optimized complex template matching or prior denoising neural network. The real life road video dataset at night time demonstrated 98% of successful detection/segmentation rate with 0% False Positive. The robust performance of proposed convolutional-recurrent neural network was also applied successfully to the tooth segmentation of dental X-ray 3D CT including the gum region. The feature extraction was based on neuromorphic visual processing filters of either hand-cut filters mimicking the visual cortex experimentation or the auto-encoder filter trained by partial X-ray images. The consistent performance of either hand-cut filters or the small auto-encoder filters demonstrated the feasibility of real-time and robust neuromorphic vision implemented by either the small embedded system or the portable computer.

Keywords: Neuromorphic visual processing · Visual cortex · Machine vision Vehicle detection · CBCT · Convolution-recurrent neural networks

1 Introduction

In this paper, the convolutional-recurrent neural networks of neuromorphic implementation are introduced for the robust abstraction of visual object recognition under limited visibility, inspired by the fundamental function of visual cortex, and the deep neural networks of multiple processing layers with the multiple levels of abstractions [1–5]. The proposed mixed neural networks of neuromorphic convolutional-recurrent neural network is investigated for enhancing the vehicle detection at night time or the

© Springer International Publishing AG 2018
Y. Bi et al. (eds.), *Intelligent Systems and Applications*,
Studies in Computational Intelligence 751,
https://doi.org/10.1007/978-3-319-69266-1_13

tooth segmentation including the noisy gum region of the dental Cone Beam Computerized Tomography (CBCT) image [6].

The proposed neuromorphic visual processing has its basis on how we, humans process visual information. Hubel and Wiesel's research on cat's striate cortex have established the concept of the simple cell within the visual cortex. The simple cell responds to the orientated edges, and various theories and algorithms in image processing, object recognition and computer vision have been developed from this experiment and its findings. The proposed system in this paper have been developed using a unique neuromorphic visual processing method, which mimics the function simple cell in the visual cortex with the reduced accuracy and orientation selectivity. For design flexibility, the real image patches are also used to train the orientation selectivity by the multiple orientations of dataset images.

The neuromorphic vision of earlier convolutional neural network demonstrated robust performance in pedestrian or cyclist detection under difficult weather conditions such as the rain, and low light condition's. This was carried out always under some variations in luminance or human objects [7, 8]. The principle was, using a few specific orientation feature extractors, based on the Head-Torso template applied as synaptic weights of neural networks. A substantial improvement from the previous versions neuromorphic network, the convolutional-recurrent neural network with Down Up scaling was introduced to offer the further accuracy and abstraction in object detection, saving the denoising process or the templates from supervised learning. There have been proposed various methods of extracting abstract features, which include the region-based convolutional neural network, customized saliency detection, or brain-inspired down-up network [9–14].

Main body of this paper explains the challenges in real life environment object detection, and how the neuromorphic visual processing is applied as a solution for different applications. In real world applications, varying level of illumination or noisy components add to the challenge of detecting target in within a complex environment. By performing robust selection of 'primitive features' plausible to the visual cortex or natural intelligence, the neuromorphic visual processing with the controlled rectifier neuron of down-up scaling is able to accomplish accurate locating of target objects for further processing.

As a test of neuromorphic convolutional-recurrent neural network, the proposed algorithm is applied on detecting human object of VRU in night-time road environment which was still the challenge in template based neuromorphic convolutional neural network.

This process is explained further with the example of segmentation for the vehicle object detection in night time condition or the tooth monitoring in the dental CBCT.

A sub-section is also devoted to describing the enhanced neuromorphic visual processing by convolutional-recurrent neural network, which includes the controlled rectifier neuron combined with down-up scaling. This particular model was applied to improve performance in urban environment and dental CBCT 3D image analysis, which frequently shows sudden strong noise signal. The feature extractors by the auto-encoder is evaluated as an alternative to the hand-cut ones inspired by Visual Cortex experimentation, illustrating the feasible aspect of one-learning algorithm.

2   Neuromorphic Visual Processing—Robustness in Object Detection

The neuromorphic visual processing of convolutional neural network in Fig. 1 is inspired by a neuromorphic neuron of simple cell, with the various orientation selective features. The system has three steps in its process which are: (1) orientation feature extraction using neuromorphic processing mimicking the simple cell of visual cortex, based on the convolution with the filter banks, (2) neural networks of template convolution is then applied to the orientation extracted image, and finally (3) the object recognition and detection is made by the evaluation of layer 5 result depending on the system's application.

The viability of real-time neuromorphic processing in real life environment has been demonstrated by the prototypes based on high performance FPGA, for its optimised computing hardware as other work [15]. In this neuromorphic research, the Xilinx device has been utilised to implement the real-time neuromorphic visual processing of 60 video frames per second without the latency, as shown in Fig. 1.

Fig. 1. An overview of neuromorphic visual processing, with the feature extraction based on repeated local convolution (top). Real-time neuromorphic system implemented on FPGA [Xilinx] (bottom)

One of the major challenges in computer vision or visual recognition in real life road condition is the nature of changing environment and luminance and its effect on the accuracy of performance. For example, most pedestrian detection algorithms have significant limitation in its detection rate at the night time or when operating under a bad climate conditions, such as rain, and blizzard. The neuromorphic vision system overcomes this problem through incorporation of the Nature's robustness inspired by Visual Cortex simple cell's orientation selectivity and the performance mimicking human intelligent function, instead of the immediate pattern matching or complex figure training. Robustness is one of the key issues in the prior research of automotive applications, especially in varying weather conditions and environmental noise. This was done through the deeper network layers to remove the incidents of larger block of saliency area to refine the realistic saliency data of the human target object on the road, which was a single human figure in the case, as demonstrated in Fig. 2 [8, 16].

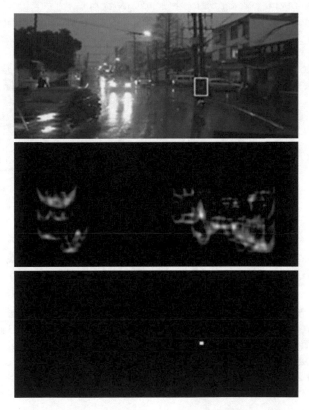

Fig. 2. The original image with the detected human figure (top); the saliency map as the output of layer 5 in Fig. 1 (middle): the saliency map after the additional denoisng layer processing (bottom)

2.1 Object Segmentation in Night-Time Road Environment

One of the challenges encountered in previous researches on neuromorphic processing is being able to account for the noise from the urban background. In particular, the strong noise at night-time has been the challenge to overcome as materialized through sudden onset and subsequent disappearance of strong signal in the image, while the object likely appears less clear. This has accounted for substantial demand of deep network of denoising purposes in earlier versions of neuromorphic visual processing, which causes the complicated learning/designing process of convolution kernels as well as extra computation resources. For some night-time road traffic environments, the abrupt changes due to the unpredictable reflection or lighting from near-by moving vehicles or changing street scene by the light from street lights or buildings. The denoising layer (exhibited in Fig. 2) enables refining the saliency data for particular object detection as shown in 'enhanced saliency map' with post filter, where the detected cyclist is represented as small white dot and red box. The optimum post filter design demands further development of learning with deep network and large data sets, though the fundamental configuration is relatively simple.

The neuromorphic convolutional-recurrent neural network is illustrated in Fig. 3, where the layer 1 and layer 2 of Fig. 1 are represented as CNN in Fig. 3 and the layer 4 of convolution neural network is removed.

The introduction of down-up resizing in Fig. 4 enables the abstract feature extraction, with the robust saliency map generation combined with the recurrent neural

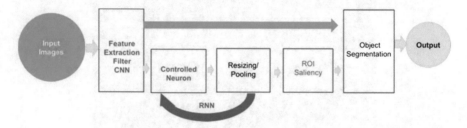

Fig. 3. The neuromorphic convolutional-recurrent neural network with down-up rescaling (resizing/pooling) to detect/segment object without template matching (or convolution)

network. The VRU was detected successfully in Fig. 5 with the neuromorphic convolutional-recurrent neural network, where the night-time video dataset in Fig. 5 was unsuccessful for the neuromorphic convolutional network of Fig. 2. The video dataset was with poorer image quality at night-time and small road condition of particular location in London, U.K.

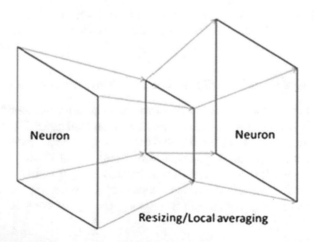

Fig. 4. The down-up rescaling layer with controlled rectifier neuron

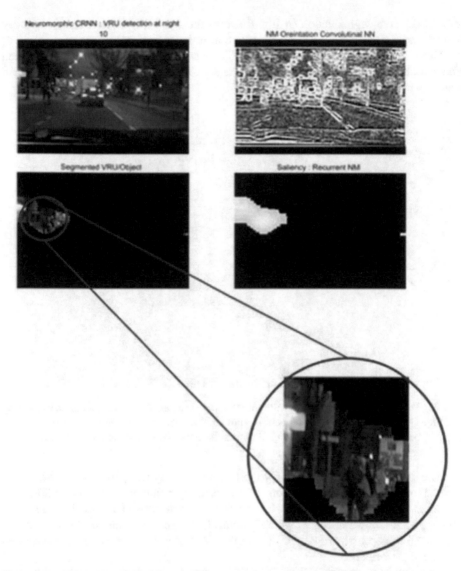

Fig. 5. The VRU detection of night-time road in London (U.K.) by object segmentation based on the proposed neuromorphic convolutional-recurrent neural network in Fig. 3

2.2 Tooth Segmentation in Dental X-ray Images

The tooth segmentation is expected to improve the dental practitioners' work process substantially, as various dental x-ray instruments are widely in use. The current status of art is the manual segmentation of tooth from x-ray image, as there are various issues including the noise. The neuromorphic visual processing was applied to the dental x-ray image analysis, and the feasibility was evaluated in Fig. 6. The demand from dental x-ray industry was the robust segmentation of individual tooth, and the individual tooth segmentation in Fig. 6 shows the feasibility of neuromorphic x-ray image analysis based on the simple evaluation of layer 2 output of Fig. 1.

However, the segmentation for 3D CBCT demands the optimum feature extraction filters, as the hand-cut feature extraction filters yield the strong signals from populated teeth, as shown in Fig. 7.

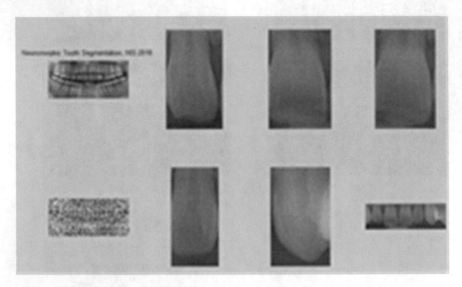

Fig. 6. Tooth segmentation by neuromorphic visual processing, 2D dental panorama image (top left); the output of neuromorphic processing (bottom right); the segmentation result (top right, 3 objects; bottom middle 2 objects); Region of Interest for tooth segmentation (bottom right)

The alternative feature extraction filters were designed based on Autoencoder (AE) trained by image patches from Dicom dataset, maintaining the compatibility to the hand-cut neuromorphic filters in size. The training dataset is based on 11 × 11 image patches shown in Fig. 8, which is equivalent to the convolution filter size. AE in MATLAB toolbox was used and the number of hidden nodes was decided based on simulated performance.

The simulated results in Figs. 9 and 10 shows the better performance with more hidden nodes, which is also more convolution filters. AE with 7 hidden odes was

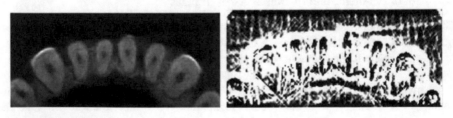

Fig. 7. Dicom image of dental x-ray CBCT (left) neuromorphic feature extraction based on hand-cut filter (right), Dicom dataset provided by Vatech in Korea

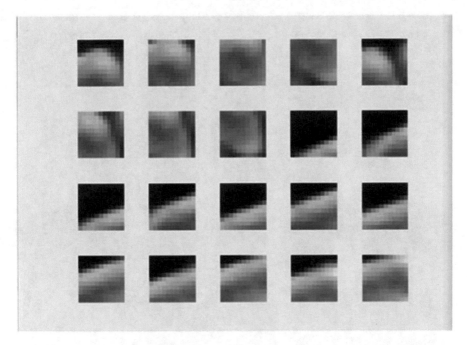

Fig. 8. Dicom image patches (11 × 11) used for training Autoencoder (AE)

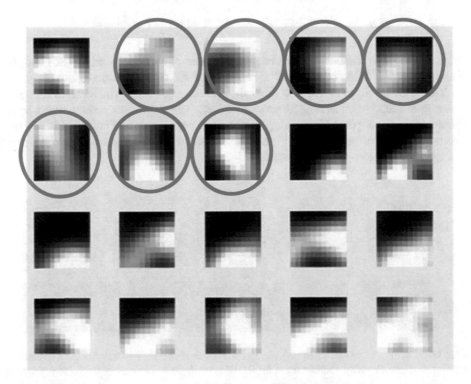

Fig. 9. Reconstructed/decoded outputs for 7 hidden nodes

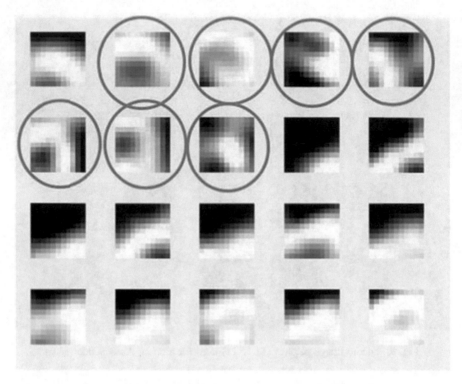

Fig. 10. Reconstructed/decoded outputs for 11 hidden nodes

chosen, and gave reasonable performance in green circles. The number of hand-cut filters was in the range of 4–6, and the number of AE filters was decided as 7.

The saliency maps in Fig. 11 show the feature extraction, which are effective in noisy gum area by enhancing the tooth objects. The feature extraction filters were trained by image patch data set, without any target value. The individual filter parameters are illustrated in Fig. 12, which were the outcome of unsupervised learning and appeared as complicated orientations.

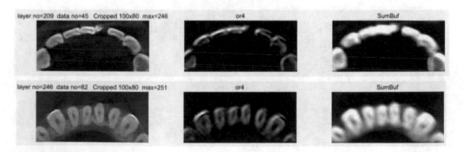

Fig. 11. Saliency maps are illustrated, though with limited accuracy of training data. 'SumBuf' represents the normalized summation of all features, while 'or4' represents one feature outcome

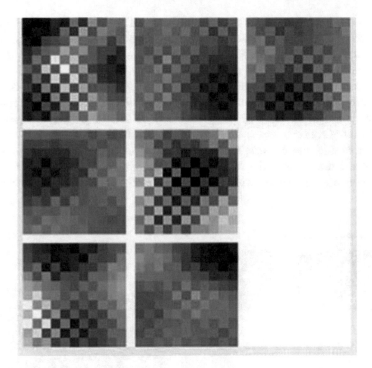

Fig. 12. Convolutional filters based on AE, trained by 11×11 images for 7 hidden nodes

3   Night-Time Vehicle Detection by Neuromorphic Convolutional-Recurrent Neural Network

Recent development allowed the earlier neuromorphic system to overcome this challenge through the down-up rescaling with controlled rectifier neuron, as the deep neural network with down-up convolution is applied to segmentation. The resizing or local processing can invoke the process of reducing the noise components of induced or processed in due course of neuromorphic processing. The controlled threshold of rectifier enables the network layers to adapt to the target functions or tasks such as object segmentation without the labelled data.

3.1   Vehicle Object Segmentation for Tracking at Night

The challenge in vehicle tracking system is from the requirements of robust operation at night time, particularly occluded with other vehicles. The neuromorphic visual processing for vehicle detection aims to deliver the robust segmentation of vehicles for enabling the further analysis and recognition of identifying individual vehicles. The performance scenario by the industry is based on the fixed video sensors and it allows the straight forward application of recurrent networks than for the mobile camera on the moving vehicle. The neuromorphic convolutional-recurrent neural network in Fig. 3 is applied to the vehicle tacking neuromorphic visual processing system, which was inspired and evaluated by the video data set from the industry. Both the hand-optimised feature filters and the AE-based feature filters are investigated with the same neuromorphic vision system.

The same neuromorphic vision system is used for moving vehicle detection in Fig. 13, with the same parameters of down-up rescaling, recurrent configuration, and rectifier neuron control. The difference is the convolutional filter set, which is from the trained AE of MATLAB or the hand-cut orientation filter. It is remarkable to analyze the difference and the consistency between the two different filters with the same algorithm and the dataset, delivering the similar outcome of vehicle detection and segmentation though with the wide difference in interim processing status. The interim

Fig. 13. Vehicle detection using neuromorphic convolutional-recurrent neural network visual processing system, hand-cut based feature extraction (top) versus AE based one (bottom), both successful at the same video frame (no 32) of dataset recorded at Beijing, China

processing based on hand-cut filter represents the visible abstract at stages of the convolutional processing and the down-up resizing/rescaling.

Vehicle detection and segmentation was successful in Figs. 13 and 14, though with the slight difference in segmented vehicle sizes. The detection/segmentation is evaluated as the successful one when the whole vehicle body or the most of frontal part is segmented. As stated earlier, the detection aims to deliver the robust segmentation of vehicles for enabling the further analysis and recognition of identifying individual vehicles.

The Vehicle detection and segmentation in Fig. 15 was unlikely successful sufficiently for the AE based on, as the detected area of vehicle is a spot. The detected spot was still the spot of vehicle's front, i.e. the location of headlight. The vehicle was detected but with the incomplete segmentation. The case of hand-cut filter was successful

Fig. 14. Vehicle detection, hand-cut based feature extraction (top) versus AE based one (bottom), both successful at the same video frame (no 38)

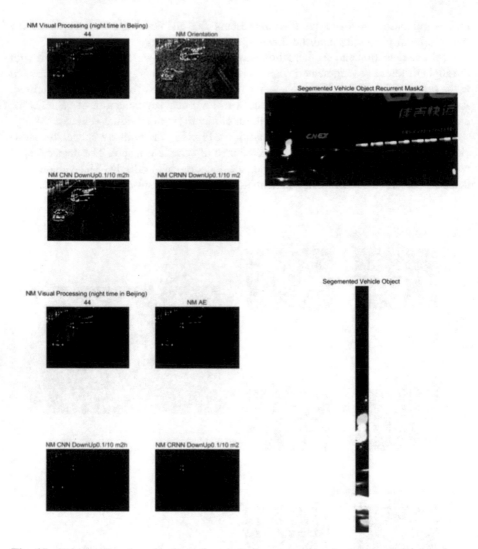

Fig. 15. Vehicle detection, hand-cut based feature extraction (top) versus AE based one (bottom), only successful for the hand-cut one (video frame no 44)

for detection and segmentation. As most of neural network based solution, it is unlikely feasible to develop the complete analysis with the quantitative reasoning for particular performance. However, the observation suggests the low level signal of AE based feature extraction caused the limitation on segmentation. The similar phenomena were observed for the signal with lower strength or lower frequency, for segmenting objects by the deep neural network either based on supervised learning or unsupervised learning.

Vehicle detection and segmentation in Fig. 15 was unlikely successful sufficiently for the AE based on, as the detected area of vehicle is a spot. The detected spot was still the spot of vehicle's front, i.e. the location of headlight. The vehicle was detected but with the incomplete segmentation. The case of hand-cut filter was successful for detection and segmentation. As most of neural network based solution, it is unlikely feasible to develop the quantitative and logical analysis thoroughly.

Down-Up rescaling was implemented with the ×10 up sizing after ×0.1 scaling down using 'imresize' in MATLAB. The vehicle segmentation was subject to the analysis of saliency map distribution after the recurrent operation and Down-Up rescaling, by controlling the threshold value of saliency map in axis. The threshold values were determined by the ratio of maximum of each axis. The threshold of row axis was applied by ×(1/2.5) and ×(1/2.2) for column axis respectively. The primitive method is applied to the segmentation process for the basic evaluation of proposed neuromorphic convolutional-recurrent neural network, though the optimum or algorithmic processing method can be applicable to the practical applications of vehicle detection/tracking.

Examples of Fig. 16 (frame no 51) and Fig. 17 (frame no 61) represent the fact that the common neuromorphic feature extractor can be extended to various applications under the natural environments, as there was no meaningful difference in performance observed with the feature extractors optimized for teeth segmentation. The overall performance is summarized in the following. The video dataset was the 42 frames video clip of cross road traffic at night-time in Beijing, China.

Type of feature extraction	Successful detection/segmentation (night-time traffic video, 42 frames)	Successful detection with poor segmentation	Failed detection (false positive)
Hand-cut orientation	41 frames (98%)	1 frame (2%)	0 frame (0%)
AE trained by teeth image	40 frames (95%)	2 frames (5%)	0 frame (0%)

3.2 3D Tooth Segmentation for x-ray CBCT Images

The 3D dental CBCT image is subject to the high level of noise due to bone structure in the gum region or low x-ray dose of head area. The neuromorphic visual system for 3D tooth segmentation is based on the neuromorphic convolutional-recurrent neural network of Fig. 3, as the same principle can be applicable to the video dataset and the 3D volume dataset as illustrated in Fig. 18.

The neuromorphic convolutional-recurrent neural network demonstrated the feasibility of segmenting object in the rough environment. The evaluation delivered the successful detection but the segmentation with some constraints. The cascaded convolutional neural network was introduced to enhance the segmentation process as in Fig. 19. In principle, the front-end convolutional neural network acts as the RoI (Region of Interest) generator by segmentation of teeth area. The orientation features of RoI is evaluated by the cascaded convolutional neural network, where RoI substantially improves the noise issue from non-tooth elements of bones or tissues in the gum

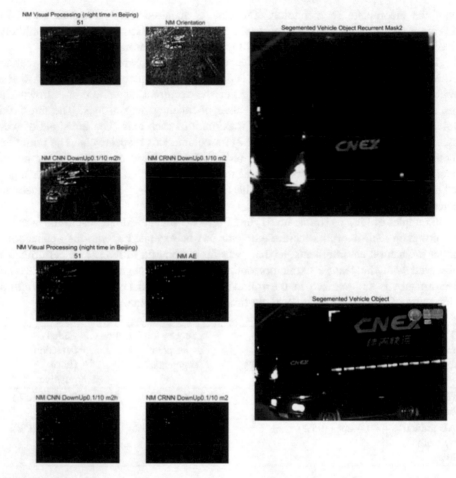

Fig. 16. Vehicle detection, hand-cut based feature extraction (top) versus AE based one (bottom), only successful for the hand-cut one (video frame no 44)

region. Gabor filter is applied as convolutional kernel as Gabor filter is an alternative to the hand-cut filter mimicking the orientation feature extracting simple cell neuron of visual cortex.

Gabor filter is a complex filter with real part and imaginary part as illustrated in the following equation, where the equation with 'cosine' term is the real part equation and that with 'sine' as the imaginary part.

$$g(x, y; \lambda, \theta, \psi, \sigma, \gamma) = \exp\left(-\frac{x'^2 + \gamma^2 y'^2}{2\sigma^2}\right) \cos\left(2\pi \frac{x'}{\lambda} + \psi\right)$$

$$g(x, y; \lambda, \theta, \psi, \sigma, \gamma) = \exp\left(-\frac{x'^2 + \gamma^2 y'^2}{2\sigma^2}\right) \sin\left(2\pi \frac{x'}{\lambda} + \psi\right)$$

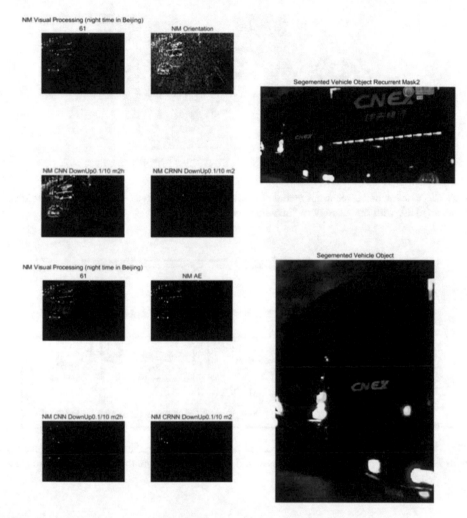

Fig. 17. Vehicle detection, hand-cut based feature extraction (top) versus AE based one (bottom), only successful for the hand-cut one (video frame no 61)

Gabor filter in Fig. 19 was implemented by two equation introduced above, and 16 angles were used to generate the convolutional filters. Gabor filters of 6 orientations were selected for the flexible interface to prior neuromorphic visual system, that is, shown in Fig. 7. The basic figure of Gabor filter is illustrated in Fig. 20, which represent the real part of filter function. The neuromorphic convolutional-recurrent neural network employs the complex Gabor filter, and the real part in the output is used as the output of convolution neural network.

In Fig. 21, the Dicom image of CBCT is applied to the neuromorphic vision system in Fig. 19. The image of top right is the outcome of convolutional-recurrent neural network after the front-end convolutional processing based on AE. Each CNN output was evaluated for the tooth segmentation, and yielded the reference positions

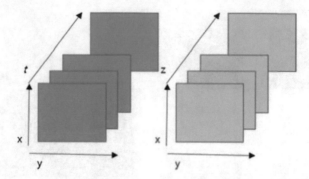

Fig. 18. Convolutional-recurrent neural network for the video dataset (left) and the 3D volume dataset (right), with the analogy of time-axis and depth—axis

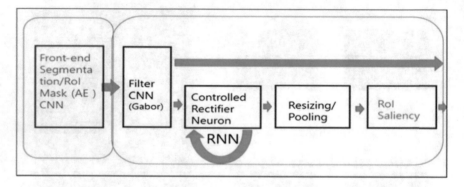

Fig. 19. Neuromorphic convolution-recurrent neural network for the 3D tooth segmentation, with the additional cascaded convolution neural network to enhance the segmentation process

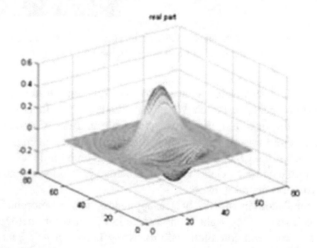

Fig. 20. Gabor complex filter illustrating the orientation feature extraction, in real-part

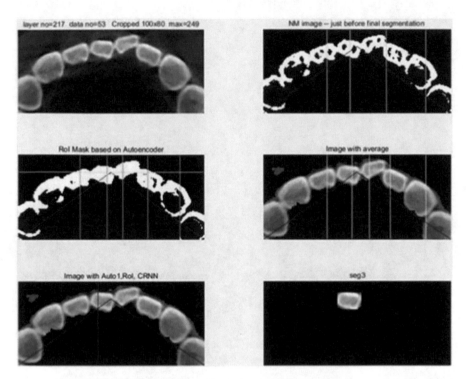

Fig. 21. 3D tooth segmentation based on the neuromorphic convolutional-recurrent neural network in Fig. 19: Dicom CBCT image of layer no 217 (top left), RoI delivered by the front-end CNN with AE (middle left), the output of CNN-RoI with Gabor filter (top right), the averaged centre (yellow lines) of each segmented tooth by AE's RoI and Gabor filter-CNN (middle right), the reference position of selected tooth (bottom left), the segmented tooth of corresponding Dicom layer (bottom right)

respectively. The averaged reference positions (corresponding to the tooth centre) were represented in the centre right figure box. The target tooth was segmented in the bottom right figure box, based on the saliency map by convolutional-recurrent process.

In Fig. 22, the dental CBCT Dicom image of gum region is applied to the neuromorphic vision system in Fig. 19. The image of middle left is the outcome of frontend CNN, which segmented tooth areas separated from the gum tissues. Contrary to the images in Fig. 21, the teeth saliency maps delivered the clear separation of each tooth. The collected Dicom layers of segmented tooth are illustrated in Fig. 23. Sub image on top right shows the segmented tooth in vertical cross section, which shows the sloped frontal tooth. The small amount of gum tissue was observed in the top right figure. Complex mixture of tissues and bones adds to the challenge of tooth segmentation. In practice, the dentists are often cautious about the operation of tooth root in the gum region, as the complicated tissue status and its adherence to the tooth root.

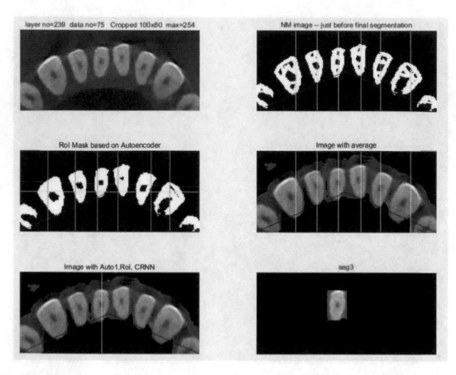

Fig. 22. 3D tooth segmentation based on the neuromorphic convolutional-recurrent neural network in Fig. 19: Dicom CBCT image of layer no 239 (top left), RoI delivered by the front-end CNN with AE (middle left), the output of CNN-RoI with Gabor filter (top right), the averaged centre (yellow lines) of each segmented tooth by AE's RoI and Gabor filter-CNN (middle right), the reference position of selected tooth (bottom left), the segmented tooth of corresponding Dicom layer (bottom right)

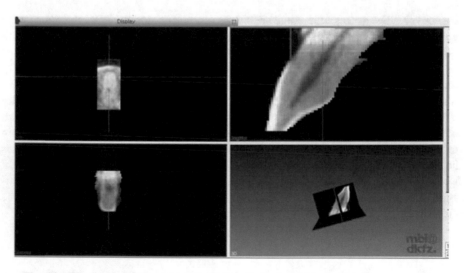

Fig. 23. 3D display of segmented frontal tooth, horizontal and vertical cross-sections

4   Conclusion

The proposed convolutional-recurrent neural network with Down-Up rescaling is evaluated successfully for its applicability and feasibility of object segmentation, by both the vehicle tracking with real video data sets of night time and the tooth segmentation with dental 3D CBCT Dicom dataset. The performance was evaluated for the robust function with over 98% successful vehicle detection and segmentation for the night time vehicle tracking. The feature extractors based on the auto-encoder as well as the hand-cut ones mimicking the neuron cell of visual cortex illustrated the insight of one-learning algorithm, as the trained by x-ray teeth images worked for the night-time street video images.

The automatic anonymous optimization of threshold control remains focus of the further work for efficient Down-Up rescaling and precise segmentation.

The neuromorphic system with new convolutional-recurrent neural network with Down-Up rescaling has been proposed for the intelligent visual sensor mimicking the visual cortex and evaluated on its applicability and feasibility for the object segmentation in the noisy environment. The research on neuromorphic and deep multi-layer network is under progress to develop the real-time sensory recognition in the diverse environment such as varying levels of sensitivity, and abrupt changing conditions, for the mobile embedded artificial intelligence solution based on real-time neuromorphic computing by FPGA or GPU.

Acknowledgements. The research of 3D tooth segmentation was sponsored by Vatech, Korea. We are thankful to Mr. Ik Kim for his cooperation to our research on tooth segmentation using Dental X-ray images, specifically on the guidance and discussion about developing the medical applications.

References

1. LeCun, Y., Bengio, Y., Hinton, G.: Deep learning. Nature **521**, 436–444 (2015)
2. Hubel, D.H., Wiesel, T.N.: Receptive fields of single neurons in the cat's striate cortex. J. Physiol. **148**, 574–591 (1959)
3. Hinton, G., Osrindero, S., Teh, Y.: A fast learning algorithm for deep belief nets. Neural Comput. **18**, 1527–1554 (2006)
4. Krizhevsky, A., Sutskever, I., Hinton, G.: ImageNet classification with deep convolutional neural networks. In: Proceedings of the Advances in Neural Information Processing Systems 25(NIPS 2012) (2012)
5. Cadieu, C., Kouh, M., Pasupathy, A., Connor, C.E., Riesenhuber, M., Poggio, T.: A model of V4 shape selectivity and invariance. J. Neurophysiol. **98**, 1733–1750 (2007)
6. Scarfe, W., Farman, A.: What is Con-Beam CT and how does it work? Dent. Clin. North Am (Elsevier) **52**, 707–730 (2008)
7. Han, W.S., Han, I.S.: Enhanced neuromorphic visual processing by segmented neuron for intelligent vehicle. In: Proceedings of the SAI, pp. 307–311, July 2016
8. Han, W.S., Han, I.S.: All Weather Human Detection Using Neuromorphic Visual Processing. Studies in Computational Intelligence, vol. 542, pp. 25–44. Springer (2014)

9. Fuster, J.: The Prefrontal cortex makes the brain a preadaptive system. Proc. IEEE **102**(4), 417–426 (2014)
10. Paisitkriangkrai, S., Shen, C., Hengel, A.: Pedestrian detection with spatially pooled features and structured ensemble learning. IEEE Trans. Pattern Anal. Mach. Intell. **38**(6), 1243–1257 (2016)
11. Girshick, R.: Fast R-CNN. In: Proceedings of the IEEE Conference Computer Vision, pp. 141–148 (2015)
12. Li, X., Flohr, F., Yang, Y., Xiong, H., Braun, M., Pan, S., Li, K., Gavrila, D.: A new benchmark for vision-based cyclist detection. In: Proceedings of the IEEE Symposium Intelligent Vehicle, pp. 1028–1033 (2016)
13. Sifibe, D., Meriaudeau, F.: Visual saliency detection in colour images based on density estimation. Electron. Lett. **53**(1), 24–25 (2017)
14. Sermanet, P., Kavukcupglu, K., Chintala, S., Lecun, Y.: Pedestrian detection with unsupervised multi-stage feature learning. In: Proceedings of the IEEE Conference Computer Vision and Pattern Recognition, pp. 3626–3633 (2013)
15. Faraber, C., Poulet, C., Han, J., Lecun, Y.: CNP: An FPGA-based processor for convolutional networks. In: Proceedings of the IEEE Conference Field Programmable Logic and Applications, pp. 32–37 (2009)
16. Hubel, D.: A big step along the visual pathway. Nature **380**, 197–198 (1996)

Weighted Multi-resource Minority Games

S.M. Mahdi Seyednezhad[1](✉), Elissa Shinseki[2], Daniel Romero[3], and
Ronaldo Menezes[1]

[1] School of Computing, Florida Institute of Technology, Melbourne, FL, USA
sseyednezhad2013@my.fit.edu, rmenezes@cs.fit.edu
[2] Department of Computer Science and Information Systems, George Fox University,
Newberg, OR, USA
eshinseki11@georgefox.edu
[3] Department of Computer Science, Weber State University, Ogden, UT, USA
danielromero@mail.weber.edu

Abstract. Game theory and its application in multi-agent systems continues to attract a considerable number of scientists and researchers
around the globe. Moreover, the need for distributed resource allocation is increasing at a high pace and multi-agent systems are known to
be suitable to deal with these problems. In this chapter, we investigate
the presence of multiple resources in minority games where each resource
can be given a weight (importance). In this context, we investigate different settings of the parameters and how they change the results of
the game. In spite of some previous works on multi-resource minority
games, we explain why they should be referred as *multi-option* games.
Through exploring various scenarios of multi-resource situations, we take
into account two important issues: *(i)* degree of freedom to choose strategy, and *(ii)* the effect of resource capacity on the different evaluation
criteria. Besides, we introduce a new criterion named *resource usage* to
understand the behavior of the system and the performance of agents in
utilizing each resource. We find that although using a single strategy may
involve less computation, using different strategies is more effective when
employing multiple resources simultaneously. In addition, we investigate
the system behavior as the importance of resources are different; we find
that by adjusting the weight of resources, it is possible to attract agents
towards a particular resource.

1 Introduction

With the expansion of computer systems, the need for distributed resources
seems to be a vital issue. Therefore, we need an effective resource allocation
system that preferably meets two important requirements: simplicity and distribution. From a different point of view, we have the concept of intelligent agents.
If a system employs several agents to reach its goal, it is called a *Multi-agent System* (MAS) [21]. Given that agents in a multi-agent systems can be designed to
coordinate or even compete with each other, they are suitable for use in distrib-

© Springer International Publishing AG 2018
Y. Bi et al. (eds.), *Intelligent Systems and Applications*,
Studies in Computational Intelligence 751,
https://doi.org/10.1007/978-3-319-69266-1_14

uted scenarios. Therefore, a resource allocation problem could involve intelligent agents with bounded rationality competing for limited resources [15]. These scenarios exist across countless disciplines. For example, sellers and buyers can be considered as agents who complete on price of items on a market. Moreover, animals can be observed as agents that are competing on their territories. In fact, these are some typical example of complex adaptive systems with self-organizing agents whose behaviors dynamically adapt over time [18]. It is arduous to model numerous different scenarios of resource allocation problems, and this makes it challenging for determining the best approach to solve them.

Latterly, the minority game has been adopted as an effective model for resource allocation using inductive reasoning [15]. We believe the literature has been incorrectly proposing multi-option games as multi-resources. In Sect. 2 we clarify this issue in more details.

1.1   The Minority Game

Back 1994, Arthur [1] introduced a resource allocation model, the *El Farol Bar problem*, that paved the way for the minority game. In the El Farol bar problem, there are a number of people who want to enjoy their evening in a bar. The bar has a limited capacity, so if people go to an overcrowded bar, they cannot enjoy their evening, and the ones who stay at home will have a better evening. Two years after the introduction of El Farol Bar problem, Challet and Zhang [6] mathematically modeled El Farol bar problem. They assumed that a population of N agents want to use a resource and the capacity of the resource is half of the population. The agents have two options: to go or not to go. If agents end up to an option chosen by less than half of the population then, they are the winners. Agents should try to be in the winner group using the previous outcome of the game and their best strategy for each round of the game.

Agents pick their favorite choice with respect to their best strategy. An strategy is a mapping table from the last m outcome of the game to a decision in a round of the game. The last m outcome is also called the history of the game which is global in a traditional classic minority game. There are 2^m possible inputs for a strategy and 2^{2^m} possible strategies. Agents reward their strategy if it leads them in the minority group of the previous round of the game, otherwise, the strategy loses points. Before choosing the best strategy in each round, agents sort their strategies based on the scores they could received.

At the end of each round, the outcome of the game is determined based on the minority rule. Consequently, winning agents are given points and the losers are taken points according to a prize-to-fine ratio. The individual strategies of the agents are reinforced in this manner as well. This payoff process is the essence of learning in the minority game, whereas agents receive feedback from environment and adjust their strategies' scores in order to detect the best strategy. Finally, the outcome of the game constructs the global memory. The game is normally repeated for a certain number of iterations. The system containing this game is preferred to have the agents' attendance close to the capacity of the resource and agents have won almost 50% of the rounds [12].

1.2 Evaluation Criteria

In order to understand the behavior of a minority game we can make use of several evaluation criteria. The conventional criteria used in the literature include attendance, variance, and winning rate. Attendance $A(t)$ is a measure of how many agents chose to attend a resource in tth iteration [15]. Usually, attendance rates are measured over many iterations. Since there is a reinforcement process for agents, the simulation should end to a state with low attendance fluctuation around the resource capacity which is considered as the threshold for the minority rule.

The variance of attendance rates are used to measure the stability of resource attendance [15]. Low variance is indicative of a stable attendance rate, while high variance is a sign of volatility. The trends in variance for complete simulations are often plotted against different memory sizes. Many works use the value called α (a function of memory size, calculated as $\alpha = 2^m/N$) to express the dynamics of the attendance rate variance (as shown in Fig. 1). The figure demonstrates the existence of memory sizes optimal for minimizing the variance for different agent population sizes. The figure also shows that after a certain value of α, increasing the memory size does not decrease the variance.

Another important criterion in minority games is the winning rate defined as a ratio between the number of times the agent was in the minority for a resource (a win) and the total number of game iterations. When resource allocation is efficient and resource capacity is 50% of the population size, winning rates among the agents would be almost 50% of the population.

The above metrics are some of the most important criteria for the minority game according to the standard literature [7,13], but how can one claim that a resource has been utilized effectively in a game or during some iterations? In our study, we want to take into account the general concept of *resource usage* [8,9] to analyze system behavior. If the number of participants in a resource is less than the resource capacity, the resource is being used effectively. Consequently, the agents that use the under-crowded resource are the winner of game. For more illustrations, we explain an example of a computer system in which some processes need to use shared resources. We would prefer the processes to hold all of the resources that they need, but the operating system may put them in a sleep mode or waiting list if they attempt to use an overcrowded resource [20,22]. Therefore, if a resource is overcrowded most of the times, it is not used properly.

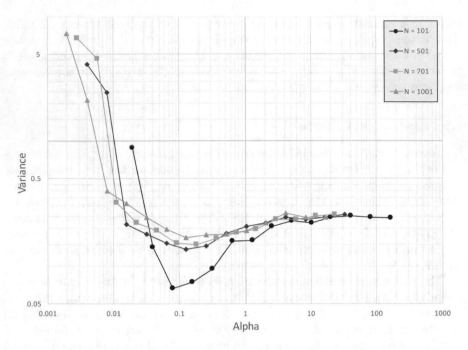

Fig. 1. Association of σ^2 and *alpha* for the classic minority game, $N = 101, 501, 701, 1001$, $s = 6$, $t = 1000$, resource capacity $= 500$

2 Multi-option Versus Multi-resource

We begin explaining the difference between multi-resource and multi-option approach with a real-world example. We consider a situation in which "Diego" wants to go to a movie theater. We assume that all movies in the theater start at the same time so he cannot watch all of them at the same night, thus, he must choose one of the movies. However, regardless of his choice, he will be using the capacity of just one single resource. If he chooses a movie playing in a movie theater that is overcrowded, he will not get in and hence not be able to watch any movie. As a result, *Diego* will not enjoy the night and will "fine" the strategy that led him to choose that movie. This situation is an example of a multi-option game in which multiple options are applied to a single resource, in this case the movie theater.

Now imagine the situation in which *Diego* wants to watch a specific movie that is being played in more than one movie theater. If one theater is overcrowded, other theaters could still be available. This situation may be considered as a multi-resource minority game. The theaters are only equivalent when *Diego* and his friends choose not to split and necessarily go all together to the same theater. In this example, if we consider *Diego* (or either of his friends) individually as an agent, we are facing a multi-option game. On the contrary, if we look at the entire group as an agent, we have a multi-resource game. In fact, the multi-resource game is a more general case of a multi-option case where agents can use more than one resource/option at each round of the game.

It should be noted that in resource allocation game studies, we can consider one single resource with two options as two resource with one option per each. If we generalize it, a game with one resource and k options for that resource is equivalent to a game with k mutual exclusive resources, i.e. agents cannot contribute to more than one resource at each round of the game. Figure 2 demonstrates the difference between multi-resource and multi-option minority games. Unlike a multi-option (MO) model in which agents make decision for one resource, in multi-resource (MR) model, each agent can decide over different resources and pick more than one of them. In other words, an agent chooses one of the k resources available in the MR model.

For more illustrations, We assume that $agent_i$ wants to pick one resource in an MR model. If the strategies of the agent suggest action a_i where $a_i = (0, 1, 1, 0, 1, \ldots, 0)$, we can say this agent decided to use the 2nd, 3rd and 5th resources (If we count the options/decisions from zero, then $a_{i,1} = 1$). Nonetheless, $agent_i$ may decide to choose the 2nd option for resource R in an MO model. In this case, the suggested action is $a_i = (0, 1, 0, 0, 0, \ldots, 0)$ or $a_i = 1$. As it can be seen, MO and MR models may be mistakingly assumed to be equal, but this is true if and only if agents are allowed to choose one resource for each round of the game. Moreover, the multi-resource model is a more generalized version of the multi-option one. Indeed, MO games can be classified as a sub-category of the MR games.

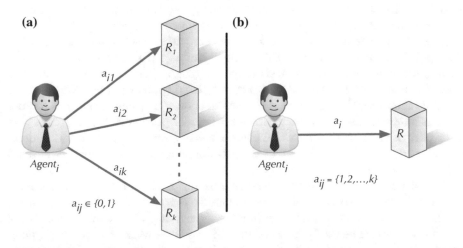

Fig. 2. Multi-option versus multi-resource: **a** Agent i should decide to choose more than one resource in the multi-resource model, **b** There are k options corresponding to the resource R and agent i should choose one of them. The action a_i can be in 1 dimension—a single value—where $a_i \in \{1, 2, 3, \ldots, k\}$

A summary of our discussion over MO and Mr models is shown in Fig. 3. As it is shown, the classic minority game—the game with one single resource—is a specific form of the multi-option (MO) model, because the agents can use one resource with two options. Furthermore, multi-option games are a specific form of multi-resource (MR) model, because the agents can choose exactly one resource/option at each iteration.

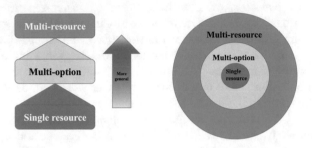

Fig. 3. From left to right: The hierarchical and Venn diagram of the relation among single resource, multi-option and multi resource model

3   Related Works

One of the earliest variations introduced for minority game is an evolutionary model in which agents' strategies are targeted to be evolved during the game in order to utilize the resources in a more efficient way [11]. In this model, instead of using binary or bipolar representation for agents' strategies, a probability p is assigned to each agent as an strategy. Agents can either refer to local information with probability p or act alone with probability of $p-1$ [10]. The system results in self-organizing agents who choose strategies in groups. The significant part of the work of Huant et al. [10] is that they actually tried to investigate a *multi-option* model; in their model, each agent is allowed to choose only one resource at a time, which is a representation of the multi-option model (as depicted in Fig. 2b). The other part of their work focuses on the observed grouping phenomenon among the agents on a lattice. They detect this phenomenon on a stock market dataset and analyze it by mean field theory.

The structure underlying the agents' relation to each other is an interesting topic, because networks allow researchers to represent the relations between agents. In real resource allocation problems, agents may often communicate over a social network. The study of structures of social networks has received much recently in the so-called field of Network Science [3]. By using different network structures in agent communication in conjunction with the minority game, agents might influence each other in groups. Remondino and Cappellini [16] connected agents using a random network. The simulation provides the agents two modes of decision-making on a resource: synchronous and asynchronous.

With asynchronous decision-making, each agent in sequence makes a decision on a resource. While in the synchronous decision-making, each agent broadcasts its proposed decision to their neighbors. Subsequently, all agents consider their neighbors' proposed decision and choose what decision to make simultaneously.

Shang and Wang [18] probed the evolutionary minority game on four different network topologies connecting agents to the neighbors: a star network, a regular network (lattice), a random network, and a scale-free network [2]. Similar to the work in [11], they assign a probability p as strategy to each agent. Then, they investigate the distribution of probability p among agents with respect to the different values of the mutation variable L and the variation of the prize-to-fine ratio ρ. With $\rho = 1$, most networks showed the agents segregating to polar probability values, while the star network showed agents with a probability close to 0.5. With $\rho = 0.9$, all network topologies showed agent probabilities clustered around 0.5, suggesting that regular, random, and scale-free networks are sensitive to the value of ρ, while the star network is sensitive to the value of L.

Caridi and Ceva [4] looked at underlying networks in the minority game by connecting agents who have similar strategies together. They found that there are different phases with different networks (e.g. different link definition). Furthermore, in the phase where the system performs like in a game of random decisions, the underlying network behaves as a random one with the same network characteristics.

Applications of the minority game consist allocating energy to rooms in smart buildings with varying energy needs from multiple energy sources [23]. Additionally, computers could use the minority game in allocating re-configurable multi-core processors to maximize efficiency [17]; the results demonstrate that the minority-game policy achieves on average 2-times higher application performance and a 5-times improved efficiency of resource utilization compared to state-of-the-art. Furthermore, the minority game may also be used in cognitive radios and other wireless networks to facilitate cooperation between devices without as much energy or communication overhead [9, 14].

Catteeuw and Manderick [5] conducted a research in which the reinforcement process of Q-learning is used in the minority game. In their version of minority game, agents decide to attend (or use) *only one* of the multiple resources, what is called here "multi-option". Agents will be punished or rewarded based on the resource that they chose to attend. The reinforcement process allows agents to balance between exploration and exploration of new strategies and their best known strategy. In fact, in their research, agents are allowed to choose just one of the resources in each round of the game. Genuinely, their approach may be considered as a minority game in which there is just one resource and more than one option to choose over that resource.

Recall that in the classic minority game, one considers the game for one resource and two mutual exclusive options. However, in distributed systems, a process may need to obtain several different resources at once (such as CPU, RAM, and bus channels) before executing [19]. Although few studies have considered multi-option instead of multiple resources in their research, none have done an in-depth exploration of the variation of the minority game parameters with multiple resources in which an agent is able to apply for more than one resource at each round.

4   The Multi-resource Minority Game Model

In our model of multi-resource minority game, we have more than two resources and two options for each one. The resources have their specific capacity and memory. The minority threshold is the capacity of each resource. Using this model, we explore scenarios that can be taken into account only in a multi-resource model. First, the way that the agents may use their strategies could be different. Here, agents can use different strategies for different resources while in another scenario they can act like simple agents in a classic single resource minority game and use the same strategy for all the resources. We use the aforementioned criteria to analyze the system behavior.

There exists many different possible scenarios one can now consider in this variation. In the following, we express a number of scenarios that deal with real world applications. A real-world example is provided with each scenario. The basis of the examples is a computer system in which a number of processes need to hold some resources to finish their task.

1. Agents need to win all the resources in order to win the game in each round. For example, a process simultaneously needs CPU, RAM, and access to the network.
2. Agents need to win at least one resource in order to win the game in each round. For example, in a multi-processor system, a process needs to win at least one of the processors.
3. Winners are the agents who win the most valuable resources. For example, holding more CPU may be better than having access to different network channels for a particular process.

There are two major variations that can be applied on the aforementioned scenarios. On the one hand, one can now *vary the memory size* independently for each resource. On the other hand, *variations on the capacity* which means we have games where the total capacity of all resources equals to N (number of agents), or are less than N, or more than N.

4.1   Weighted Resource Model

After we explain the concept of multi-resource, one may ask: what if the resources are valued different by the agents? We examine the effect of two different kinds of resource-weighting methods, linear and exponential. In fact, weights could be used to prioritize the acquisition of one resource over another. In these models, each resource has a weight that is applied to strategy scores during the payoff phase of the game. In the linear method, in every iteration, the weight of every resource is either added to the old score when the agent is the winner, or it is subtracted from the old score when the agent is not in the minority for the resource. This is accomplished with:

$$S(t+1) = S(t) + w \times \beta, \qquad (1)$$

where $S(t)$ is the score of the agent at iteration t, w is the resource weight, and β is 1 when the agent is in the minority, and it is -1 when the agent is not in the minority. Increasingly larger values for resource weights alter agent and strategy scores at a rate linearly proportional to the value of the changing weight itself.

With the exponential method, updating the score is a multi-step process. First, for every resource, the old agent and strategy scores are multiplied by the resource weight to the power of β (Eq. 2). Second, the value obtained by this equation for each resource is averaged to compute the new score. In other words, the new agent and strategy score is calculated by Eq. 3.

$$S(t+1) = S(t) \times w^{\beta}, \qquad (2)$$

where $S(t)$ is the score of the agent at iteration t, w is the resource weight, and β determines positive or negative payoff.

$$S_k(t+1) = \frac{\sum_{j=1}^{R} S_j(t) \times w_j^{\beta}}{K}, \qquad (3)$$

where K is the number of resources and w_j is the base/weight of the rth resource, where w must be greater than 1. More details on resource weighting are found in Sect. 5.

5   Experimental Results

In this section, we try to compare the situation in which agents are allowed to use one strategy for all resources with the situation where agents can use different strategies for resources. Further, the different weighting methods will be compared with each other. In the same strategy simulation, agents use their best strategy to make decision over all the resources in the system. On the other hand, in the different-strategies simulation, each agent refers to its best strategies with respect to the resource on which it makes decision. As it is mentioned before, we can have different minority rules, here we consider an agent a winner if it is in the minority for all resources. If we use other types of the minority rule we will mention them. In all simulations, $N = 501$ is the population size, $S = 3$ is the number of strategy, $m = 4$ is the memory size and $T = 1000$ is the maximum iteration of the game. We consider $k = 3$ resources for our simulations because having 3 resources is enough to satisfy our assumption about the multi-resource model.

5.1   Multi-resource Simulation

The first experiment is about the attendance of the agents which use the same strategy, i.e. $A(t)$. In this experiment, the capacity distribution is $c = (\frac{N}{4}, \frac{N}{2}, \frac{3N}{4})$, i.e. capacity of resource 1 is 25% of the population (125), resource 2 is 50% (250), and resource 3 is 75% (375). Figure 4 (left) shows the impact of the resource capacity on the attendance in that resource. The obvious trends are steady over-attendance for resource 1 and steady under-attendance for resource 3. This is because the capacity is the lowest one in resource 1 and agents have a difficult time winning by attending in it. Meanwhile, it is easy to win by attending resource 3 because its capacity is the highest (375). This phenomenon causes the history to consistently suggest agents avoid attending resource 1, and always attend resource 3. In this simulation, the strategies which have zero in their first element and 1 in the last elements can lead their corresponding agents to win the rounds of the game.[1]

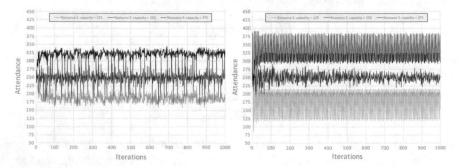

Fig. 4. Attendance versus iterations for $N = 501$, $T = 1000$, $S = 3$, and $m = 4$. Left: Agents are allowed to use just one strategy for all resources (SS approach). Right: Agents are allowed to use different strategies for different resources (DS)

In order to start the analysis, we look at $A(t)$ when agents are allowed to use different strategies in Fig. 4 (right). The attendance rate is still heavily under the influence of the resource's capacity; but by allowing agents to change their strategy we can observe variation in the outcomes. The dashed lines in Fig. 5 imply that losing agents tend to change their strategies in order to avoid attending resource 1. That is to say, the number of agents using resource 1 is decreased and that allows those agents to win. Therefore, a considerable number of agents reward the strategy that helps them avoid this resource. Similarly, we can observe a repetitive pattern with resource 3 where agents that have not

---

[1] The first element of the strategy taken into account corresponds to the situation where all previous outcomes of the game (or the recent ones in memory) should be zero (i.e. no agent chooses to use or go for the resource). Similarly, the last element represents the situation where all of the previous outcomes of the game is one (i.e. all agents choose to use the resource).

attempt for resource 3 try to switch their strategy to a new strategy in order to be in the resource 3. As a result, a periodic over-attendance happens for resource 3. Figure 5 also allows us to observe the differences between the use of different strategies (DS) (dashed lines) and the same strategy (SS) (solid lines). One main difference between the DS and the SS approach is that if agents use SS, then resource 1 is mostly over-utilized while resource 3 is mostly under-utilized. Another key fact to remember is that the variance in the utilization of resource 2 in DS approach is lower than the SS approach.

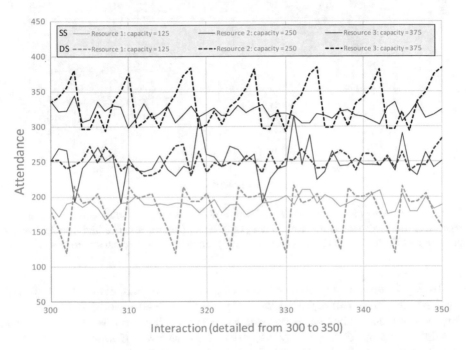

Fig. 5. A zoomed in view of what is happening from iteration 300 to iteration 350. Dashed lines and solid lines show the attendance for different strategy and same strategy approaches receptively

Furthermore, the impact of memory size in attendance for the SS and DS approaches is investigated in Fig. 6. As it is shown in Fig. 6, using different strategies let the agents have consistent variances over different memory size. Additionally, it is observed that in the same strategy approach, there are more fluctuations on the resource with capacity of 50% of the population. The reason behind this phenomenon is that agents are using just one strategy, and they try to use the strategy that gives them the most for all three resources.

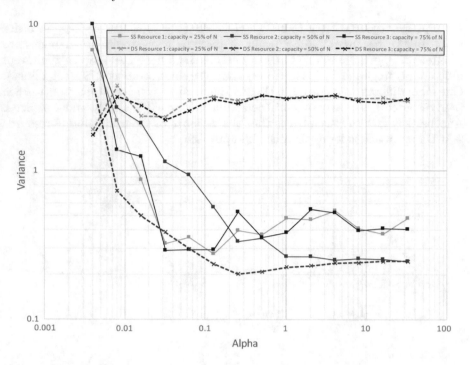

Fig. 6. σ^2 versus α for $N = 501$, $T = 1000$, and $S = 3$ for the same-strategy simulation (SS) and different-strategies simulation (DS). The variance is quite high for 25% of the capacity (125) and 75% of the capacity (375) when using different strategies

In Fig. 7 (left), we explore the resource usage for the same strategy simulation. As a result, the resource with lowest capacity has been never successfully used during the game. That happens because it is mostly overcrowded and difficult to win; so, agents decide to not attend to this resource. However, the resource with highest capacity has been 100% successfully utilized.

The story is different when agents are allowed to use different strategies for different resources. Evidently, it is more possible for agents to obtain all resources successfully. Since the different strategies used for resources are independent, agents have the opportunity to use their best strategy for individual resources. Therefore, resource 1 can be periodically under-capacity, which was never the case when using the same strategy. As shown in Table 1 the number of iterations in which agents successfully utilize resource 1 increases to about 14%. Figure 7 (right) develops the claim that if agents need all resources in order to accomplish their tasks, then using different strategies is beneficial for them.

Another mentioned criterion is the average resource usage that is stated in Table 2. The results from SS analysis shows that if you want your agents to

Table 1. Percentage of iterations with under-capacity attendance for $N = 501$, $T = 1000$, $S = 3$, and $m = 4$

Same strategy			Different strategies		
Resource #			Resource #		
1	2	3	1	2	3
0.000	0.520	1.000	0.140	0.500	0.870

use the resources, and the resources are the same, it is better to use the same strategy. However, the results suggest to consider different strategies in the case that you need to utilize different resources. The average resource usage (ARU) is calculated based on Eq. 4.

$$ARU_i = \frac{\sum_{t=1}^{T} A_i(t) \cdot \delta_i(t)}{T.C_i}, \qquad (4)$$

where i and t represents the resource index and iteration number, $A_i(t)$ is the number of agents using resource i at iteration t (aka attendance rate), C_i is the capacity of resource i, T is the total number of iterations, $\delta_i(t) = 0$ if resource i is overcrowded at iteration t, and $\delta_i(t) = 1$ if resource i is not overcrowded at iteration t.

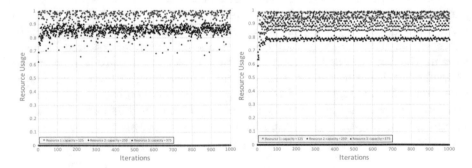

Fig. 7. Resource usage versus iterations for $N = 501$, $T = 1000$, $S = 3$, and $m = 4$. Left: Same strategy Method. Right: Different strategy method. unlike in the same-strategy simulation, resource 1 is successfully utilized

When agents use the same strategy for each resource, they have a higher average winning rate, as seen in Table 3, The pitfall is the fact that some agents cannot win even one single round of the game. In order to understand this phenomenon, we should take a closer look into the strategies for each resource separately. Since most of the times th resource number 1 in over crowded, the one which decide not to use it are the winners. As a result, most of the times the

Table 2. Average resource usage for $N = 501$, $T = 1000$, $S = 3$, and $m = 4$. Resource 1 is never successfully used in the same-strategy simulation. Note that here we do care about the number of agents (attendance rate) in a resource that is not overcrowded

Same strategy			Different strategies		
Resource #			Resource #		
1	2	3	1	2	3
0.000	0.494	0.858	0.137	0.479	0.748

memory is as $m = (0, 0, \ldots, 0)$ for resource 1. So agents need to have the value 0 in the element number 0 in their strategy to be able to win the next round with a high probability.

Whats more, we consider the resource 3 (the resource with capacity of 75% of the population). In this case, it is more likely that the agents that use this resource are the winners. So the outcome will be 1 and it should happens frequently. Consequently, the agent's memory will be $m = (1, 1, \ldots, 1)$ for the third resource. Then, the winners will be the agents which have 1 in the last element of their strategy. All facts considered, the agents that have a strategy $s = (0, \ldots, 1)$ are guarantied to make the correct decision for two resources. They just need to adjust their strategy for the resource 2 (with capacity of 50% of the population). In other words, if agents have strategies that meet these requirements (i.e. having 0 and 1 in their first and last element of strategy respectively), the deciding factor is their attendance in resource 2. The advantage of using the same strategy for each resource happens when agents have strategies meeting this criterion given that they only need to make a decision for resource 2 in order to win. The disadvantage of using the same strategy is the considerable number of agents which never win a round of game (almost one-third of the population based on Table 3). Agents tend to have lower average winning rates when they can use different strategies for each resource. Although it is more difficult for agents to become winner, we found that every agent is able to win at least one round of the game.

Table 3. Average winning rate of agents and the average percentage of agents with no wins in all simulation iterations

	Avg. winning rate	Agents w/o wins
Same strategy	19.8	29.1
Different strategies	15.8	0.0

The resource usage percentage is shown in Table 4. We considered two other types of capacity distribution over the resources. In one of them, all the resources have different capacity but all at the level that is less than the half of the population. In this situation, agents are expected to have more chance to win all resources. In the other distribution, resources have the same capacity (33%) at a level less than half of the population size. In a general case, the capacity of resource c_i can be considered as $c_i = \frac{N}{K}$, where N is the population size and K is the number of resources. We also cover a scenario with a different winning rule in which agents have only to win two resources out of three available ones. In this scenario, we consider the same distributions we use in the last simulation.

Table 4. Resource usage for different capacity distribution. Terms "DS" and "SS" are refer to different strategy and same strategy respectively

Need to win 3 out of 3 resources								
	20	30	40	25	50	75	33	33	33
SS	0	19.9	19.9	0	49.4	85.8	14.3	14.3	14.29
DS	12.5	14.2	35.5	13.7	47.9	74.8	14.3	14.3	14.3

Need to win 2 out of 3 resources								
	20	30	40	25	50	75	33	33	33
SS	5.8	19.9	37.6	11.79	53.0	88.3	14.3	14.3	14.3
DS	12.5	14.2	35.5	14.2	50.2	87.5	12.5	14.2	35.5

First we discuss other types of the distribution in the scenario of "win 3 out of 3" situation. As it can be seen in Table 4, using different strategies (DS) can improve the resource usage significantly where resources have different capacity distributions and the capacities are less than half of the population. However, when the capacity is uniformly distributed, there is no significant difference between SS and DS. Since the same strategy approach requires less computational power, it is better to follow it. This situation can be modeled as assigning similar resources to the processes in a computer system.

In the "winning two out of three resources" scenario, we expect that agents will be biased to the resources with higher capacity. The results in Table 4 show that using SS works better in this scenario. Nonetheless, using the DS approach helps the system to have a better load balance over the resources. However, we will prefer to use the SS approach in the systems that the capacity distribution is $c = (\frac{N}{4}, \frac{N}{2}, \frac{3N}{4})$, because it is as effective as the DS approach.

The lower part of Table 4 refers to the scenario in which agents should win two out of three resources. Since there is a higher chance for them to win or utilize resources with higher capacity, we anticipate that the agents tend to participate in those resources. Furthermore, we expect that simulations with DS approach may show much better results, because agents have more degree of freedom to choose a suitable strategy. Surprisingly, the results show that in this experiment, using the same strategy (SS) performs better than DS approach. Particularly, using SS seems to be more efficient where the capacities are less, equal and more than half of the population size. Notwithstanding, it is better to use different strategy (DS) when a balanced load on the resources is more important than the computational complexity of DS approach. Similar to the previous scenario, there is no significant difference between SS and DS when the capacity is uniformly distributed.

5.2   Multiple Weighted Resources

In this section we investigate the linear and exponential methods of weighting resources. In the weighted model, either the high capacity resource has a greater weight or the low capacity resource has the greater weight. We refer to the former as heavier high capacity resource and the latter as heavier low capacity. Note that resource weighting simulations were initially performed with three resources, two resources with a capacity of 250 and one resource with a capacity of 166 1/2 and 1/3 of the agent population of 501, respectively). Weights were chosen to sum to 1, with the two identical resources having identical weight values. For clarity of the description, only two resources are presented in the figures.

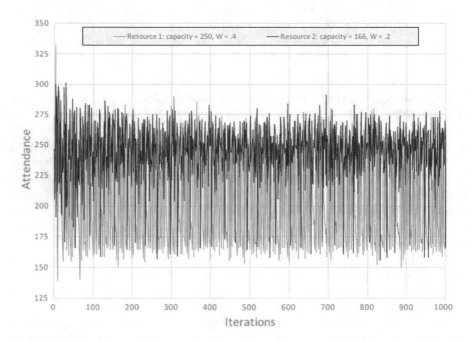

Fig. 8. Attendance rate for resources with different capacities with heavier high capacity resource using a linear resource weighting method, N = 501, T = 1000, S = 6, and m = 3

Figure 8 shows a situation in which the higher capacity resource is weighted higher (0.4) than the lower capacity resource (0.2). As expected, the attendance rates for both resources stay close to the capacity of the more heavily weighted

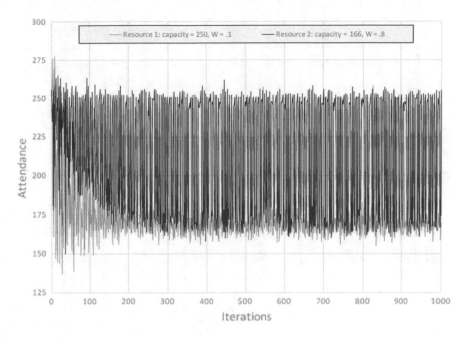

Fig. 9. Attendance rate for differing capacity resources with heavier low capacity resource using a linear resource weighting method, N = 501, T = 1000, S = 6, and m = 3. Note that attendances are influenced by the heavily weighted low capacity resource

higher capacity resource 1. This causes significant and consistent over-attendance for resource 2. In contrast, Fig. 9 shows a situation in which the lower capacity resource is weighted higher (0.8) than the higher capacity resource (0.1). The attendance rate of resource 2 quickly drops to its capacity of 166 over the course of the simulation, while attendance for Resource 1, with a capacity of 250, swings back and forth from around 250 to around 166. The attendance rate is more predictable for both resources and more importantly, both resources enjoy efficient utilization, as attendance more frequently is at or below capacity.

Figure 10 shows a situation in which the low capacity resource is weighted at 2 and the higher capacity resource is weighted at 1.5. The plot looks remarkably similar to the baseline plot, however the point at which the attendance spikes meet rise from approximately 220–230. This is unexpected, as higher weights for the low capacity resource should have resulted in a stronger pull towards the low capacity. It is unclear why this occurred. However, when the low capacity resource is weighted at 1.01 and the higher capacity resource is weighted at 1.5, there is a dramatic change in plot appearance, as seen in Fig. 11. The attendance of both plots approach equilibrium about highly weighted resource 1's capacity of 250. Variance around the attendance mean minimizes very rapidly. This suggests that the effect of a lower capacity resource on the attendance rate of a higher capacity resource can be minimized if the weight of the low capacity resource is reduced.

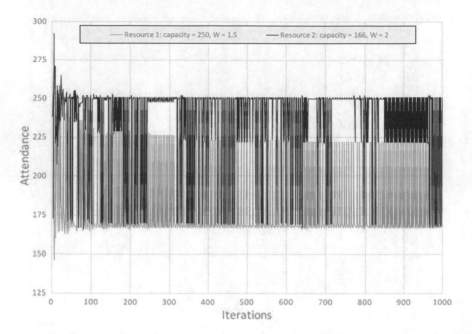

Fig. 10. Attendance rate for differing capacity resource resources with heavier low capacity resource using resource weighting method 2, N = 501, T = 1000, S = 6, and m = 3

Based on these experiments, the weighting methods appear to have different strengths. The linear weighting method can increase resource utilization for all resources in the system by weighting low capacity resources more heavily. The exponential weighting method can dramatically decrease the attendance variance of a high capacity resource by weighting low capacity resource less heavily.

Another interesting result for the weighted multi-resource minority game simulation has been observed when we try two other weight distributions. In this simulation, we investigate a different game with a uniform capacity distribution. In one system, the capacity of each resource is $\frac{1}{3}$ of the population, i.e. $c_i = \frac{N}{3}$; in the other one, the capacity is half of the population, i.e. $c_i = \frac{N}{2}$. Surprisingly, the resource usage is the same for SS and DS approaches in the case that resource capacity is $\frac{N}{2}$, and neither using different weights for resources nor different weighting methods can make difference. Apparently, it is more important for agents to be a winner than taking the risk to win a resource with higher weight.

In the simulation with the capacity $c_i = \frac{N}{3}$, the resource usage is almost the same for the resources with different weights in each simulation. Table 5 shows the results for this experiment. In the simulation with 3 strategies per agent and memory size of 4, the resource usage is 0.14 for all conditions. While changing the number of strategy per agent does not make any significant difference, increasing the memory size causes dropping down in resource usage except the situation in

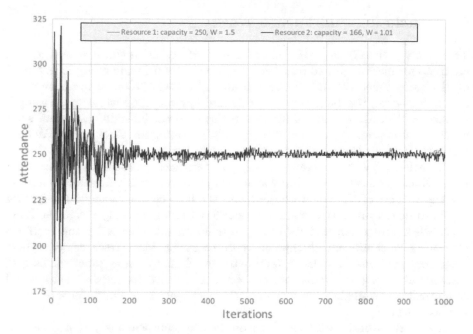

Fig. 11. Attendance rate for differing capacity resource resources with heavier high capacity resource using resource weighting method 2, N = 501, T = 1000, S = 6, and m = 3. Shows attendance of low capacity resource pulled to capacity of resource with heavier weight

which we use weighting method 2 and different strategy approach. One reason behind this phenomenon is the fact that agents avoid utilizing resources because they can easily be overcrowded. However, increasing the number of strategies per agent can slightly improve the resource usage.

Table 5. Resource usage for weighting methods 1 and 2 with SS and DS approaches. Each number in the table represents the resource usage for all three resources, because they have the same resource usage. Capacity of each resource is $\frac{N}{3}$, $N = 501, T = 1000$, and $k = 3$; the weight distribution is $w = (0.4, 0.35, 0.25)$ for weighting method 1, and $w = (2, 1.5, 1.01)$ for weighting method 2

Strategy per agent	3	5	7	3	3
Memory size	4	4	4	6	8
Weighting method 1 + SS	0.14	0.27	0.28	0.11	0.08
Weighting method 1 + DS	0.14	0.26	0.29	0.11	0.09
Weighting method 2 + SS	0.14	0.26	0.28	0.11	0.09
Weighting method 2 + DS	0.14	0.26	0.28	0.16	0.11

6   Conclusion and Future Work

This study proposed an extension of the minority game in which agents need to compete for more than two resources and are able of utilizing multiple resources at the same time. We introduced the idea of a multi-resource minority game and investigated the behavior of agents and systems for several cases using 3 resources. We clarified that multi-resource as we describe is different than what we call multi-option (but other authors insist in calling multi-resource). We have made the case for the naming as we use in this text.

Additionally, we investigate the weighted resources approach and we observed that a linear method is potentially able to improve resource utilization by adjusting the weights of the resources. Further, the impact of exponential weighting seems to be crucial to the attendance variance of high capacity resources. However, when the resource distribution is uniform, different weighting methods cannot make significant changes in resource usage. More research with these weighting methods has the potential to unveil many more practical uses of resource weighting with multiple resources. For instance, having different method of rewarding in a single close system can change agent behavior in favor of a particular resource.

A further potential work can focus on the mathematical analysis of the evolutionary multi-resource minority game. In this case, instead of using a traditional binary/bipolar representation of strategies, agents can use a probability to chose the resource. Additionally, analyzing the significance of different memory size with respect to the umber of strategies needs more explorations. Besides, analyzing the influence of connected agents over different networks may be useful for the situation in which the agents share information.

Acknowledgments. The authors acknowledge support from National Science Foundation (NSF) grant No. 1263011. Any opinions, findings, and conclusions or recommendations expressed in this material are those of the authors and do not necessarily reflect the views of the NSF.

References

1. Arthur, W.B.: Inductive reasoning and bounded rationality. Am. Econ. Rev. **84**(2), 406–411 (1994)
2. Barabási, A.-L., Albert, R.: Emergence of scaling in random networks. Science **286**(5439), 509–512 (1999)
3. Börner, K., Sanyal, S., Vespignani, A.: Network science. Ann. Rev. Inf. Sci. Technol. **41**(1), 537–607 (2007)
4. Caridi, I., Ceva, H.: The underlying complex network of the minority game (2008). arXiv:0802.0672
5. Catteeuw, D., Manderick, B.: Learning in minority games with multiple resources. **5778**, 326–333 (2011)
6. Challet, D., Zhang, Y.-C.: Emergence of cooperation and organization in an evolutionary game. Physica A: Stat. Mech. Appl. **246**(3), 407–418 (1997)

7. Challet, D., Zhang, Y.-C.: On the minority game: analytical and numerical studies. Physica A: Stat. Mech. Appl. **256**(3), 514–532 (1998)

8. Galstyan, A., Kolar, S., Lerman, K.: Resource allocation games with changing resource capacities. In: Proceedings of the Second International Joint Conference on Autonomous Agents and Multiagent Systems, pp. 145–152 (2003)

9. Galstyan, A., Krishnamachari, B., Lerman, K.: Resource allocation and emergent coordination in wireless sensor networks (2004)

10. Huang, Z.-G., Zhang, J.-Q., Dong, J.-Q., Huang, L., Lai, Y.-C.: Emergence of grouping in multi-resource minority game dynamics. Sci. Rep. **2** (2012)

11. Johnson, N.F., Hui, P.M., Jonson, R., Lo, T.S.: Self-organized segregation within an evolving population. Phys. Rev. Lett. **82**(16), 3360 (1999)

12. Kets, W., Voorneveld, M.: Congestion, equilibrium and learning: the minority game. *Available at SSRN 1012271* (2007)

13. Lo, T.S., Hui, P.M., Johnson, N.F.: Theory of the evolutionary minority game. Phys. Rev. E **62**(3), 4393 (2000)

14. Mähönen, P., Petrova, M.: Minority game for cognitive radios: cooperating without cooperation. Phys. Commun. **1**(2), 94–102 (2008)

15. Moro, E.: The minority game: an introductory guide (2004). arXiv:cond-mat/0402651

16. Remondino, M., Cappellini, A.: Minority game with communication: an agent based model. *Simulation in Industry* (2004)

17. Shafique, M., Bauer, L., Ahmed, W., Henkel, J.: Minority-game-based resource allocation for run-time reconfigurable multi-core processors, pp. 1–6 (2011)

18. Shang, L., Wang, X.F.: Evolutionary minority game on complex networks. Physica A: Stat. Mech. Appl. **377**(2), 616–624 (2007)

19. Tanenbaum, A.S., Van Steen, M.: *Distributed Systems*. Prentice-Hall (2007)

20. Tanenbaum, A.S., Woodhull, A.S., Tanenbaum, A.S., Tanenbaum, A.S.: Operating Systems: Design and Implementation, vol. 2. Prentice-Hall Englewood Cliffs, NJ (1987)

21. Van der Hoek, W., Wooldridge, M.: Multi-agent systems. Found. Artif. Intell. **3**, 887–928 (2008)

22. Woodhull, A.S., Tanenbaum, A.S.: *Operating Systems Design and Implementation*. Prentice-Hall (1997)

23. Zhang, C., Wu, W., Huang, H., Yu, H.: Fair energy resource allocation by minority game algorithm for smart buildings, pp. 63–68 (2012)

M2M Routing Protocol for Energy Efficient and Delay Constrained in IoT Based on an Adaptive Sleep Mode

Wasan Twayej[✉] and H. S. Al-Raweshidy

Department of Electronic and Computer Engineering, College of Engineering, Design and Physical Sciences, Brunel University London, London, UK
{wasan.twayej, hamd.al-rewsdy}@brunel.ac.uk

Abstract. In recent years, the number of machine-to-machine (M2M) networks that do not require direct human intervention has been increasing at a rapid pace. However, the need for a wireless platform to control and monitor these M2M networks, one with both a vast coverage area and a low network deployment cost, continues to be unmet. Wireless Sensor Networks (WSNs) with energy efficiency routing protocols in M2M environments are emerging to meet the challenges of such communication through network convergence. M2M communication is considered as the core of the Internet of Things (IoT). IoT refers to a network of billions of objects that can send and receive data. Energy efficiency, delay are a critical issue in M2M and there is a shortfall in IP addresses in IoT. In this chapter, an energy efficient routing protocol for Wireless Sensor Networks (WSN) is presented, which provides a platform to control and M2M networks. Inefficient energy consumption caused by nodes being active all the time is tackled using an adaptive sleep mode solution to maintain high levels of Network Performance (N.P). Firstly, a Multilevel Clustering Multiple Sink (MLCMS) with IPv6 protocol over Low Wireless Personal Area Networks (6LoWPAN) is promoted using a sophisticated mathematical equation for electing cluster heads (CH) for each level, so as to prolong network lifetime. Secondly, enhanced N.P that prolongs the life time of the system and maximises the reduction of delay is achieved through an adaptive sleep mode scheme. The sensor field is divided into quarters with different levels of cluster heads (CHs) and two optimal location sinks. The performance of the MLCMS protocol is evaluated and compared with the multi-hop low-energy adaptive clustering hierarchy (M-LEACH) protocol. MLCMS performs 62% better than M-LEACH and 147% more effectively regarding energy efficiency. Next, 6LoWPAN for the proposed model is constructed, and its impact on the performance of MLCMS by Network Simulator (NS3) simulation is evaluated. This increases the packets received by the system by 7% more than using MLCMS without 6LoWPAN and it improves the flexibility of the proposed model. Subsequently, an adaptive sleep mode scheme, based on CH's residual energy for the active period time, is introduced for MLCMS and a comparative analysis establishes that it extends the lifetime of the system twice as much as the evaluated MLCMS without the adaptive sleep mode algorithm. Furthermore, with the sleep mode algorithm, this reduces the delay by a half and increases the delivery by 10%.

© Springer International Publishing AG 2018
Y. Bi et al. (eds.), *Intelligent Systems and Applications*,
Studies in Computational Intelligence 751,
https://doi.org/10.1007/978-3-319-69266-1_15

Keywords: IoT · WSN · M2M · MLCMS · LEACH
M-LEACH 6LoWPAN

1 Introduction

M2M is the key to enabling the IoT to achieve its core vision, i.e. enhancing the communication of real things in the world, transmitting information and executing smart tasks [1, 2]. Many solutions in the M2M domain have been provided, and it is expected that billions of devices will be connected using M2M in the future [3, 4]. Intensive research has focused in recent years on M2M networks that allow networked nodes (sensors) to exchange information without human intervention for monitoring and surveillance. Specifically, they have been used in the development of wireless sensor networks (WSNs), helped by the IPv6 technique, to improve the quality of data transmission by reducing packet losses and providing more flexibility. Moreover, most appliances have been equipped with capabilities for sensing people's daily needs [5, 6]. As a result, M2M wireless communication is the key to implementing sophisticated connections between machines.

Energy efficiency is critical when designing a wireless M2M sensor network; the sensor nodes need to be self-governing with small batteries that last for several months or even years, since replacing batteries for a large number of devices is impractical in distant or unfavourable environments [7]. Furthermore, sink node isolation has been noted as a direct result of energy consumption imbalances in WSNs, referred to as a hotspot or energy hole problem, and energy is the most crucial resource in WSNs. Consequently, minimising energy consumption is a key focus for researchers, and this has been extensively investigated.

One method for doing so is energy efficient transmission protocols for wireless M2M sensor networks, which are categorised into routing and clustering types. Clustering refers to organising the nodes into groups in accordance with the network's needs. Each group is assigned a leader, namely the cluster head (CH), and the others are simply normal nodes. The sharing communication channel of an M2M network in a multilevel arrangement is efficient for energy consumption. This clustering protocol form lessens energy consumption by aggregating multiple sensed data, which are then transmitted to the sink node [8, 9]. Furthermore, multiple sinks allow for more uniform energy consumption traffic configurations across the network.

Other issues, apart from energy efficiency, are the demand for more IPs and a lack of flexibility in system management, both of which can be addressed by using 6LoWPAN. This protocol, 6LoWPAN, is the use of the IPv6 protocol over low wireless personal area networks, and its application has been rapidly increasing. In particular, it has been playing an important role in extending throughput, which has been somewhat limited for IEEE 802.15.4. The main reason for designing 6LoWPAN has been to pave the way for IoT. As a result, IoT can support different applications like home automation, smart lighting and addressing traffic congestion [10]. However, energy efficiency is an important consideration with 6LoWPAN, also it is crucial to make the nodes being active all the time for consuming unnecessary power. This is achieved by keeping some nodes active and the others inactive, based on their role in the network. A sleeping node is defined as one that goes to sleep voluntarily to save power. With this model, an efficient structure to specify the time periods during which the nodes go to sleep and when they become active again is proposed. All the issues related to M2M WSN and try to be tackled in this chapter are summarised in Fig. 1.

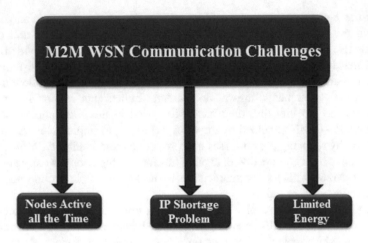

Fig. 1. Problem statement

Efficient ways to conserve power are presented in this chapter in three parts. In the first evaluation section, the performance of MLCMS will be compared with LEACH and M-LEACH protocols. Secondly, this algorithm will be implemented using a 6LoWPAN network interface and then compared to a system without it. These procedures impact positively on the performance of the system by maximising the packet received. Finally, the third part will introduce and evaluate its performance an adaptive sleep mode on MLCMS with the proposed sleep mode approach in MLCMS, architecture. The rest of the chapter is as follows: Sect. 2 discusses related work, whilst Sect. 3 explains the MLCMS protocol in detail. Section 4 includes the simulation results of the proposed protocol and Sect. 5 contains the conclusion.

2   Related Work

2.1   Hierarchical M2M Networks with Clustered Nodes

Over the years, developments in WSN technology have enabled reliable monitoring and analysis of unknown and untested environments. Nevertheless, some fields that offer immense scope still need further research and development. One of them, as mentioned before, is energy conservation. Network energy waste can be minimised by several routing protocols. The first cluster-based energy efficient protocol, in which both the sensor nodes and the sink are static, is LEACH [11]. Other protocols include M-LEACH [11], PEGASIS [12], TEEN [13] and APTEEN [14]. LEACH and M-LEACH will be considered in this chapter, as they form the base hierarchy clustering protocols. LEACH depends on a threshold regarding CH formation for every next round. That is, if an existing CH has spent little energy during its operation and has more than the required threshold remaining, it will stay in this role for the next round. Thus, this shows how energy in routing packets for a new CH and cluster formation can be saved. However, if the converse is the case and the CH has less energy than the required threshold, it will be replaced in accordance with the LEACH algorithm. This is considered the main issue concerning improvements. Multi-hop transmission was subsequently made possible through Multi-hop LEACH or M-LEACH, which is a modification of the original. For this protocol, distant CHs send information to the BS through CHs nearer to it, which avoids high energy expenditure of faraway CHs, if the data has to be directly sent to the BS, as in LEACH. By sharing the load through multi-hops, distant clusters and their CHs from the BS can substantially reduce energy expenditure, thereby extending those nodes' lifetimes [11]. During the development process for MLCMS, it became apparent that this commonly deployed technique of electing the CH can be changed according to the requirements of the system model. From previous studies, it can be concluded that some models are more stable than others, but not energy efficient, while the others are more energy efficient, but not stable. The goal in this chapter is to improve routing quality in M2M MLCMS by reducing the overall energy consumption, increasing the stability period of the network and prolonging the network lifetime.

2.2   IETF Standards

In 2003, the standardisation bodies of the IEEE and the Internet Engineering Task Force (IETF) began constructing a framework for the communication protocols of IoT systems [15, 16]. Recently, the 6LoWPAN working group (WG) invented an adaptation layer, which delivers compression of the header of IPv6 and fragmentation for the datagram. In addition, for low power packet transmission over 802.15.4, the WG has focused on providing better IPv6 connectivity, rather than constrained node networks [10]. Most hierarchical networks use a routing protocol based on IEEE 802.15.4, but this lacks the extension of IPs. Low power and short range wireless communication technologies have an important role in the efficiency of network communication, and these technologies are commonly adopted in WSNs and M2M systems. Given the lack of IP addresses and the need to overcome this, it has been demonstrated that 6LoWPAN can lessen power consumption and compress the size of the packets [17].

2.3   Sleeping Mode

In spite of the advantages of 6LoWPAN, the issue of all the nodes being active all the time still needs to be addressed. This can be solved by developing a model for scheduling the times for active and sleeping nodes [18]. There are many sleep mode models, but the adapted sleep mode in the proposed system supports a reduction in energy consumption. The proposed adaptive sleep approach actually doubles the efficiency of energy consumption.

3 Proposed Work

3.1 System Architecture and Energy Model for M2M WSN

We modelled the network by a graph $G = (V, E)$ where V is the set of sensor nodes, and $E \subseteq V^2$ is the set of link between two nodes when these nodes are within same transmission range represents the wireless connections between nodes. Here, the hierarchical network is considered in which a sensor field is logically divided into quarters, with levels of CHs. Each cluster has a sink and a cluster head (CH), and the remaining components are the ordinary sensor nodes. It is assumed that (n), a sensor node in an N*N sensing field, is used with a homogeneous distribution, as shown in Fig. 2.

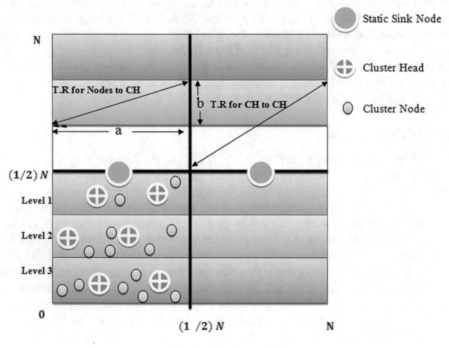

Fig. 2. System model

The proposed network model contains the following elements.
   The nodes: Distributed randomly.

- Two stationary sinks: One sink is located in the middle of the left side ($[1/4]N, [1/2]N$) and the other one is in the middle of the right side ($[3/4]N, [1/2]N$).
- Four quarters of the sensing field: Each quarter is divided into three levels horizontally. At each level, there is a maximum transmission range (TR), which is the diagonal length of the level. Nodes that are beyond this are unable to connect with

the CHs, whereas those within this range use this to transmit their parameters so that they might themselves become the new CH and once chosen, the others will send data to it.

The TRs are formulated below depending on the levels, as shown in Fig. 2. The TR of CH-to-Node will be:

$$H_{CH-Node} = \sqrt{(a)^2 + (b)^2} \tag{1}$$

The TR of CH-to-CH will be:

$$H_{CH-CH} = \sqrt{(a)^2 + (2 * b)^2} \tag{2}$$

The TR of CH- or Nodes-to-Sink will be:

$$H_{CH-Sink} = \sqrt{(0.5 * a)^2 + (3 * b)^2} \tag{3}$$

The operation of the MLCMS routing protocol is divided into two parts in each round, which are the set-up phase and the steady-state phase. This approach involves building a hierarchical structure with a sophisticated CH distribution. The selection of the CHs in each quarter is considered as the set-up phase, which involves the central node or that nearest the centre for each of the three levels in each quarter, initially sending a Hello message to all the nodes at the same level. Only those within transmission range can receive the message and hence, send information regarding residual energy, which is the remaining energy divided by the initial energy and their position (x, y). In this phase, the node with the highest weight is elected as the CH. The weight of the node is calculated by Eq. (4), which takes the residual energy, number of neighbours and the distance to the sink into account. Thus, we can generate more balanced clusters in terms of energy and position in that the two nodes with the maximum W(n) are then selected as the CHs of this level, as shown in Fig. 3. These CHs are uniformly distributed to prevent them from being too close to each other. That is, if the two elected CHs in the level are close to each other then only one of them will be chosen to be CH, as shown in Fig. 4. Then, the elected CHs broadcast their election to the rest of the nodes by sending an INVITE message. The nodes within their transmission range will receive the INVITE message and will measure the signal strength to join the CH that has the largest. Furthermore, the CH's energy is checked regularly and if it is less than 25% of its residual energy, it should be changed. The new CHs for each level will be selected by the old CH sending a request to all related nodes' about their residual energy, number of neighbours and their (x, y) position, so as to measure their distance from their sink. The one that has maximum, W(n) as in Algorithm 1, will be awarded the role of new CH for that level, depending on the equation below:

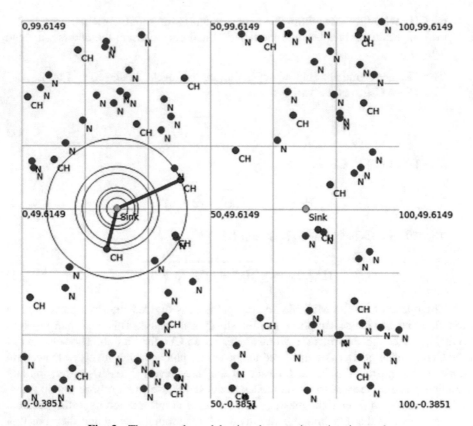

Fig. 3. The network model using the netanim animation tool

$$W(n) = \varpi_1 E(n) + \varpi_2 N(n) - \varpi_3 D(n) \qquad (4)$$

$$0 \le \varpi_1, \varpi_2, \varpi_3 \le 1$$
$$\varpi_1 + \varpi_2 + \varpi_3 = 1$$

where, $\varpi_1, \varpi_2, \varpi_3$ are the effect factors, $E(n)$ is the residual energy of node n, $N(n)$ is the number of neighbours of node n and the $D(n)$ is the distance between node n and the sink.

$$E(n) = \frac{R^n(t)}{E_{in}} \qquad (5)$$

$R^n(t)$ is the remaining energy of node n at time (t) and E_{in} is the initial energy of node n.

$$D(n) = \sqrt{(X_2 - X_1)^2 + (Y_2 - Y_1)^2} \qquad (6)$$

The priority levels or weights selected for (4) are based on the Rank Order Centroid (ROC) method [19]. The ROC is a simple method of assigning weights to a number of functions, ranked according to their priority or importance. The priorities of each

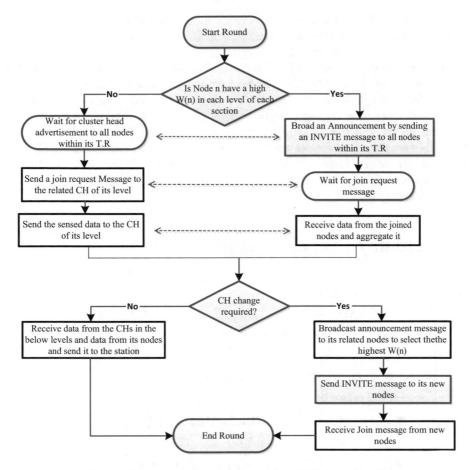

Fig. 4. Round procedure during selection CHS flow diagram

function are taken as an input and converted into the weight for it. The conversion is performed using the following formula:

$$\varpi_i = \left(\frac{1}{F}\right) \sum_{n=i}^{F} \frac{1}{n} \tag{7}$$

where, F is the number of functions $(E(n), N(n)$ and $D(n))$ and ϖ_i is the weight of the ith function. E(n), ranked first, is weighted as $(1/1 + 1/2 + 1/3)/3 = 0.6$, N(n), ranked second, is weighted as $(1/2 + 1/3)/3 = 0.3$ and D(n), ranked third, is weighted as $(1/3)/3 = 0.1$. Figure 4 shows a general flow diagram of the CH selection procedure. The second phase is the steady-state phase, which is responsible for forwarding the information to the sink. It is the crucial factor of the model, which is the CH at level three in each quarter and cannot connect directly to a sink. It should be attached to that at level two, but if there is no CH at this level, it can connect directly to a sink. Hence, this is a proposal for multi-level clustering (CH to CH and then to a sink), which involves some CHs being far from a sink, but still being able to send their data to one, thereby linking CHs on multiple levels without losing excessive amounts of energy. In addition, there is a particular technique for sending data from the CHs to the sinks, which depends on the priority of the farthest CH.

To preserve energy, a transmission mechanism for communicating with a CH is implemented that brings the distance between it and the nodes into consideration. According to this protocol, each cluster has only one CH and sends data to it, which then sends them in an aggregated form to its relevant sink. The clustering method proposed in the MLCMS protocol has many advantages. An important benefit is that the CHs are located in a more uniform way and this is, in fact, more suitable for large-scale networks. Moreover, it can prolong the network lifetime. Also, there is the flexibility of scalability in extending the size of the network by increasing the number of levels in the same way.

Algorithm 1 CHs election phase

for each Level L **do**
  for each node n **do**
      send 'hello' message
      find the degree
      $E = \{(n, v) \subset V/D(n, v) \leq R\}$;
      $Deg(n) = |E|$;
      calculate the residual energy $E(n)$;
      $E(n) = E_R(n)/E_{in}(n)$;
      calculate the distance to the sink;
      $D(n) = \sqrt{(X_2 - X_1)^2 + (Y_2 - Y_1)^2}$;
      calculate the weight
      $W(n) = \varpi_1 E_i(n) + \varpi_2 Deg(n) - \varpi_3 D(n)$
  end
  if W (n) = biggest W (L) **then**
      state(n) = CH
      send 'clusterhead elected'
  else
      state(n) = ME
      send 'cluster head accepted'
  end
end

3.2   Energy Model

An energy model similar to the one proposed in [20, 21] has been used, but with many changes according to the MLCMS algorithm, which is based on the free space (power loss) or multi-path fading (power loss) channels. A free space model is calculated based on the distance between a transmitter and receiver. In the given model, E_{TX} of energy is consumed by each sensor node to transmit an L-bits packet over a distance d. In this model, the energy consumed per bit is E_{elec} and is used to run the transmitter or receiver circuit, where ε_{fs} and ε_{mp} represent the transmitter and amplifier's efficiency of channel conditions [9]. The proposed MLCMS model relies on two levels of power to amplify signals, depending on the nature of the transmission, which is dependent on the transmission range of the proposed model, as in Eqs. (1), (2) and (3). The cluster-based network can take three forms: inter-cluster transmission (CH-to-CH); intra-cluster transmission (cluster member-to-CH); and CH-to-sink transmission, also known as the base station. Moreover, in the proposed model, multi–power levels also diminish the packet drop ratio, collisions and/or interference with other signals. When a node is a CH, the routing protocol informs it to use high power amplification, and during the next round, when it reverts to being a mere cluster member, it is switched to low-level power amplification.

The energy dissipated for transmission between nodes is formulated by:

$$E_{TX}(L, d) = \begin{cases} L * E_{elec} + L * \varepsilon_{fs} * d^2, d \leq d_o \\ L * E_{elec} + L * \varepsilon_{mp} * d^4, d > d_o \end{cases} \tag{8}$$

E_{Tx} is the transmission energy.where,

$$d_o = \sqrt{\frac{\varepsilon_{fs}}{\varepsilon_{mp}}} \tag{9}$$

The energy that is consumed by the receiving packet is:

$$E_{RX}(L) = L * E_{elec} \tag{10}$$

The value of ε_{fs} and ε_{mp} in the equation, according to the transmission forms; if it is between (CH-to-sink), these values will be the same as shown in Table 1.

When transmitting between (cluster member-to-CH), the ε_{fs1} and ε_{mp1} will be used as below:

where,

$$\varepsilon_{fs1} = \frac{\varepsilon_{fs}}{K} \tag{11}$$

$$\varepsilon_{mp1} = \frac{\varepsilon_{mp}}{K} \tag{12}$$

Table 1. Parameters definition

Parameters	Definition
N	Number of nodes
D(n)	Distance between n and its sink
Deg(n)	Number of neighbours of one hop
W(n)	Wight of node n
State of n	State of n: "CH" or "ME"
CH	Cluster head
ME	Member as a normal node
$E(n)$	Residual energy
$E_{in}(n)$	Initial energy
$E_{cu}(n)$	Current energy

where,

$$K = \frac{H_{CH-Sink}}{H_{CH-Node}} \tag{13}$$

While transmitting between (CH-to-CH), the ε_{fs2} and ε_{mp2} will be used as below:

$$\varepsilon_{fs2} = \frac{\varepsilon_{fs}}{C} \tag{14}$$

$$\varepsilon_{mp2} = \frac{\varepsilon_{mp}}{C} \tag{15}$$

where

$$C = \frac{H_{CH-Sink}}{H_{CH-CH}} \tag{16}$$

In the proposed work, when the transmission between nodes is initiated, then the energy equation will be used depending on the conditions above.

3.3   Construction of 6LoWPAN Over MLCMS

The aim of using 6LoWPAN is to overcome the gap in relation to shortages of IPs. Moreover, the dynamic short addresses capabilities, by making the assumption that the routing protocol occurs in the adaptation layer. Owing to the limitation of the payload size of the link layer in 6LoWPAN networks, the adaptation layer in the 6LoWPAN standard covers the compression of the packet header, fragmentation and reassembly of the datagram. For instance, the IEEE 802.15.4 frame size may go beyond the maximum transmission unit (MTU) size of 127 bytes for big application data, whereas that for

6LoWPAN is 1,280 bytes, in which case fragmentation is needed. 6LoWPAN networks are connected to the Internet through the 6BR (6LoWPAN Border Router) [16]. Using 6LoWPAN allows for more opportunities for M2M services with better packet transmission quality and reduced loss. The main advantage of 6LoWPAN is the adaptation layer. This involves using stacking of headers as well as adding the definition of the encapsulation header stack in front of the datagram of IPv6.

3.4 Sleep Mode Coordination

In this part, the aim is to extend the lifetime of an M2M WSN by improving its energy efficiency. To prolong the network lifetime, some sleep scheduling methods are always employed in WSNs. There are many ways to meet the requirement of an effective sleep mode in terms of how and when the nodes are scheduled to be active or asleep. The proposed hierarchy clustering topology architecture with 6LoWPAN includes the addition of an adaptive sleep mode to extend the lifetime of the system. The proposed sleeping mode technique relates to the structure of the model. During the entire functional cycle, the MLCMS model with 6LoWPAN divides the time period for (T_a), which is the active time period according to the multilevel clustering. This makes the CHs at each level active for specific periods of time (T_a), and these overlap so that one or two are awake at the same. This adaptive sleep method extends the lifetime of the system, which, as stated previously, is the main goal of the proposed model.

Energy conservation is generally addressed in two ways:

(1) Alternating the nodes between sleep and active modes in an optimal manner;
(2) Efficient trade-off between energy consumption and connectivity through effective control of transmission power;

An effective process of adaptive scheduling to maximise energy saving is effective use of the sleep and active modes amongst the sensor nodes. Scheduling of these two modes is initiated immediately after a cluster has been has been created one or more CHs, i.e. those with the greatest residual energy, are kept active and the rest are put in the sleep mode so as to conserve energy. Any active CH needs to identify the nodes with the highest residual energy in its cluster for the sensing task. Once chosen, a WORK message is sent by the CH to those nodes to be active. Also, the CH sends a SLEEP message to all the other nodes, telling them to be inactive in the next period, which is scheduled in adaptive sleep mode scheme.

The protocol is for data collection applications (e.g. for monitoring purposes, where the sensor nodes need to report to a sink node periodically. Compared to other protocols, it has two distinct advantages. First, because it is not linked to specific medium access control (MAC) protocol, it is applicable to a range of sensor platforms. Second, it can quickly adjust the sleep/active pattern to suit variations in network conditions, e.g. changes in traffic demand, network congestion etc.), thus leading to improved deployment of the energy resources. Thorough simulation analysis has been undertaken for a range of contexts and operating conditions to ascertain the level of performance. The outcomes show that owing to the injected flexibility, adaptive sleep mode performs better than the original scheme of MLCMS without it regarding energy efficiency, message latency, prolonging system life span and in terms of the delivery ratio. Consequently, the sleep adapted protocol emerges as offering a substantial improvement monitoring purposes, thereby making the long-term deployment of WSNs more likely.

Stepwise Algorithm: These steps are followed if there are two CHs in the level, but if there is just one, then this CH will be active all the specified time for this level.

Step 1: Initialise the number of parameters, such as time (T_a), which is the time of activation for each level, the number of CHs (n) and number of levels (m). $E(n)$ is the residual energy, $R^n(t)$ is the remaining energy of node n at time (t) and E_{in} is the initial energy of node n.

Step 2: Set variables to compute CH residual energy separately, as in Eq. (5), for computing α, using the formula:

$$\alpha = (E(n)) / \sum_{n=1}^{CH} E(n) \qquad (17)$$

Step 3: Using the resultant α, cluster activation time (i.e. CH active time for a given time interval) is computed using the formula:

$$CH - Time = \alpha * T\_a * 100; \quad //CH\,Activation\,time\,in\,\% \qquad (18)$$

4   Simulation Studies

The NS3 is a new simulator for modeling different WSNs. It has been used to simulate the proposed M2M MLCMS algorithm by specifying its characteristics, as shown in Table 2. In this section, the proposed algorithm is compared with LEACH and M-LEACH, commonly deployed algorithms, in terms of their performance. Our proposal is found to outperform both of the latter two algorithms in relation to energy conservation and hence, prolonging network life time. In addition, simulations are undertaken for MLCMS over 6LoWPAN with and without the adaptive sleep mode algorithm, thereby also addressing the efficiency of the method proposed. The parameters network lifetime, residual energy at the end of each turn and total packet received by sink are compared for all three algorithms. It should be noted that the node arrangement for the 6LoWPAN model is randomly chosen, so too their positioning in the square area, which implies that the results are independent of node location. All the parameters pertaining to the LEACH and M-LEACH algorithms are the same as those employed in Table 1, which also describes the parameters of the scenarios.

Table 2. Simulation parameters and values

Parameter name	Value
Number of the sensor nodes (n)	100
Base station locations	(25, 50), (75, 50)
Length of the packet (L)	4,000 bits
Initial energy of the sensor nodes (E_{init})	0.5 J
Energy consumption on circuit (E_{elect})	50 nJ/bit
Channel parameter in free space model (E_{fs})	10 pJ/bit/ m^2
Channel parameter in two-ray model (E_{mp})	0.0013 pJ/bit/ m^4
Network size (N * N)	100 m × 100 m

4.1 Performance Evaluation of MLCMS Based on the 6LoWPAN Model

The number of "alive" nodes over the simulation time is illustrated in Fig. 5. The life time of the system refers to the period between its first operation and the demise of the last node. It can be seen that MLCMs lasted for 3716 rounds, whilst LEACH and M-LEACH achieved 1500 and 1900 rounds, respectively. This is observed in the shape of the graphs, whereby a steep decrease in LEACH and M-LEACH shows where energy equalisation is attained. When the last node is reached, all nodes are depleted of their energy resources, leading to this quick decline. On the other hand, MLCMS shows a gradual decrease in the number of alive nodes, which is due to the fact that energy equalisation is not coordinated and some nodes run out of battery energy much more quickly than others. We can also observe an early jump for LEACH and M-LEACH in Fig. 5, which is seen at the instance when many sensors have depleted their energy resources.

In addition, the performance of the proposed model improves the lifetime of the system over that of the LEACH and M-LEACH architecture, as given by:

$$W = \frac{PA - PB}{PB} \times 100\% \tag{19}$$

PA is the performance of MLCMS, and PB is the performance of the M-LEACH architecture. From Eq. (19), the improvement of the MLCMS model over that of the LEACH and M-LEACH architecture is 147 and 62%, respectively. Clearly, the number of active nodes in MLCMS can extend the lifetime of the system more than the others algorithms. This MLCMS algorithm is superior to other clustering routing protocols, CH allocates appropriate places for the nodes belonging to its CH. A noticeable difference between MLCMS and both LEACH and M-LEACH is that when the latter two lose all nodes, MLCMS still has some that are still functioning, but with low residual energy. To address the question as to which protocol is best for collecting data, the total number of messages each delivers throughout the lifetime of the system is calculated and MLCMS sends the most for various node density arrangements.

The aim of this performance evaluation for the packet received is to test the basic operations and features in the MLCMS model when using 6LoWPAN. This is achieved by installing the 6LoWPAN structure model in each sensor node of an M2M system

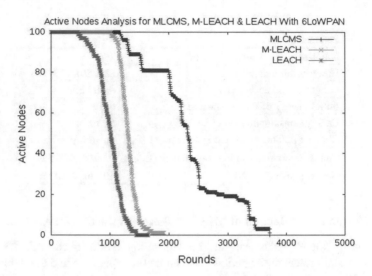

Fig. 5. Number of active nodes

and testing how this enhances performance. Moreover, NS-3 packet capturing (pcap) is used to show whether the bit streams of the packet headers are compliant with their corresponding standards and the module follows RFC 4944. Figure 6 below shows that the number of packets received by the sink is improved 7% by using 6LoWPAN than without using it in the data rate of the system.

The amount of data sent to the base station is an important factor for quality analysis of any routing protocol. MLCMS with 6LOWPAN is marginally better without it, because 6LOWPAN deploys header compression which increases the packet rate at the destination.

4.2 Performance Evaluation of MLCMS Based on the 6LoWPAN Model with an Adaptive Sleep Mode

The comparative performance is assessed when using adaptive sleep mode with the MLCMS of 6LoWPAN routing protocol and comparing between MLCMS with adaptive sleep mode and without using the proposed adaptive sleep mode. Clearly, the results obtained from using the sleep mode technique on MLCMS with 6LoWPAN enhances the lifetime of the system. Furthermore, implementing the adaptive sleep mode leads to increases in the number of alive nodes and in the remaining energy. In all aspects, the proposed adaptive sleep mode increased performance better than the MLCMS without using the sleep mode scheme, as can be seen in Figs. 7 and 8.

It is evident from these figures that the lifetime of the network is extended further and the network energy consumption is greatly reduced when adding the adaptive sleep mode rather than the alternative. Most notably, the extension of the proposed model lifetime is double that when using MLCMS without adaptive sleep mode. Figure 8 shows the remaining energy in both models, which reveals that the lifetime of the

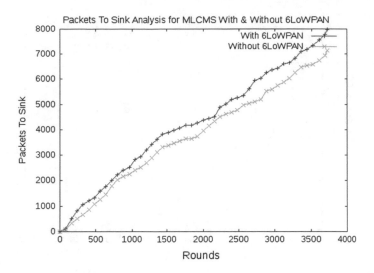

Fig. 6. number of packets received by the sink

proposed model with adaptive sleep mode ends in round 6,100, whereas using it without it results in active nodes by round 3,716. The proposed architecture can extend the lifetime of wireless sensor nodes, because the MLCMS with 6LoWPAN and adaptive sleep is a sophisticated algorithm aimed at balancing energy consumption and reducing the incidence of node death during transmission and reception. In Fig. 7, the coefficient of variation of active nodes levels is plotted with respect to round time. MLCMS with adaptive sleep mode shows minimal variation in energy levels, while

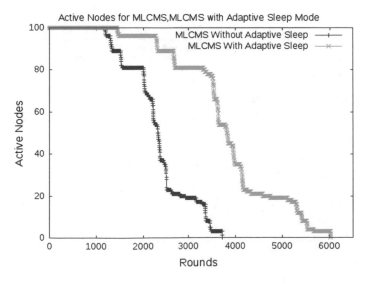

Fig. 7. Number of active nodes

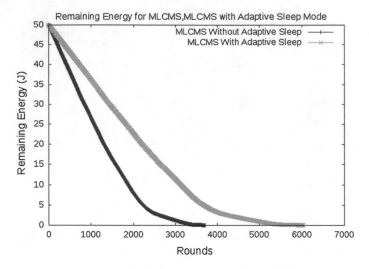

Fig. 8. Remaining energy

without it there are larger fluctuations. Figure 8 shows the residual energy levels of the sensor nodes for the MLCMS with the proposed adaptive sleep mode and for the protocol without considering the period of time for active and sleep nodes. It indicates that the sensor nodes in MLCMS with adaptive sleep are used for improving the remaining energy level. As can be clearly observed, it achieves energy equalisation with adaptive sleep mode showing a better performance regarding energy conservation. Furthermore, after 1,500 rounds, MLCMS with adaptive sleep mode nodes has significantly higher residual energy stocks compared to sensors in MLCMS without it.

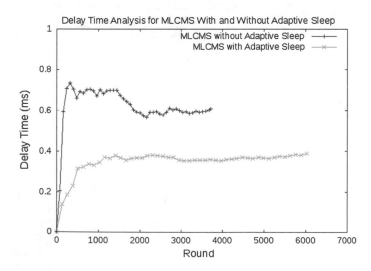

Fig. 9. End-to-end delay

In Fig. 9, we plot the expected end-to-end delay associated with the rounds. As can be seen, the end-to-end delay decreases when the traffic load of the system is minimised, by taking into account the sleep mode periods for the CHs and their nodes in an adaptive way based on the residual energy of its and achieve the overlapping between CHs to keep the performance of the multi-level clustering. The delay is linearly proportional to the traffic load. Indeed, as the traffic between the source and destination of consecutive forwarders increases, the time taken to forward a data packet increases and hence, the end-to-end delay increases. That is, the more the conjunction increases, the more end-to-end delay there is.

5 Conclusion

In this chapter, energy efficient MLCMS of the M2M routing protocol has been proposed. Furthermore, the time period of active and asleep CHs has been considered by adding an adaptive sleep mode scheme to the MLCMS algorithm. The evaluation has demonstrated that the energy consumption when applying the MLCMS protocol is reduced when compared with the LEACH and M-LEACH algorithms. Simulation results have shown that the approach consumes much less energy than M-LEACH and LEACH, with a 62% and 147%, respectively enhancement in energy efficiency and stability. Moreover, deployment of the sink nodes with an optimal multiple number has been introduced based on a number of experiments. A global-coverage M2M MLCMS model system has been successfully implemented by using sensors based on the 6LoWPAN protocol. By combining 6LoWPAN and an IP network using IPv6 addresses, the system's data rate is increased by 7%, and higher accessibility to the M2M nodes and a substantial extension of the network are realised. Moreover, using MLCMS with 6LoWPAN and the proposed adaptive sleep mode, with consideration of the residual energy in scheduling the timing of the sleep mode, doubles the lifetime of the system, increase the reduction of end to end delay by a half and improves the packet delivery ratio by 10%, when compared to using the MLCMS model without this mode.

References

1. Ekbatanifard, G., et al.: An energy efficient data dissemination scheme for distributed storage in the internet of things. Comput. Knowl. Eng. **1**(2), 1–8 (2017)
2. Razzaque, M.A., Milojevic-Jevric, M., Palade, A., Clarke, S.: Middleware for internet of things: a survey. IEEE Internet Things J. **3**(1), 70–95 (2016)
3. Palattella, M.R, Accettura, N., Vilajosana, X., Watteyne, T., Grieco, L.A., Boggia, G., Dohler, M.: Standardized protocol stack for the internet of (important) things. IEEE Commun. Surv. Tutorials **15**(3), 1389–1406 (2013)
4. Jung, S.-J., Chung, W.-Y.: Non-Intrusive Healthcare System in Global Machine-to-Machine Networks. Sensors Journal, IEEE **13**(12), 4824–4830 (2013)
5. Miao, G., Azari, A., Hwang, T.: E^2-MAC: energy efficient medium access for massive M2M communications. IEEE Trans. Commun. **64**(11) 4720–4735 (2016)

6. Al-Khatib, O., Hardjawana, W., Vucetic, B.: Traffic modeling for Machine-to-Machine (M2M) last mile wireless access networks. In: 2014 IEEE Global Communications Conference (GLOBECOM), pp. 1199–1204 (2014)
7. Huang, P., Xiao, L., Soltani, S., Mutka, M.W., Xi, N.: The evolution of MAC protocols in wireless sensor networks: a survey. IEEE Commun. Surv. Tutor. 15(1), 101–120 (2013)
8. Park, I., Kim, D., Har, D.: MAC achieving low latency and energy efficiency in hierarchical M2M networks with clustered nodes. Sens. J. IEEE 15(3), 1657–1661 (2015)
9. Farouk, F., Rizk, R., Zaki, F.W.: Multi-level stable and energy-efficient clustering protocol in heterogeneous wireless sensor networks. Wirel. Sens. Syst. IET 4(4), 159–169 (2014)
10. Wang, X.: Multicast for 6LoWPAN wireless sensor networks. Sens. J. IEEE 15(5), 3076–3083 (2015)
11. Kodali, R.K., Aravapalli, N.K.: Multi-level LEACH protocol model using NS-3. In: 2014 IEEE International Advance Computing Conference (IACC), pp. 375–380 (2014)
12. Lindsey, S., Raghavendra, C.S.: PEGASIS: power-efficient gathering in sensor information systems. In: 2002 Aerospace Conference Proceedings, vol. 3, pp. 3–1125–3–1130. IEEE (2002)
13. Manjeshwar, A., Agrawal, D.P.: TEEN: a routing protocol for enhanced efficiency in wireless sensor networks. In: Parallel and Distributed Processing Symposium., Proceedings 15th International, pp. 2009–2015 (2001)
14. Manjeshwar, A., Agrawal, D.P.: APTEEN: a hybrid protocol for efficient routing and comprehensive information retrieval in wireless. In: Parallel and Distributed Processing Symposium., Proceedings International, IPDPS 2002, Abstracts and CD-ROM, p. 8 (2002)
15. Lee, D., Chung, J.M., Garcia, R.C.: Machine-to-machine communication standardization trends and end-to-end service enhancements through vertical handover technology. In: Midwest Symposium on Circuits and Systems, pp. 840–844 (2012)
16. Le, A., Loo, J., Lasebae, A., Vinel, A., Chen, Y., Chai, M.: The impact of rank attack on network topology of routing protocol for low-power and lossy networks. Sens. J. IEEE 13(10), 3685–3692 (2013)
17. Buratti, C., Stajkic, A., Gardasevic, G., Milardo, S., Abrignani, M.D., Mijovic, S., Morabito, G., Verdone, R.: Testing protocols for the internet of things on the EuWIn platform. IEEE Internet Things J. 3(1), 124–133 (2016)
18. Kumar, N.H., Karthikeyan, P., Deeksha, B., Mohandas, T.: Enhanced routing over sleeping nodes in 6LoWPAN network. In: 2014 International Conference on Future Internet of Things and Cloud (FiCloud), pp. 272–279 (2014)
19. Roberts, R., Goodwin, P.: Weight approximations in multiattribute decision models. J. Multi-Criteria Decis. Anal. 303(June), 291–303 (2002)
20. Mahmood, D., Javaid, N., Mahmood, S., Qureshi, S., Memon, A.M., Zaman, T.: MODLEACH: a variant of LEACH for WSNs. In: 2013 Eighth International Conference on Broadband and Wireless Computing, Communication and Applications (BWCCA), pp. 158–163 (2013)
21. Chen, Y.-L., Shih, Y.-N., Lin, J.-S.: A four-layers hierarchical clustering topology architecture with sleep mode in a wireless sensor network. In: 2013 Seventh International Conference on Complex, Intelligent, and Software Intensive Systems (CISIS), pp. 335–339 (2013)

Integration of Fuzzy C-Means and Artificial Neural Network for Short-Term Localized Rainfall Forecasting in Tropical Climate

Noor Zuraidin Mohd-Safar[1]([✉]), David Ndzi[1], David Sanders[1],
Hassanuddin Mohamed Noor[1], and Latifah Munirah Kamarudin[2]

[1] School of Engineering, University of Portsmouth, Portsmouth, UK
UP680382@myport.ac.uk, {david.ndzi,david.sanders,
hassanuddin.mohamednoor}@port.ac.uk
[2] School of Computer and Communication Engineering, Universiti Malaysia
Perlis, Arau, Malaysia
latifahmunirah@unimap.edu.my

Abstract. This paper proposes and analyses the applicability of integrating Fuzzy C-Means (FCM) and artificial neural network (ANN) in rainfall forecasting. The algorithm of ANN and FCM clustering are integrated and applied to forecast short-term localized rainfall in tropical weather. Rainfall forecasting in this paper is divided into state forecast (raining or not raining) and rainfall rate forecast. Various type of back propagation extended network with hidden layers of ANN structured were trained. Training algorithm of Levenberg-Marquardt, Bayesian Regularization and Scaled Conjugate Gradient are used and trained. Transfer function in each neuron uses linear, logistic sigmoid and hyperbolic tangent sigmoid. Initial statistical analysis of weather parameter, data pre-processing approach and FCM clustering method were used to organize input data for the ANN forecast model. Input parameters such as atmospheric pressure, temperature, dew point, humidity and wind speed have been used. One to six hour predicted rainfall forecast are compared and analyzed. The result indicates that the integrated of FCM-ANN forecast model yield 80% for 1 h forecast.

Keywords: Fuzzy c-means · FCM · ANN · Rainfall forecast · Rainfall prediction · Neural network · Artificial neural network · Soft computing · Meteorology · Tropics · Tropical climate · Soft clustering

1 Introduction

Meteorological processes are highly non-linear and complicated to predict at high spatial resolutions. Weather forecasting provides critical information about future weather that is important for flood prediction systems and disaster management. This

The original version of this chapter was revised: Email ID was updated. The erratum to this chapter is available at https://doi.org/10.1007/978-3-319-69266-1_23

© Springer International Publishing AG 2018
Y. Bi et al. (eds.), *Intelligent Systems and Applications*,
Studies in Computational Intelligence 751,
https://doi.org/10.1007/978-3-319-69266-1_16

information is also important to businesses, industry, agricultural sector, government and local authorities for a wide range of reasons. Rainfall event is one of the weather's non-linear parameters for which the relationships between meteorological parameters are dynamic. The uncertainty of a future occurrence and rain intensity can have a negative impact on many sectors which depend on weather conditions. Therefore, having an accurate rainfall prediction is important to be useful in human decision. Rain prediction is among the most challenging issues in hydrological process as a result of various spatial and temporal of rainfall dissemination, vigorous behaviour of tropical climate and extremely non-linear rainfall events. The analysis of rain data and meteorological data of North-West Malaysia is used to develop an applicable forecast model. Numerical Weather Prediction (NWP) technique is normally used for weather forecasting in Malaysia (Shahi 2009). NWP method was established to serve vast coverage regions and it is therefore not suitable for localized tropical rain prediction that typically involves a small-scale of rain cell size distribution. The characteristic of NWP process demands complex scientific calculation and a reliable computer with a size of gigantic storage capability. Additionally, the NWP method that implemented in Malaysia is came from Europe and Japan which is contradictory with the tropical climate (Shahi 2009). European and Japan rainfall condition is widespread which typical phenomenon in temperate climate, due to this differences therefore adapting Europe and Japan NWP is not applicable in tropical rainfall prediction. Artificial Neural Network (ANN) approaching is susceptible of acquisition the correlation between rainfall condition and related meteorological data. Supervised learning methods are applied throughout the training to find appropriate weights that reduce the inaccuracies between real and predicted rate. The recommended ANN is utilizing intensify meteorological parameters like temperature, air pressure, dew point, humidness and wind speed data at the same point of measured rain quantities. Integrating pre-process and soft clustering method on those parameters possess the potential of increasing forecast accuracy. The algorithm of FCM soft clustering is applied to discover the correlation between the meteorological variables and rain intensity. Integrating additional processes or technique in ANN recently attracted more attention in meteorological study. Other similar study of adapting additional technique in ANN implementation is by integrating FCM clustering in daily rainfall forecast model. FCM clustering has been used to cluster rainfall into three clusters in training dataset (Chau and Wu 2010). Both of the mentioned methods were applied in non-tropical climate. Furthermore, the integration of FCM was used for rainfall-runoff modelling and daily rainfall prediction. This study's emphasis is on smaller lead time forecasting that makes use of localized environmental data and is applicable for tropical climates. Additionally, this study evaluates the reliability of applying soft clustering method in rain intensity forecast by studying the combination of FCM soft clustering and ANN. The success results of the proposed model are calculated by assessing the degree of the error between the measured data and forecasted outcome. Results exhibited that 1–4 h forecasts produce more than 75% for state forecast (raining or not raining) and more than 50% for rainfall intensity forecast. The results indicate that the integration of FCM and ANN models able to forecast rain events and intensities.

2 Literature Review

2.1 Related Works

ANN is an algorithm originally motivated by the goal of replicating human neurons patterns using machine learning. Ever since the ANN has been established, it has been numbers of study in weather condition forecasting with ANN. Maqsood et al. (2004) used the application of ANN in hourly meteorological prediction for temperature, humidity and wind speed. The study area is conducted for south Saskatchewan region in Canada. The outcome of the study indicate that ANN method together with radial basis function is better than statistical model (Maqsood et al. 2004). The application of ANN in short term rainfall and weather forecasting have also been described in Hu and Root (1964), Hung et al. (2008) and Klent Gomez Abistado and Maravillas (2014). Hu and Root (1964) used ANN to forecast whether it is raining or not raining irrespective of rainfall amount. The outcome from the study were analysed with the U.S Weather Bureau data (Hu and Root 1964). Various patterns of sea level pressure were trained to forecast 12 and 24 h rainfall for San Francisco bay. The prediction results show that adapting ANN pattern recognition system has the potential in forecasting raining and not raining events without inferred in details the vigorous of the weather pattern. A research of tropical climate using ANN approach to improve the rainfall forecast was setup in Bangkok, Thailand (Hung et al. 2008). Forecast outcome for 1–3 h obtained greater than 70% accuracy (Hung et al. 2008). A variation of weather parameters from 75 weather stations were used in network training. Feed-forward network and Multi-layer Perceptron (MLP) network model were tested in the study (Hung et al. 2008). A related study was organized in the Philippines by utilizing day-to-day meteorological data during training (Klent Gomez Abistado and Maravillas 2014). The experimentation showed that forecast results attained greater than 90% when the combination of ANN and Bayesian network were used (Klent Gomez Abistado and Maravillas 2014). All of these methods of weather and rainfall forecasting achievement are benefit from high spatial and temporal data. For instance, data obtainability in Hung et al. (2008) came from more than one hundred weather stations with the measured data more than 15 years. The current leading approaches of ANN technique in modelling non-linear and dynamic condition of application is one of the interest to explore the uses of ANN in rainfall forecasting of a tropical area.

2.2 Fuzzy C-Means Clustering

FCM soft clustering is an algorithm originally established by Dunn (1973) and Bezdek (1984). FCM uses fuzzy partitioning by assigning membership to each data point corresponding to each cluster centre on the basis of distance between the cluster centre and the data point. The data point is the property of several cluster groups with belongs to fuzzy truth value between 0 and 1. Then, the summation of membership of each data point is equal to one (Lu et al. 2013). The interactive algorithm of FCM clustering will define the cluster centre known as centroids from the data reciprocal distances in order to minimize differences. The aim of using FCM is to decrease the non-linearity of the raw data by grouping similar segments of the training data. It attempts to organize uncategorized input into clusters where data points within a cluster are similar rather than those belonging to different clusters (Sabit and Al-Anbuky 2014). In this study, the integration of FCM and ANN in the proposed FCM-ANN forecasting model consists of supervised FCM and ANN algorithm. ANN inputs selection is one of the important aspects of designing ANN forecasting model. The forecasting model will perform poorly if the inputs are not chosen carefully (Srivastava et al. 2010). An appropriate number of FCM clusters for a given input data are pre-determined before the input is applied to ANN. Non-linear relationship between identified inputs and output (rainfall) can be captured using FCM clustering algorithm to achieve better prediction (Irwin et al. 2011).

In the proposed FCM-ANN rainfall forecasting model, training data is portioned into five clusters. This approach has been used to identify the substantial input (environmental parameters) for hourly rainfall prediction. At the beginning of the implementation, pre-processing of data is used to validate the raw data, and then followed by dividing the data into two groups representing raining and not raining conditions. Figure 1 shows the graphic representation of integrating FCM and ANN application where FCM clustering is used individually for zero rain and non-zero rain data to create input and output clustered vectors. Denoised data and cascaded implementation of FCM of input vectors before the forecasting model will determine the correct range. The process of clustering rainfall dataset is important to determine a correlated class pattern between input parameters for raining and not raining events. FCM clustering will improve the input-output relationship in ANN model and reduce overfitting problem in training. Overfitting arises when network training strongly success on the training set but not on the test set (Srivastava et al. 2014). FCM is algorithm that works iteratively until the termination criterion is satisfied (Karlik 2016).

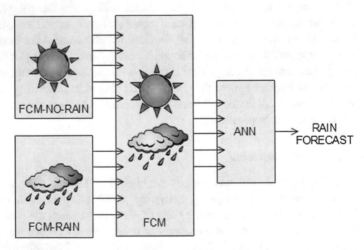

Fig. 1. Schematic illustration of FCM clustering integration into ANN rainfall forecasting model

Initially, FCM arbitrarily sets the associated matrix using Eq. (1)

$$\sum_{i=1}^{c} u_{ij} = 1, \forall j = 1, \ldots, n \tag{1}$$

The distinction function which is used in FCM is given by (2) with matrix dataset is U

$$J(U, c_1, c_2, \ldots, c_c) = \sum_{i=1}^{c} J_i = \sum_{i=1}^{c} \sum_{j=1}^{n} u_{ij}^m d_{ij}^2 \tag{2}$$

where u_{ij} is a degree of membership of x_i that has a range element from 0 to 1. In given data point j, the summative of the range element for all clusters is finally equal to 1. c_i is the cluster centre of i, n is the clusters amount, d_{ij} is the Euclidian distance between ith cluster centre (c_i) and jth data point, and m is the exponential values of weight (where $1 \le m \le \infty$) (Vega-corona 2012). An iterative optimization algorithm is applied to reduce the distinction functions using Eq. (3) where x_i is the ith data point. Updated U^k matrix to U^{k+1} matrix is the iteration phase by using Eq. (4) where k is the number of iteration step. The iteration step will end when $\{||u_{ij}^{k+1} - u_{ij}^k||\} \le \varepsilon$ where ε is a termination criterion that has been defined.

$$c_i = \frac{\sum_{j=1}^{n} u_{ij}^m x_i}{\sum_{j=1}^{n} u_{ij}^m} \tag{3}$$

$$u_{ij} = \frac{1}{\sum_{k=1}^{c} \left(\frac{d_{ij}}{d_{kj}}\right)^{2/(m-1)}} \tag{4}$$

The following steps are the implemented of the FCM algorithm:

1. Firstly, random initialization of the membership matrix ($U = [u_{ij}]$) using Eq. (1), initialization of clusters (c), weighting exponent (m), termination criterion (ε) and the number of maximum iteration.
2. Calculate the cluster centre (c_i) using Eq. (3).
3. Calculate distinction between cluster centre and data points using Eq. (2)
4. Update U^k to U^{k+1} using (4).
5. If $\{||u_{ij}^{k+1} - u_{ij}^k||\} \le \varepsilon$ then the iteration will stop then else it recalculates the cluster centre (c_i) using Eq. (3) in step 2.

FCM iteratively updates the cluster centre and the membership value for each data point in a group of dataset. However, updates done by FCM technique not guarantee to a finest result since the cluster centre are initialized in matrix U in random order. The achievement of the FCM performance is subject to the initial selected of clustered centre. For a strong implementation, FCM should be executed for a number of times using different initialization value. FCM method requires users to specify the number of clusters. The cluster validity criterion is usually performed to find the optimal number of clusters for the dataset (Rezaeianzadeh et al. 2013). Before input parameters are prepared to an ANN initial layer, soft clustering was applied to gain better input-output correlation and pre-classification for data reduction (Vega-corona 2012).

2.3   Artificial Neural Network

ANN is a mathematical data processing systems model. The key function of ANN is to predict or classify appropriately the particular input instances. Similarly to biological neurons, learning in ANN is achieved by adapting the synaptic weights to the input data (Gardner and Dorling 1998). ANN structure has numbers of interconnected artificial neurons connected together. It is a mathematical model which functions like the neuron patterns of the human brain. A supervised learning ANN must be trained through a training data set and it creates the patterns and the rules governing the network (Reed and Marks 1998). In early model of ANN, feed forward ANN was introduced as perceptron that uses a single input perceptron layer for targeting single output (Hagan et al. 1996). The drawback of this technique make the perceptron is not able to train and recognize many types of patterns. Each of the artificial neuron is composed of two main parts; weighted coefficient and transfer function. Figure 2 shows a single neuron that consists of a weighted input, summing function and transfer function.

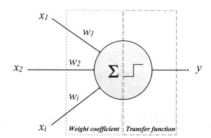

Fig. 2. Single neuron

For a single neuron that accept n number of input, in every input x_i (where $i = 1 \dots n$) is weighted by the product of the weight w_i. The amount of the $w_i x_i$ multiplication provides the net activation of the neuron. Then the activation rate is depending on a transfer function of (f) to create the neuron's output y. A single neuron can be described by Eq. (5):

$$y = f\left(\sum_{i=1}^{n} w_i x_i\right)$$

(5)

Extending the number of perceptron layer in a network topology will improve the capability of the ANN to identify various classes of patterns. This additional layer of nodes between input and output nodes is known as MLP. The neurons are organized into an input layer and an output layer where one or more hidden layers were placed in between. MLP is trained with backpropagation algorithm derived initially from first hidden layer of MLP. The optimization of weight from the uses of backpropagation algorithm will make the network correctly map the arbitrary input-output. The processes continually calculate an error and regulate the weight with smaller error. A step by step process of regulating the weights in the network structure can be described below:

1. Each training stage start with imposing input signal from training data set.
2. Determine the output arbitrary of neuron for every network layer.
3. Lastly, optimum weight with less error is selected to map input-output arbitrary.

Figure 3 shows the transmission of signals for every neuron in the form of input-hidden-output layer. Supposedly there are L amount of layers including input, hidden and output layer, where l is the input layer and a hidden layers that has N numbers of nodes in the form of $N(l)$, where $l = (0, 1,..., L; l = 0$ *is the input parameter)* and $i = 1,..., N(l)$ is the node that hold the output came from preceding layer. Then the output is the input for the succeeding layer. $y_{l,i}$ is the output that rely on the inbound values of $x_{l,i}$ and variables of α, β, γ. Thus the Eq. (6) is the simplification of the output from every node:

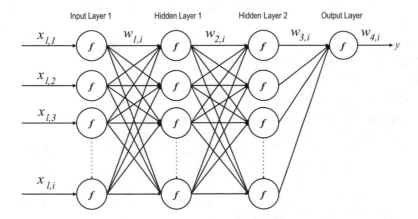

Fig. 3. ANN layer representation

$$y_{l,i} = f_{l,i}(x_{l-1,1}w_{l-1,1}, \ldots, x_{l-1,N(l-1)}, \alpha, \beta, \gamma, \ldots) \qquad (6)$$

When the signals were propagated, the next pace is to match the output signal from the node and the predicted output (z) from training data. The differences $\delta_{l,i}$ between the output and predicted output can be generalized in Eq. (7):

$$\delta_{l,i} = z - y_{l,i} \tag{7}$$

Suppose that the training data set has P number of records, the calculated error of the Pth record is the summation of the squared error:

$$E_P = \sum_{k=1}^{N(L)} (z - y_{l,i})^2 \tag{8}$$

where k is the amount of z component. Adjusted of the individual node's weight are scaled to the error of the connected neurons using backpropagation Gradient descent algorithm in order to find minimized weighted error. The algorithm can be generalized by:

1. Get the gradient vector.
2. Compute the signal error of $\varepsilon_{l,i}$ derived from the error measure of E_p with accordance to the output of node i in layer l for direct and indirect paths of the neurons navigation. The derivation is represented in Eq. (9):

$$\varepsilon_{l,i} = \frac{\partial^+ E_p}{\partial y_{l,i}} \tag{9}$$

For each input data for network training, it should be related to the expected output form. The ANN models uses supervised learning MLP trained with backpropagation algorithm (Abraham et al. 1992) and the step by step training procedure can be summarized below:

1. Set the weights by mapping the arbitrary input-output data.
2. Randomly set bias by selecting data originated in training data.
3. Calculate neurons error and updating weights.
4. Repeat step 3 until the error is minimized.

Three training algorithms were applied to find the best accuracy, each of them having a variety of different computation and storage requirement.

Levenberg-Marquardt (LM) gives a good conversion for a moderate size of the network and capable to reduce memory size. LM uses estimation technique to change network weights and biases. LM learning approach is used in Foresee and Hagan (1997) and Pellakuri et al. (2015). It is capable of reducing the number of chances that it converges to local minima (Goss 1993).

Bayesian Regularization (BR) method increases generality and decreases the struggle of determining finest network design. BR use functions such as Mean Squared Error (MSE) to increase generalization. Detail discussion of BR can be found in Moré (2015), MacKay (1992a, b).

Scaled Conjugate Gradient (SCG) training in a neural network minimizes global error function, which is a multivariate function that depends on the weights in the network. The Conjugate Gradient Method is a numerical analysis method based on optimization technique of both linear and non-linear systems (Shewchuk 1994).

Contrasting with other conjugate gradient approaches SCG was designed to reduce line search process to make it faster for large amount of weight (Okut et al. 2013). SCG does not demand any detailed parameter and its mathematical calculation is faster. Detail explanation of the SCG can be found in Møller (1993). In ANN implementation transfer function is used to compute net weight between input-output layers. The finest prediction outcome are originated by the combination setting of ANN training algorithm and transfer function (Shamseldin 1997; Lohani et al. 2011; Hung et al. 2008). Numerous arrangements of the transfer functions are used in order to gain better performance.

Dataset are divided into training dataset and validation dataset. It is important that they are representative of the same population because ANNs is unable to extrapolate beyond the range of the data used for training (Toth et al. 2000). Poor forecast results will occur if the validation dataset contains values outside the range of those used for training. However, it might be difficult to form a representative validation dataset when limited number of data is available to be explored. Methods for maximizing the available data for training have been reviewed by Maier and Dandy (2000). Other issues related to model input data and network training is overfitting. It occurs when the network has not learned to generalize to new situations (Sheela and Deepa 2013). To resolve overfitting problems, the available dataset need to be divided into three, training data, testing set and validation sets (Hung et al. 2008; Yuan et al. 2007). The testing set is a cross-validation dataset in the network implementation. The training set is used to train a number of different ANN model configurations. The test set is used to decide when to stop training and also to determine which of the networks is provides good results. Finally, the validation set is used to evaluate the chosen model against independent data.

3 Study Area and Meteorological Data

The study focused on meteorological data available for Chuping, Alor Setar and MARDI Bukit Tangga weather stations. Those stations are located in Perlis and Kedah with similar climatic characteristic and topology. The climate tropical has long hours of sunshine, uniform temperature and high humidity. The average annual relative humidity is between 70 and 90% (Lim and Samah 2004) and most of the places exceed 80% (Nor and Rakhecha 2008). Malaysia has an average annual surface temperature of 26.7 °C (Hamidi et al. 2014) and the study area, the yearly typical temperature is 27.5 °C. The average day light is between 7 a.m. to 7 p.m. Meteorological parameter such as air mass pressure, temperature, dew point, humidity, wind speed, wind direction, rainfall amount and rainfall rate are available. Figure 4 shows the study area and the distances between the stations. Three datasets were created; each of the dataset was derived from all weather stations.

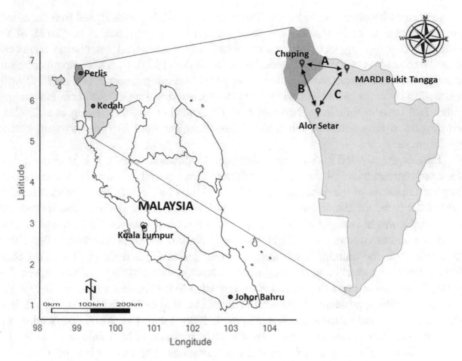

Fig. 4. Study area and weather stations (A = 26 km, B = 35 km, C = 32 km)

4   FCM-ANN Rainfall Forecasting Model

To develop the model, the data was pre-processed to eliminate noise. FCM was then used to create complete datasets by estimating missing data (imputation). The imputation result was evaluated to ascertain the reliability of new dataset. After that, FCM clustering was applied to raining and non-raining condition data. Environmental parameters and rainfall rate were clustered into several subsets. Input vectors of clustered dataset together with pre-processed environmental data will be used in FCM-ANN forecasting model. Figure 5 show the flowchart of the FCM-ANN forecasting model.

The ANN forecasting model is built on MLP with backpropagation algorithm. Diverse neural network configurations have been tried. Figure 6 is an illustration of the ANN forecasting model architecture. Model input selection will determine the input data that will feed into network. In standard ANN rainfall forecasting model, all meteorological parameters are selected and put into the network without any sanitation process. The proposed FCM-ANN rainfall forecasting model takes advantage of pre-processing method in model input selection. Additional inputs that improves the forecast result can also be selected such as clustered (Lu et al. 2013; Krzhizhanovskaya et al. 2011).

The input-output parameters (P_{input} and P_{output}) for the neural network structure can be simplified as follows:

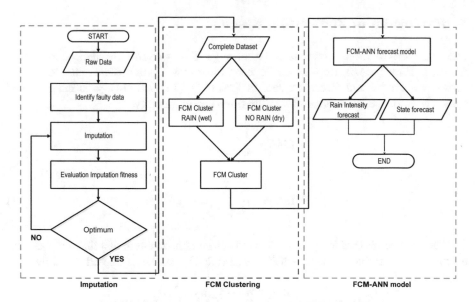

Fig. 5. Flowchart of forecasting for modular FCM-ANN implementation

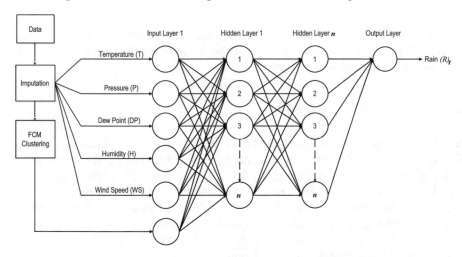

Fig. 6. FCM-ANN architecture for rainfall forecasting model

$$P_{input} = [(P, T, D, H, WS, R)_{FCM}, P_{t-n}, T_{t-n}, D_{t-n}, H_{t-n}, WS_{t-n}, R_{t-n}] \qquad (10)$$

$$P_{output} = [R_t] \qquad (11)$$

where R is the rain intensity at time t and n is the number of hours before the rain, P, T, D, H and WS is the environmental parameters which is pressure, temperature, dew point, humidity and wind speed respectively. $(P, T, D, H, WS, R)_{FCM}$ is the cluster value for each environmental parameters.

5 Evaluation of Forecasting Results

The FCM-ANN model was assessed through the performance indices of Mean Absolute Error (*MAE*), Root Mean Squared Error (*RMSE*) and correlation coefficient between observed and forecast value. In Eqs. (12), (13) and (14), y_t is the observed value, $\hat{y}_t$ is forecasted value and n is the observation quantity.

$$MAE = \frac{1}{n}\sum_{t=1}^{n}|y_t - \hat{y}_t| \tag{12}$$

$$RMSE = \sqrt{\frac{1}{n}\sum_{t=1}^{n}(y_t - \hat{y}_t)^2} \tag{13}$$

The correlation coefficient (*R*) is used to measure the strength of a linear association between two variables (Hauke and Kossowski 2011) can be expressed as follows:

$$R = \frac{n\sum y_t\hat{y}_t - (\sum y_t)(\sum \hat{y}_t)}{\sqrt{n(\sum y_t^2) - (\sum y_t)^2} - \sqrt{n(\sum \hat{y}_t^2) - (\sum \hat{y}_t)^2}} \tag{14}$$

The value of *R* will determine the linear association between the forecast and actual value. If *R = 1*, this imply that there is a perfect linear correlation but when *R* is nearly to zero, then it present no linear correlation.

6 Results and Discussions

In this study, forecasting of raining and not raining occurrences and rain intensity were tested. The input vectors of meteorological parameters between 2 and 7 h were utilized in ANN forecasting model. Predicted value for state forecast is a binary number of 0 and 1 (where 0 is not raining and 1 is raining) and target vectors for rain intensity forecast is the amount of rainfall rate. Meteorological data from January 2012 to December 2014 were placed into the input-output matrix. Each of the ANN forecast structured will search the finest result, assesses the error magnitude and correlation coefficient between forecast and actual value.

6.1 FCM Clustering

Data clustering, by definition, is the grouping of data into similar categories to identify natural groupings of data from a large dataset. The model proposed in this study uses FCM clustering and ANN on the interpreted historical data and the relationship between parameters. Based on these interpretations, datasets that are required for analysis are generated. By using FCM clustering technique, the whole training set can be divided into subsets with lower complexity. Therefore, based on these subsets, the stability of individual ANN can be improved and the precision can also be enhanced

(Sugumar et al. 2015). Data clustering can be performed using supervised or unsupervised learning. The FCM algorithm implemented in this study is a supervised algorithm because a priori knowledge of the number of clusters required. The proposed FCM clustering for optimal partition of data was applied to all dataset (Chuping, Alor Setar and MARDI Bukit Tangga). Each dataset consists of 26304 instances comprising weather parameters (with six attributes: pressure, temperature, dew point, humidity, wind speed and rainfall rate) are clustered using FCM algorithm. The initial number of clusters is set to five clusters. The maximum number of iterations is set to 100. The termination criterion is fixed to 10^{-4}. Training data are clustered into five and then the data in each group are used to train the network. Since the data could be divided into rain (wet) and not raining (dry) condition, the proposed method partitions the data into two FCM cluster group according to the rain condition (refer to Fig. 5 FCM Clustering flowchart). Once the training data are prepared, the next step is to utilize supervised ANN to capture the relationship between these input-output parameters.

Figures 7, 8, 9 and 10 show the results for each cluster and represent the FCM clustering for raining condition. Five clusters are labelled and represent homogenous group for every input parameters. These labels will be used as input vectors in the proposed FCM-ANN forecasting model described later. Most of the cluster group are overlapping because the nature of environmental parameters in tropical climate is dynamic and uncertain (Deser et al. 2012). For example, the same temperature value can be obtained during wet and dry conditions. There are no distinctive attributes that will specifically determine the rainfall condition. Therefore, the extension of the relationship among others parameters is crucial in forecast determination.

Dry condition or not raining condition is the dominant condition. Out of 78912 samples in the dataset, 5611 (7%) of them were raining events. In contrast, 93% of the data is biased towards not raining event. Environmental parameters commonly have similar characteristic in raining and not raining conditions. Reducing dataset pattern will increase the relationship between input and output parameters for zero rain condition. During raining events, humidity varied from 65 to 100%. Thus; if the humidity is less than 65% most probably it is not raining. For temperature greater than 33 °C and dew point greater than 28 °C, no rain event was recorded. Unfortunately, for pressure

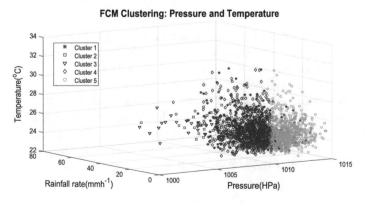

Fig. 7. FCM clustering for pressure and temperature during raining condition

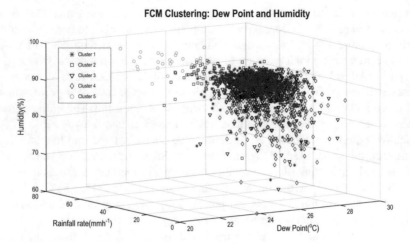

Fig. 8. FCM clustering for dew point and humidity during raining condition

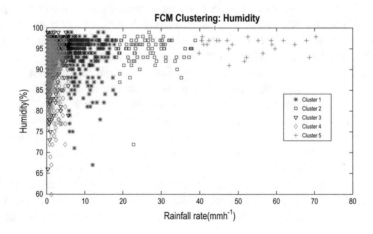

Fig. 9. FCM clustering for humidity during raining condition

and wind speed there is no clear distinction between rain and not raining events. Only specific range of parameters values that contribute to dry condition were selected and clustered. Figures 11 and 12 show the results of FCM clustering for pressure, temperature, humidity and dew point for zero rain condition. Each of the cluster group has its own cluster centre, also known as centroid. Cluster centre or cluster centroid is a vector containing one number for each variable, where each number is the mean of a variable for the observations in that cluster (Bora 2014).

After successfully clustering the environmental parameters for rain and not raining conditions, the next step is to apply the results as inputs to the rainfall forecasting model.

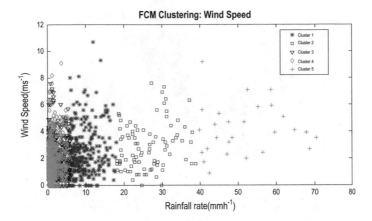

Fig. 10. FCM clustering for wind speed during raining condition

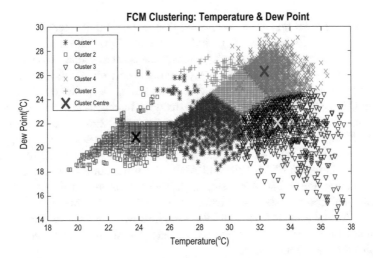

Fig. 11. Dew point and temperature clustering for dry condition

6.2   Forecast Result

FCM clustering was applied to all three weather stations (Chuping, Alor Setar and MARDI Bukit Tangga). The main purpose for testing on dataset from all weather stations was to investigate the generalization and capability of proposed model in different location with similar climatological condition. Conventional ANN forecasting model were compared with the proposed FCM-ANN model in Tables 1 and 2. This section presents the comparison between basic ANN and FCM-ANN forecasting model. Results presented in this section are for 1–6 h forecasts, state forecast (comparing raining versus not raining prediction—a binary forecast) and rain intensity forecast (rainfall rate).

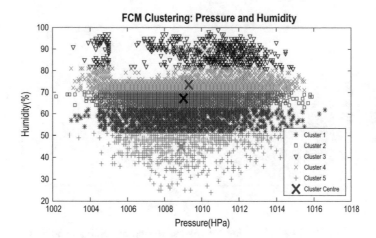

Fig. 12. Humidity and pressure clustering for dry condition

In the ANN model that are not clustered (Table 1), rainfall intensity forecast is less than 10% accuracy for the shorter period of forecast (2 h). The result for FCM-ANN model is 68% accuracy for 3 h rainfall intensity forecast and more than 75% accuracy for 3 h state forecast. Results show that the rainfall prediction meets the projected value when FCM clustering applied to model input vectors. Magnitude of error index of RMSE values are 1.1–2.5 for FCM-ANN structure. In contrary, ANN implementation without applying soft clustering yields more than 2.5. From all of these forecasts results, FCM-ANN model shows a better solution for short-term forecast.

In Tables 1, 2, 3 and 4 show the differences of MAE, RMSE and correlation coefficient (R) between FCM-ANN and ANN. From the comparison, mainly based on R value it shows that FCM-ANN structure produces improved forecast for localized short-term prediction. Binary prediction in state forecast achieve 0.92 correlations because ANN was trained from multiple input into two possible outputs, it is either 0 (not raining) or 1 (raining). Figures 13, 15 and 16, show forecast comparison and scatter plot for Alor Setar dataset.

Variations of rain intensity from 0.3 to 70 mmh^{-1} make forecasting model challenging. Table 5 shows rain intensity category for all weather stations. There were 279 recorded events when the rain intensity exceeds 30 mmh^{-1}. Therefore the convergence of environmental parameters towards high intensity rain forecasting is poor (Fig. 14).

One hour FCM-ANN prediction model for raining or not raining occurrences gives 0.92 correlations coefficient and rainfall rate prediction gives 0.80. Rainfall rate forecast using ANN model without applying FCM produced lower correlation value. The results show that increasing the forecast time will reduce the convergence result. Six hour prediction can reach 0.67 correlations and 0.42 correlations using FCM-ANN for state and rain intensity forecast respectively. State forecast gained higher correlations value as a result of ANN learning process simply adapts to binary prediction of rain or no raining events. Figure 15 shows the contrast between actual and forecasted rainfall rate and Fig. 16 the regression plots for 1–6 h forecasting for Chuping weather station. Figures 17 and 18 show the rainfall rate forecast for Alor Setar.

Table 1. Rain prediction for basic ANN forecasting model for Chuping weather station

	Model	Transfer function	Training function	MAE	RMSE	R
Rainfall rate forecast	1-h	HT, HT, HT	Bayesian	0.373	2.408	0.334
	2-h	HT, HT, LN	Bayesian	0.542	2.426	0.305
	3-h	HT, HT, LN	Bayesian	0.608	2.536	0.094
	4-h	HT, HT, HT	Bayesian	0.598	2.543	0.057
State Forecast	1-h	HT, HT, HT	Bayesian	0.089	0.212	0.589
	2-h	HT, HT, LN	Bayesian	0.117	0.240	0.407
	3-h	HT, HT, LN	Bayesian	0.157	0.284	0.341
	4-h	HT, HT, LN	Bayesian	0.309	0.288	0.310
	5-h	HT, HT, HT	Bayesian	0.163	0.291	0.276
	6-h	HT, HT, HT	Bayesian	0.225	0.339	0.269

Bayesian = Bayesian Regularization, Levenberg = Levenberg Marquardt,
Conjugate = Scaled Conjugate Gradient, HT = Hyperbolic Tangent Sigmoid, LN = Linear,
SG = Sigmoid

Table 2. FCM-ANN rainfall forecasting model for Chuping

	Model	Transfer function	Training function	MAE	RMSE	R
Rainfall rate forecast	1-h	HT, HT, HT	Bayesian	0.412	1.116	0.808
	2-h	HT, IIT, LN	Bayesian	0.812	1.691	0.705
	3-h	HT, HT, LN	Bayesian	0.971	2.004	0.680
	4-h	HT, HT, HT	Bayesian	1.212	2.238	0.527
	5-h	HT, HT, HT	Bayesian	1.397	2.443	0.491
	6-h	HT, HT, HT	Bayesian	1.554	2.591	0.422
State forecast	1-h	HT, HT, HT	Lavenberg	0.076	0.194	0.920
	2-h	HT, HT, HT	Bayesian	0.133	0.257	0.856
	3-h	HT, HT, HT	Bayesian	0.176	0.297	0.802
	4-h	HT, HT, HT	Bayesian	0.214	0.326	0.754
	5-h	HT, HT, HT	Bayesian	0.254	0.356	0.697
	6-h	HT, HT, HT	Bayesian	0.271	0.368	0.671

Bayesian = Bayesian Regularization, Levenberg = Levenberg Marquardt,
Conjugate = Scaled Conjugate Gradient, HT = Hyperbolic Tangent Sigmoid,
LN = Linear, SG = Sigmoid

MAE and RMSE are used to evaluate quantitative error in every model (Basic ANN and FCM-ANN) outcome. Figure 19 shows the performances of the correlation coefficient between observed and forecast value for state and rainfall rate forecasting. State forecasting yields better result compare to rainfall rate forecasting. One hour forecast in both state and rainfall intensity yields good results. As the forecast time increases, the forecast performances decrease.

Table 3. FCM-ANN rainfall forecasting model for Alor Setar

	Model	Transfer function	Training function	MAE	RMSE	R
Rainfall rate forecast	1-h	HT, HT, HT	Bayesian	1.425	3.655	0.826
	2-h	HT, HT, HT	Bayesian	1.468	3.956	0.799
	3-h	HT, HT, LN	Bayesian	1.861	4.740	0.661
	4-h	HT, HT, HT	Bayesian	2.110	4.888	0.520
	5-h	HT, HT, HT	Bayesian	2.355	5.674	0.468
	6-h	HT, HT, HT	Bayesian	2.543	5.855	0.438
State Forecast	1-h	HT, HT, HT	Bayesian	0.007	0.026	0.995
	2-h	HT, HT, HT	Bayesian	0.074	0.163	0.921
	3-h	HT, HT, HT	Bayesian	0.150	0.264	0.846
	4-h	HT, HT, HT	Bayesian	0.196	0.307	0.784
	5-h	HT, HT, HT	Bayesian	0.231	0.337	0.732
	6-h	HT, HT, HT	Bayesian	0.254	0.355	0.695

Bayesian = Bayesian Regularization, Levenberg = Levenberg Marquardt, Conjugate = Scaled Conjugate Gradient, HT = Hyperbolic Tangent Sigmoid, LN = Linear, SG = Sigmoid

Table 4. FCM-ANN rainfall forecasting model for MARDI Bukit Tangga

	Model	Transfer function	Training function	MAE	RMSE	R
Rainfall rate forecast	1-h	HT, HT, HT	Bayesian	0.722	2.870	0.835
	2-h	HT, HT, HT	Bayesian	1.233	3.826	0.701
	3-h	HT, HT, HT	Bayesian	1.207	3.882	0.671
	4-h	HT, HT, HT	Bayesian	1.607	4.295	0.600
	5-h	HT, HT, HT	Bayesian	1.976	5.277	0.520
	6-h	HT, HT, HT	Bayesian	2.178	5.474	0.435
State Forecast	1-h	HT, HT, HT	Lavenberg	0.114	0.195	0.899
	2-h	HT, HT, HT	Bayesian	0.118	0.241	0.843
	3-h	HT, HT, HT	Bayesian	0.141	0.256	0.826
	4-h	HT, HT, HT	Bayesian	0.152	0.277	0.790
	5-h	HT, HT, HT	Bayesian	0.185	0.323	0.723
	6-h	HT, HT, HT	Bayesian	0.213	0.328	0.673

Bayesian = Bayesian Regularization, Levenberg = Levenberg Marquardt, Conjugate = Scaled Conjugate Gradient, HT = Hyperbolic Tangent Sigmoid, LN = Linear, SG = Sigmoid

The MAE and RMSE values of each of the prediction models are plotted in Fig. 20. Surprisingly, some of the forecast results using basic ANN forecasting model outperformed the proposed FCM-ANN model. For 1–3 h forecast, both MAE and RMSE yield smaller error for FCM-ANN model than ANN. Therefore, for all performance validation, most of the more accurate forecast results were obtain from the FCM-ANN model.

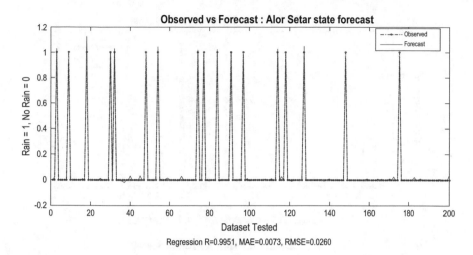

Regression R=0.9951, MAE=0.0073, RMSE=0.0260

Fig. 13. Actual and forecast evaluation for 1 h state prediction (FCM-ANN) for Alor Setar dataset

Table 5. Categorization of rainfall

Category[a]	Rainfall rate at any given time	Number of rain incidences
Light	1–10 mmh^{-1}	3887
Moderate	11–30 mmh^{-1}	955
Heavy	31–60 mmh^{-1}	252
Very heavy	Greater than 60 mmh^{-1}	27

[a]Categorization of the rainfall in Malaysia (Department of Irrigation and Drainage (DID) Manual (Volume 1—Flood Management) 2009)

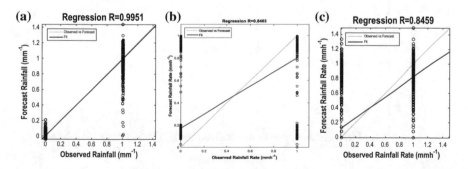

Fig. 14. Scatter plot for **a** 1 h, **b** 3 h and **c** 6 h state forecast using FCM-ANN for Alor Setar dataset

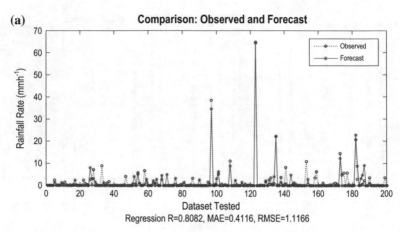

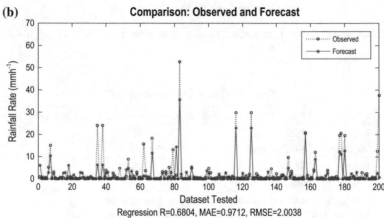

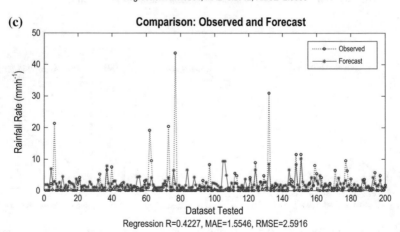

Fig. 15. Rainfall forecast for Chuping weather station **a** 1 h, **b** 3 h and **c** 6 h

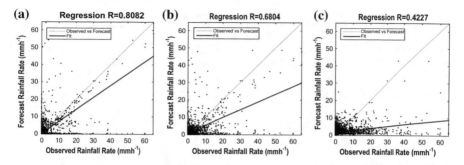

Fig. 16. Rainfall rate intensity regression plot for Chuping weather station **a** 1 h, **b** 3 h and **c** 6 h

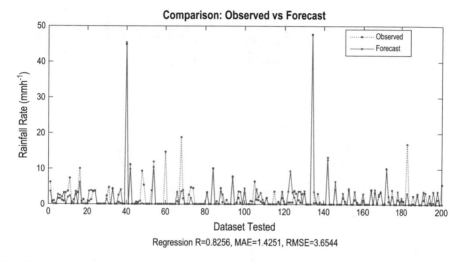

Fig. 17. Rainfall rate for FCM-ANN 1 h rain intensity forecast for Alor Setar weather station

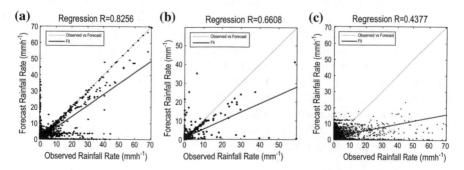

Fig. 18. Rainfall rate intensity regression plot for Alor Setar weather station **a** 1 h, **b** 3 h and **c** 6 h

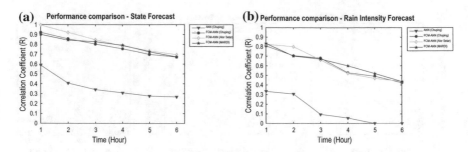

Fig. 19. Performance comparison of forecasting model for **a** state and **b** rainfall rate prediction

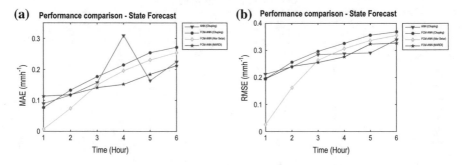

Fig. 20. **a** MAE and **b** RSME performance comparison

7 Summary

Rain characteristic is a very challenging weather condition to forecast, even a large datasets covering many decades unable to make better forecast (Yuan et al. 2007). In tropical area, the dynamic conditions of rainfall make it an additional challenging to predict because of their localized characteristic. The geo-location of the covering area has higher rain intensity within a short duration that is convective. The extremely dynamic and non-linear of this weather phenomenon is a reflection of the complicated relationship between the numerous processes at diverse scales. Therefore, the process of designing an efficient forecast system is challenging to comprehend. In this study, it utilized ANN to improve a rainfall forecast. The available data for training the network for higher rainfall rate are small (only 27 incidences were recorded) that make the process of developing an optimum model is challenging. The best prediction is for 1 h forecast for state prediction using FCM-ANN that able to reach 80% accuracy.

References

Abraham, A., Philip, N.S., Joseph, K.B. In: Will We Have a Wet Summer? Soft Computing Models for Long-term Rainfall Forecasting (1992)

Bezdek, J.C.: FCM: the fuzzy c-means clustering algorithm. Comput. Geosci. **10**(2), 191–203 (1984)

Bora, D.J.: A comparative study between fuzzy clustering algorithm and hard clustering algorithm **10**(2), 108–113 (2014)

Chau, K.W., Wu, C.L.: A hybrid model coupled with singular spectrum analysis for daily rainfall prediction. J. Hydroinformatics **12**(4), 458–473 (2010)

Department of Irrigation and Drainage (DID) Manual (Volume 1—Flood Management). Department of Irrigation and Drainage, Malaysia, 1 (2009)

Deser, C., Phillips, A., Bourdette, V.: Uncertainty in climate change projections: the role of internal variability 527–546 (2012)

Dunn, J.C.: A fuzzy relative of the process and its use in detecting compact well-separated clusters. J. Cybern. **3**(3), 32–57 (1973)

Foresee, F.D., Hagan, M.T.: Guass-Newton approximation to bayesian learning. In: Proceedings of the International Conference on Neural Networks, Houston, Texas, pp. 1930–1935. IEEE (1997)

Gardner, M.W., Dorling, S.R.: Artificial neural networks (the multilayer perceptron) a review of applications in the atmospheric sciences. Atmos. Environ. **32**(14), 2627–2636 (1998)

Goss, D.F.E.: Forecasting with neural networks: an application using bankruptcy data. Inf. Manage. **24**(3), 159–167 (1993)

Hagan, M.T., Demuth, H.B., Beale, M.H., De Jesús, O.: Neural Network Design, vol. 20. PWS Publishing Company, Boston (1996)

Hamidi, Z.S., Shariff, N.N.M., Monstein, C.: Understanding climate changes in Malaysia through space weather study. Int. Lett. Nat. Sci. **13**, 9–16 (2014)

Hauke, J., Kossowski, T.: Comparison of values of pearson's and spearman's correlation coefficients on the same sets of data. Quaestiones Geographicae **30**(2), 87–93 (2011)

Hu, M.J.C., Root, H.E.: An adaptive data processing system for weather forecasting. J. Appl. Meteorol. **3**(5), 513–523 (1964)

Hung, N.Q., Babel, M.S., Weesakul, S., Tripathi, N.K.: An artificial neural network model for rainfall forecasting in Bangkok. Thai. Hydrol. Earth Syst. Sci. **13**(8), 1413–1425 (2008)

Irwin, S., Srivastav, R., Slobodan, P.S.: Instructions for Operating the Proposed Regionalization Tool Cluster-FCM Using Fuzzy C-Means Clustering and L-Moment Statistics (2011)

Karlik, B.: The positive effects of fuzzy c-means clustering on supervised learning classifiers. Int. J. Artif. Intell. Expert Syst. (IJAE) **7**, 1–8 (2016)

Klent Gomez Abistado, C.N.A., Maravillas, E.A.: Weather forecasting using artificial neural network and Bayesian network. J. Adv. Comput. Intell. Intell. Inf. **18**(5), 812–817 (2014)

Krzhizhanovskaya, V.V., Shirshov, G.S., Melnikova, N.B., Belleman, R.G., Rusadi, F.I., Broekhuijsen, B.J., Meijer, R.J.: Flood early warning system: design, implementation and computational modules. Procedia Comput. Sci. **4**, 106–115 (2011)

Lim, J.T., Samah, A.A.: Weather and Climate of Malaysia. University of Malaya Press (2004)

Lohani, A.K., Goel, N.K., Bhatia, K.K.S.: Comparative study of neural network, fuzzy logic and linear transfer function techniques in daily rainfall-runoff modelling under different input domains. Hydrol. Process. **25**(2), 175–193 (2011)

Lu, Y., Ma, T., Yin, C., Xie, X., Tian, W., Zhong, S.: Implementation of the fuzzy c-means clustering algorithm in meteorological data. Int. J. Database Theory Appl. **6**(6), 1–18 (2013)

MacKay, D.J.C.: A practical Bayesian framework for backpropagation networks. Neural Comput. **4**(3), 448–472 (1992a)

MacKay, D.J.C.: Bayesian interpolation. Neural Comput. **4**(3), 415–447 (1992b)

Maier, H.R., Dandy, G.C.: Neural networks for the prediction and forecasting of water resources variables : a review of modelling issues and applications **15,** 101–124 (2000)

Maqsood, I., Khan, M., Abraham, A.: An ensemble of neural networks for weather forecasting. Neural Comput. Appl. **13,** 112–122 (2004)

Møller, M.F.: A scaled conjugate gradient algorithm for fast supervised learning. Neural Netw. **6** (4), 525–533 (1993)

Moré, J.J.: The levenberg-marquardt algorithm: implementation and theory. In: Lecture Notes in Mathematics, pp. 105–116. Springer (2015)

Nor, M., Rakhecha, P.: Analysis of a severe tropical urban storm in Kuala Lumpur, Malaysia. In: 11th International Conference on Urban Drainage, pp. 1–9 (2008)

Okut, H., Wu, X.-L., Rosa, G.J.M., Bauck, S., Woodward, B.W., Schnabel, R.D., Gianola, D.: Predicting expected progeny difference for marbling score in Angus cattle using artificial neural networks and Bayesian regression models. Genet. Sel. Evol.: GSE **45**(1), 34 (2013)

Pellakuri, V., Rajeswara Rao, D., Lakshmi Prasanna, P., Santhi, M.V.B.T.: A conceptual framework for approaching predictive modeling using multivariate regression analysis vs artificial neural network. J. Theor. Appl. Inform. Technol. **77**(2), 287–290 (2015)

Reed, R.D., Marks, R.J.: Neural Smithing: Supervised Learning in Feedforward Artificial Neural Networks. Mit Press (1998)

Rezaeianzadeh, M., Tabari, H., Arabi Yazdi, A., Isik, S., Kalin, L.: Flood flow forecasting using ANN, ANFIS and regression models. Neural Comput. Appl. 14–16 (2013)

Sabit, H., Al-Anbuky, A.: Multivariate spatial condition mapping using subtractive fuzzy cluster means. Sensors (Switzerland) **14**(10), 18960–18981 (2014)

Shahi, A.: An effective fuzzy c-mean and type-2 fuzzy. J. Theor. Appl. Inf. Technol. 556–567 (2009)

Shamseldin, A.Y.: Application of a neural network technique to rainfall-runoff modelling. J. Hydrol. **199**(3–4), 272–294 (1997)

Sheela, K.G., Deepa, S.N.: Review on methods to fix number of hidden neurons in neural networks. Math. Probl. Eng. (2013)

Shewchuk, J.R.: An introduction to the conjugate gradient method without the agonizing pain. Science **49**(CS-94–125), 64 (1994)

Srivastava, G., Panda, S.N., Mondal, P., Liu, J.: Forecasting of rainfall using ocean-atmospheric indices with a fuzzy neural technique. J. Hydrol. **395**(3–4) (2010)

Srivastava, N., Hinton, G., Krizhevsky, A., Sutskever, I., Salakhutdinov, R.: Dropout: a simple way to prevent neural networks from overfitting. J. Mach. Learn. Res. **15,** 1929–1958 (2014)

Sugumar, R., Rengarajan, A., Jayakumar, C.: A technique to stock market prediction using fuzzy clustering and artificial neural networks. Comput. Inf. **33**(5), 992–1024 (2015)

Toth, E., Brath, A., Montanari, A.: Comparison of short-term rainfall prediction models for real-time flood forecasting. J. Hydrol. **239**(1–4), 132–147 (2000)

Vega-corona, A.: ANN and Fuzzy c-Means Applied to Environmental Pollution Prediction, 1–6 (2012)

Yuan, H.L., Gao, X.G., Mullen, S.L., Sorooshian, S., Du, J., Juang, H.M.H.: Calibration of probabilistic quantitative precipitation forecasts with an artificial neural network. Weather Forecast. **22**(6), 1287–1303 (2007)

Neural Network Configurations Analysis for Identification of Speech Pattern with Low Order Parameters

Priscila Lima[1($\boxtimes$)], Allan Barros[1], and Washington Silva[2]

[1] Federal University of Maranhão, Portugueses Avenue, 1966, Bacanga Village, São Luís-Ma, Brazil
priscilalima_rocha@hotmail.com, akduailibe@gmail.com
[2] Federal Institute of Maranhão, Getúlio Vargas Avenue, 4, Monte Castelo, São Luís-Ma, Brazil
washington.silva@ifma.edu.br

Abstract. This work proposes the analysis between two neural network configurations for development a intelligent recognition system of speech signal patterns of numerical commands in Brazilian Portuguese. Thus, the Multilayer Perceptron (MLP) and Learning Vector Quantization (LVQ) networks are evaluated their performance in the course of training, validation and testing in speech signal recognition, whose pattern of speech signal is given by a two-dimensional time matrix, resulting of the encoding of the mel-cepstral coefficients (MFCC) through application of discrete cosine transform (DCT). These patterns have reduced set of parameters and the configurations of neural network in analysis use few examples for each pattern through training. It was carried out many simulations for network topologies and some selected learning algorithms to determine the network structures with best hit and generalization results. The potential this proposed approach is shown by check up on obtained outcomes with others classifiers, represented by Gaussian Mixture Models (GMM) and Support Vector Machines (SVM).

1 Introduction

The growing researches in pattern recognitions show the intention in improve the efficiency of systems based on this field due to practical applications that are motivated by necessity of people in communicate with computational machines through natural language and the interest in idea of designing and building smart machines that can carry out some tasks with abilities comparable to human abilities [1].

Among these applications, it has the systems based on pattern of speech signal recognition. These systems bring benefits on several scopes, such as telephony; on automobilistic industry; on computer and robotic systems, beside help individuals with disabling health condition through automation of the domestic and hospital environments [2–8].

© Springer International Publishing AG 2018
Y. Bi et al. (eds.), *Intelligent Systems and Applications*,
Studies in Computational Intelligence 751,
https://doi.org/10.1007/978-3-319-69266-1_17

The task of pattern classification consists in classify or categorize the patterns through a specific set of properties and characteristics. The classes may be previously defined by designer of system (supervised classification) or they are determined by similarities among characteristics presented by patterns (unsupervised patter).

In spite of be a task of high computational cost, the design and use of elaborate methods of pattern analysis and classification got more accessible due to advance and availability of several computation resort. This way, the necessary stages for implementation of pattern recognition system can be defined as follow [9]:

1. Acquisition of data, preprocessing and extraction of feature more discriminative;
2. Representation of data;
3. Definition of classifier for made decision.

The first and second stages have many approaches. In case of the recognition of speech signal patterns, the improve efficiency on recognition was possible because the progress in the digital signal processing techniques. These techniques allowed withdraw from speech signal the necessary elements for carrying out the recognition either speaker or locution [10,11]. Thus, during theses stages, the techniques of digital encoding worry in to obtain representations with few parameters of speech signal with purpose decrease computational cost and highlighting significant feature of speech signal for maximize the performance in recognition. This way, it is necessary remove information of the speech signal that do not contribute to acoustic features and accentuate the phoneme combinations that compound the speech signal to obtain an appropriate encode [12–15].

The thirst stage in pattern recognition process, that constitutes the intelligent core of system in charge of the making decision, is given by classical linear discriminant analysis by the most of classic recognizers, as Hidden Markov Models. This is still the dominant project on theoretic referential of pattern recognition, that also called Bayes' theorem [9,10].

The Bayes' theorem is a method of probabilistic classification based on the characteristics and classes of a problem may be modeled as random variables. In this context, if a pattern is chosen among patterns to be classified, the class which it belong would be a holding of a discrete random variable. The same way, the values of the characteristic of randomly chosen pattern can be as holding of random variable, that are normally continuous.

So, it have the samples of the population of patters to be classified, the estimates of the prior probabilities of the classes $\{\Pr(C_i)\}$ and the densities of conditional probability $p_X(\mathbf{x}|C_i)$ of their characteristics are available. Thus, the solution for classification problem, that is, the posterior probability of a class given unknown pattern, is the solving Bayes' formula, that is given by (1):

$$\Pr(C_i|\mathbf{x}) = \frac{p_X(\mathbf{x}|C_i)\Pr(C_i)}{\sum_j p_X(\mathbf{x}|C_j)\Pr(C_j)} \tag{1}$$

Even so Bayes mathematical formality, there is a huge difficult in practical applications due to estimation of quantities of right-hand side of Eq. (1). The obtaining of a good estimation of prior probabilities of the classes $\{\Pr(C_i)\}$ is often a easy task, carried out through a simple frequency counting of each class in the sample. Conversely, the estimation of the likelihoods $p_X(\mathbf{x}|C_i)$ is subject to problem called *curse of dimensionality*, that means if is necessary a reliable estimate of likelihood, the number of pattern would rise exponentially with dimension of feature vector.

Thus, when low level representation of patterns are used, the number of features may be large. In this context, the approach for classifier represented by neural networks are an alternative interesting in place of Bayes' classification because they can provide a direct estimate of the posterior probabilities of the classes and likelihoods [16–18].

The potential of neural networks in pattern classification tasks is derived of analogy that their computation structure make based on connection among neuronal synapses of the human brain. As a result of this, neural networks provide robust solutions due to capability learn complex information and patterns merely through examples that characterize the environment which they are working. Other attractive of the neural networks is that they can adjust the transformations of data provide during learn process just adjustment synaptic weight, allowing their application in time variant systems [19,20].

Thus, to the end learn process through reduced number of input-output dates from system in analysis, the neural network are also able to generalize the information learned. In doing so, when new information originating from environment are presented to network, it will present adequate result according to learned features. There are some kinds of neural network configurations that can be applied in solution of several tasks. In special case of pattern classification, the literature point out Multilayer Perceptron-MLP and Learning Vector Quantization-LVQ due to obtained results [21–23].

For these reasons, it is presented the development of an identification system of speech signal given by numerical commands. Multilayer Perceptron and Learning Vector Quantization will be assessed by their carrying out during training, validation and testing in speech signal recognition, whose pattern of speech signal is given by a two-dimensional time matrix, resulting of the encoding of the mel-cepstral coefficients (MFCC) through application of discrete cosine transform (DCT). These patterns have reduced set of parameters and the MLP and LVQ networks are trained with a few examples of each pattern.

2   Neural Network Configurations Analysis

The propose this work is analyze two neural network configurations, Multilayer Perceptron and Learning Vector Quantization during training and testing stages to verify the potential of these configurations as classifiers of speech signal recognition system. For this application, samples of spoken numeric commands from Brazilian Portuguese ('0' to '9') are encoded on low order parameters set, generated through bi-dimensional DCT time matrix formed by 2×2, 3×3 and 4×4 elements. The parameters enclosed into bi-dimensional DCT time matrix preserve temporal information of short-term and long-term beside frequency information, given by spectral envelope.

This way, the performance analysis of the MLP and LVQ configurations as speech signal pattern classifiers, where the patterns are defined by few parameters provide from bi-dimensional DCT time matrix, and verification of relevance of the results by means of comparative with other approaches are important contributions this research.

Thus, after speech signal encoding, it was adopted some procedures for evaluate neural networks behavior in research study. So, it was considered two stages: training jointly validation of pre-establishment topologies and check generalization through testing with new samples of speech signal.

During training and validation phase, the proposed topologies are submitted to patterns composed by 4, 9 and 16 parameters in a effort to verify network response when the number of input is increased. It is training phase that all specified topological elements are combined. At least, as selection criteria of the topologies to testing phase, it was determined only configurations that presented validation global hit result superior to 80% will be testing.

Finally, testing are realized with different speech signal parameters that were used in training process and obtained results by topologies are evaluated taken into consideration both hit rate and quantity of neurons used.

Therefore, in face of methodology applied on preparation this work, it can defined between neural networks configurations studied, the configurations that are better adequate to speech recognition system for classification of speech signal pattern determined by reduced number of parameters. The performance study of the LVQ network in this work for low order predictive models of speech signal provide an optional approach for classifier, since MLP configuration is neural network most carried out in pattern classification problems.

3   Neural Networks for Speech Identification System

The correct classification of the numeric commands in Portuguese by proposed recognition system is performed through some stages presented by block diagram in Fig. 1.

3.1   Codification of Speech Signal

General problem of the processing and manipulation of the information contained in speech signal consists in obtaining a representation for this signal through a model and then, to apply a transformation in order to leave it in a more convenient form. At last, it is made the extraction and utilization of the

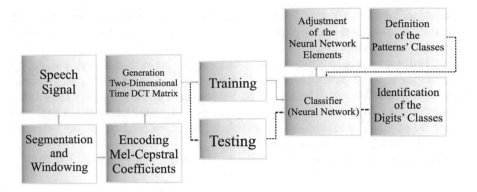

Fig. 1. Scheme of neural networks applied to speech identification system

characteristic contained in speech signal. However, before phase of processing of signal, it is necessary to carry out the right acquisition this signal. In this section, it is described all these procedures.

3.1.1   Database of Speech Signal

The voice bank with samples of the Portuguese locutions of the digits from 0 to 9 was obtained through three other voice banks. For obtain each them were made procedures of acquisition with particular characteristics, such as noise control or pause between the word pronunciation.

Thus, this shows that the system can work in different situations of the acquisition process and generate good results. It was used three bank voice: Voice Bank from University of São Paulo (EPUSP), Institute of Telecommunications (Inatel) and Federal Institute of Maranhão (IFMA). These voice banks are constituted by male and female speakers, all in the age group 18–30 years old.

Although three voice bank adopt some different procedures during the acquisition, all them used the same sampling frequency, $f_a = 22050$ Hz, with 16-bits resolution.

3.1.2   Speech Signal Preprocessing and Mel-Cepstral Extraction

The phase of signal preprocessing of includes to make the segmentation and windowing of the samples of speech signal from constructed database. The segmentation is necessary for limit the time interval that certain number of parameters are considered valid for Fast Fourier Transform calculation. After the segment size defined, the window function is applied in each these segments, making overlap between windows to avoid the attenuation of the signal samples that happen in the end of the window.

Thus, developed speech signal preprocessing algorithm used Hamming function for carry out the windowing of the segments, since in speech recognition systems, the Hamming function is almost unique used. The adopted overlap percentage between window functions was 50%. It was also calculated the window size in samples by multiplication length of time of the window $T_\omega = 20$ ms by sampling frequency f_a.

At last speech signal preprocessing, the specific characteristics of speech signal are extracted to use in recognition phase. These characteristics are given by mel-cepstral coefficients. These coefficients translate the no linearity that sound is perceived by human ear. This is possible due to power of speech signal necessary to obtain the mel-cepstral coefficients be calculated in frequency bands positioned on logarithm scale called *mel*.

Therefore, the mel-cepstral coefficients can be obtained through Eq. (2):

$$\text{mfcc}[k] = \sum_{i=1}^{N_F} E[i] \cos \left[\frac{i(k - 0.5)\pi}{N_F} \right] \tag{2}$$

where k represents the number of mel-cepstral coefficients; N_F is the quantity of filters and $E[i]$ is the log power output of the i-th frequency band.

3.1.3   Determination of Bi-dimensional Time Matrix

The mel-cepstral coefficients obtained from speech signal samples were encoded by application of the Discrete Cosine Transform (DCT) that condense the alternation of the spectral envelope of the speech signal over time [24]. So, this encoding resulted in formation of a two-dimensional DCT time matrix, given by (3):

$$C_k(n, T) = \frac{1}{N} \sum_{t=1}^{T} \text{mfcc}_k(t) \cos \left[\frac{(2t - 1)n\pi}{2T} \right] \tag{3}$$

where k, that varies of $1 \le k \le K$, is the k-th line component of t-th segment of the matrix. K is the number of mel-cepstral coefficients; n, that varies of $1 \le n \le N$, is the n-th column. n is the order of the matrix DCT; T is the number of vectors of observation of the mel-cepstral coefficients in time axis; $\text{mfcc}_k(t)$ represents the mel-cepstral coefficients.

Consequently, all examples for a given numeric command **D** have a two-dimensional DCT time matrix C_{kn}^{jm}, being $j = 0, 1, 2, \ldots, 9$ the number to be identified and the samples taken for each numeric command given by $m = 0, 1, 2, \ldots, 9$.

The parameters of two-dimensional matrix C_{kn}^{jm} were adequate in form the column vectors C_N^{jm}, being that N represents the quantity of elements in C_{kn}^{jm}. The organization of the parameters in C_N^{jm} maintains the temporal order of mel-cepstral coefficients and its formula is specified by (4):

$$C_N^{jm} = \left[c_{11}^{jm}, \ c_{12}^{jm}, \ \ldots, \ c_{1n}^{jm} c_{21}^{jm}, \ c_{22}^{jm}, \ \ldots, \ c_{2n}^{jm}, \ \ldots, \ c_{kn}^{jm} \right]' \tag{4}$$

In doing so:

$$C_N^{00} = \begin{bmatrix} c_{11}^{01}, & c_{12}^{01}, & \dots, & c_{1n}^{01}, c_{21}^{01}, & c_{22}^{01}, & \dots, & c_{2n}^{01}, & \dots, & c_{kn}^{01} \end{bmatrix}'$$

$$C_N^{01} = \begin{bmatrix} c_{11}^{02}, & c_{12}^{02}, & \dots, & c_{1n}^{02}, c_{21}^{02}, & c_{22}^{02}, & \dots, & c_{2n}^{02}, & \dots, & c_{kn}^{02} \end{bmatrix}'$$

$$\vdots$$

$$C_N^{09} = \begin{bmatrix} c_{11}^{09}, & c_{12}^{09}, & \dots, & c_{1n}^{09}, c_{21}^{09}, & c_{22}^{09}, & \dots, & c_{2n}^{09}, & \dots, & c_{kn}^{09} \end{bmatrix}'$$

$$C_N^{10} = \begin{bmatrix} c_{11}^{10}, & c_{12}^{10}, & \dots, & c_{1n}^{10}, c_{21}^{10}, & c_{22}^{10}, & \dots, & c_{2n}^{10}, & \dots, & c_{kn}^{10} \end{bmatrix}'$$

$$C_N^{11} = \begin{bmatrix} c_{11}^{11}, & c_{12}^{11}, & \dots, & c_{1n}^{11}, c_{21}^{11}, & c_{22}^{11}, & \dots, & c_{2n}^{11}, & \dots, & c_{kn}^{11} \end{bmatrix}'$$

$$\vdots$$

$$C_N^{19} = \begin{bmatrix} c_{11}^{19}, & c_{12}^{19}, & \dots, & c_{1n}^{19}, c_{21}^{19}, & c_{22}^{19}, & \dots, & c_{2n}^{19}, & \dots, & c_{kn}^{19} \end{bmatrix}'$$

$$\vdots$$

$$C_N^{99} = \begin{bmatrix} c_{11}^{99}, & c_{12}^{99}, & \dots, & c_{1n}^{99}, c_{21}^{99}, & c_{22}^{99}, & \dots, & c_{2n}^{99}, & \dots, & c_{kn}^{99} \end{bmatrix}'$$

Therefore, the vector C_N^{jm} was defined as the pattern for each digit **D** to be recognized. The conversion between the two-dimensional matrix C_{kn}^{jm} and the column vector C_N^{jm} was made to turn appropriate the parameters of speech signal pattern with the inputs of the neural network. This way, the number of neural network inputs was determined according to dimension of C_N^{jm}. Thus, the performance analysis of the neural networks was also carried out in order to increase of number of parameters that compose input patterns. Because this, two-dimensional matrices C_{kn}^{jm} of order $n = 2, 3$ and 4 were generated and obtained three kind of patterns, represented by column vector C_N^{jm} with $N = 4$, 9 and 16, respectively.

3.1.4 Training and Testing Sets

Regarding training and testing phase of the MLP and LVQ neural networks, the voice banks EPUSP, INATEL and IFMA provide necessary samples for get the digit patterns. In order to verify the neural network behavior with increase number of parameters, it was formed sets with patterns C_N^{jm} where N was changed to 4, 9 and 16. The structure of training and test sets is explain as follow:

1. Training Sets Ω_{NL}^{Tr}: It is composed by total of 200 locutions. This set is characterized by Tr, L is the quantity of samples and N informs which number of parameters that form the pattern is being used. For each digit, it was used 20 samples ($m = 20$). The selection of the locutions among voice banks was made as follow: 60 examples from EPUSP bank; 40 samples from INATEL bank and finally, 100 locutions from IFMA bank. Half of all samples that compose the set is from female speaker and the other, from male speakers. There are two subsets from the training set. The first subset, called estimation subset Ω_N^E is used for modify the synaptic weights of the network and the second, called validation subset Ω_N^V, is deployed for prevent overfitting and consequently ensure the generalization of the trained topologies. Ω_N^E subset has 80% of all samples of the training set and the remaining is to validation subset ($\Omega_{N200}^{Tr} = \{\Omega_N^E \cup \Omega_N^V\}$).

2. Testing Set Ω^T_{NL}: For this set, 20 speakers were selected, where 10 speakers are female (Ω^{TF}_{NL}) and 10 speakers are male (Ω^{TM}_{NL}). All speakers belong to IFMA bank, however these persons are different from these individuals that participate of the training. The number of samples for each digit spoken by each speaker was 10 ($m = 10$), So, the whole test set has 2000 exemplar for applying during testing ($\Omega^T_{N2000} = \{\Omega^{TM}_{N1000} \cup \Omega^{TF}_{N1000}\}$).

3.1.5   Neural Network Structure

The configurations presented in literature have a set of changeable topology elements that, chosen in appropriate form, allow neural network presents a minimum error of response in its output for problem to be solve and present a good generalization when different samples are used in testing phase.

The MLP and LVQ networks were chosen for performance analysis in recognition pattern of speech signal in this work because both configurations of neural networks have excellent solutions in pattern classification applications, how demonstrate specialized bibliography [22, 23].

Thereby, for obtaining the adequate topology of both configurations, it was made many association among learning algorithms chosen for training and topological units. It was also taken in consideration the adopted procedures in others related works. The details about MLP and LVQ design are shown afterward:

LVQ Design

LVQ network structure was define by learning rate η and number of neurons n in competitive layer. The amount of variable elements of the LVQ network topology are few and first, it was proposed the set $\eta = 0.01, 0.1, 0.5, 0.9$ and the set $n = 20, 25, 30, 35, 40, 45, 50, 55, 60$ for simulating the different topologies and allowing the choice of the best among them. The values chosen to η are frequently used in literature [20] and the n set was defined with number of neurons superior to input vector dimension and concurrent, they also must be superior to quantity of output or classes.

The inputs of neural network are specified by length of C^{jm}_N, that is, the simulated neural networks have 4, 9 and 16 inputs. The outputs of neural network are given by number of numeric command from '0' to '9' to be classified, that is, the output layer of neural network have 10 neurons, one for each class. This procedure is known as one c-classes method [25]. It is observed an increase of 5 among the number of neurons in n for verify the neural network behavior in recognizing of patterns with increase of the number of neurons in hidden layer.

This way, all the combinations among the elements of topology were made and, during the simulation, it was checked that the increase in learning rate value η led to fast convergence of the algorithm, but in unstable form, presenting mean squared error (MSE) greater than the project error of 10^{-3}. Thus, it was only used the topologies that have had $\eta = 0.01$. The number of epoch was 1000 that was enough for convergence of algorithm.

In Table 1 there are the elements of topology and training algorithm simulated in the course of training stage for LVQ neural networks.

Table 1. Component of LVQ configuration for training

Component	Representation	Specified interval
Training algorithms	–	LVQ-1
Number of hidden layers	n_1	20, 25, 30, 35, 40, 45, 50, 55, 60
Learning rate	η	0.01
Number of epoch	–	1000

MLP Design

In contrast the LVQ neural networks, the MLP neural networks have many elements of topology and training algorithms that may be associated in the simulations during training. The choice of number of inputs and outputs, η value, n_1 set and number of epoch for MLP networks have the same reasons given for the LVQ networks.

MLP networks topologies were simulated with 1 and 2 hidden layers because for pattern classification problems, the use until two hidden layers is enough for this application [19,21]. Consequently, it was possible to verify an improvement or not in learning and generalization of MLP networks with increase of number of hidden layers. For simulations relate to the MLP networks with two hidden layers, it was defined that the second hidden layer has 15 neurons. This value was specified in order to be the lowest value among all values belong to n_1 set and, at the same time, it must be superior to class number. The quantity of neurons on second hidden layer was the same for all simulations with the values of n_1 set.

The other important element in a MLP network is activation function. The hyperbolic tangent function was used in all neurons because it is a no-linear function that can map any kind of problem and it presents a continuous range between $[-1, 1]$ which is the same range of normalization of the patters of input of the neural network [21]. Gradient Descent (GD), Gradient Descent with Momentum (GDM), Resilient Backpropagation (RP) and Levenberg-Marquardt (LM) are four different learning algorithms for training of the MLP topologies that were analyzed. This allowed to check which algorithm shows the best results for set of parameters applied to network input.

A important factor of performance for MLP network that should be taken into consideration is the appropriate attribution of values for weight matrices in beginning of each training. This specification, many times realized random form, it will produce different responses that can be adequate or not for application.

So that, for each quantity of neurons belonging to n set was made for simulations (TR_1, TR_2, TR_3, TR_4) with distinct initial values for weights, nevertheless these values are included in the uniform distribution range from -0.01 to 0.01. This interval of random initialization of weights is justified by the fact to be smaller than interval of values that comprehend the parameters of patterns from training set, avoiding saturation of activation function used by neurons, preventing neural network from convergence.

Therefore, with these procedures during training phase, it examine the training duration in addition to increase of capacity of extrapolating for data that were not used in the course of learned process, since weight initialization influence on training speed and quality of obtained solution after convergence, providing a greater rate of correct classification for new data [26].

In summary, the topological units and leaning algorithms defined for simulations of MLP configuration are presented in Table 2.

Table 2. Components of MLP configuration for training

Component	Representation	Specified interval
Training algorithms	–	GD, GDM, RP, LM
Number of hidden layers	–	1 and 2
Number of hidden neuron 1 hidden layer	n_1	20, 25, 30, 35, 40, 45, 50, 55, 60
Number of hidden neuron 2 hidden layers	n_1 and n_2	$n_2 = 15$
Learning rate	η	0.01
Momentum term	α	0.8
Number of epoch	–	1000
Activation function	–	Hyperbolic tangent

4 Experimental Results

After the simulations of all possible topological combinations for MLP and LVQ neural networks, the obtained results on training and testing phase for the two configurations in analysis are presented in the next subsections. It is important to remember that all simulations of the training phase and performed testing were made with each three kind of patterns of speech signal.

4.1 LVQ Training and Validation

In Figs. 2, 3 and 4, respectively, are shown achieved results, in percentage, of the global hit of digits on training and validation in order to n neurons set simulated, using the patterns C_4^{jm}, C_9^{jm} and C_{16}^{jm}.

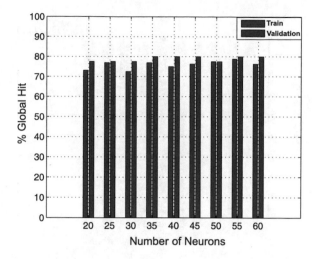

Fig. 2. LVQ C_4^{jm}: Global hit after training and validation stages

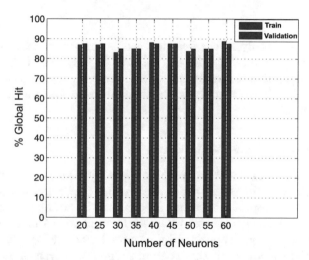

Fig. 3. LVQ C_9^{jm}: Global hit after training and validation stages

4.2   LVQ Testing

For carrying out testing, just the trained simulations that presented global validation hit greater than 80% were used. Finishing tests, the topology that presented the highest global mean hit in relation to the testing set was chosen as the best topology for speech recognition. However, based on established criterion, if more than one topology present the same global mean hit, the topology with smaller number of neurons has priority in the choice.

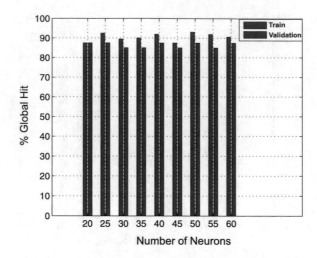

Fig. 4. LVQ C_{16}^{jm}: Global hit after training and validation stages

Thus, after this analysis, the two best scores (in percentage) for application of testing in Ω_{N1000}^{TM} and Ω_{N1000}^{TF} for the patterns C_4^{jm}, C_9^{jm} and C_16^{jm} are shown in Table 3. The highlighted values in Table 3 for each tested pattern represent the topologies that have not only highest global mean hit, but also the smaller number of neurons.

4.3 MLP Training and Validation

In the end of all simulations that combine topology elements and training algorithms, the achieved outcomes made possible the performance analysis as well as the choice of appropriate topology. Regarding the training algorithms used, two of them, Gradient Descent and Gradient Descent with Momentum exhibited results with levels below to specified by proposed methodology. It was observed that this algorithms conduced the training to high error values, indicating that simulated topologies did not extract of the presented patterns the necessary informations for right classification of the numeric commands. For this reason, the responses theses algorithms are not exposed to reader.

The results obtained with RP algorithm were good, but almost always less than the results obtained with LM algorithm. Therefore, this work presents only the results for MLP configuration, using LM training algorithm in topologies with one hidden layer, where the inputs are C_4^{jm}, C_9^{jm} and C_{16}^{jm}.

From Figs. 5, 6 and 7 are displayed the training responses and from Figs. 8, 9 and 10 are shown validation results obtained to MLP topologies with one hidden layer.

In accordance with Figs. 5, 6, 7, 8, 9 and 10, it is observed that the behavior of four simulations made for each value belonging to neuron set was very different. Initial weights used in each of theses training determined convergence of a same

Table 3. LVQ networks: female and male speakers tests

	C_4^{jm}		C_9^{jm}		C_{16}^{jm}	
	40	60	20	45	25	45
Loc_F1	91	94	88	89	94	95
Loc_F2	90	88	86	87	95	95
Loc_F3	80	78	81	83	90	90
Loc_F4	84	76	87	87	85	84
Loc_F5	79	87	82	87	96	97
Loc_F6	90	89	90	90	99	99
Loc_F7	57	55	72	75	73	74
Loc_F8	76	79	87	86	95	93
Loc_F9	69	74	73	85	80	80
Loc_F10	78	77	64	64	80	80
Loc_M1	62	62	69	68	80	78
Loc_M2	77	85	81	82	87	88
Loc_M3	77	78	80	79	85	86
Loc_M4	70	79	80	80	88	89
Loc_M5	64	67	64	61	74	79
Loc_M6	65	69	90	86	78	86
Loc_M7	72	68	82	75	84	82
Loc_M8	79	75	81	71	84	82
Loc_M9	63	65	69	73	79	79
Loc_M10	81	83	77	77	90	88
Mean	75.2	76.4	79.15	79.25	85.8	86.2

structure for adequate and in addition unsatisfactory results. This demonstrates that weight initialization impact on directly neural network performance.

4.4 MLP Testing

By the same token LVQ neural networks, after training and validation of proposed topologies, it was used the same procedures on MLP networks based on adopted testing criterion.

During testing, it was possible to verify the effect that the networks of one and two hidden layer caused in recognition result. As each proposed topology was carried out four times (T_1, T_2, T_3, T_4), only the training these topologies that presented validation results up 80% were chosen.

Due to the MLP networks topologies trained with LM learning algorithm present the best results in relation to RP algorithm, just the results of the networks simulated with LM algorithms are shown. The greatest response (in percentage) obtained in testing carried out, taking the MLP neural networks structured with 1 and 2 hidden layer by LM algorithm are verified in Table 4.

By result analysis present in Table 4 and based on established criterion for choice of the best topology for recognition problem, the highlighted results show that these topologies have not only less topological complexity but also a good global mean hit compared with the other topologies.

5 Analysis of Experimental Results

According to obtained results during the carrying out of proposed experiments, some considerations about the analysis of results are pointed out afterward:

1. The obtained results during training and validation phase for LVQ and MLP networks in three carried out experiments show that there was not significant increase in global hit with increase of the quantity of neurons in intermediate layer. It was established for MLP network that the increase in the number of hidden layer did not improve the results in recognition, showing that the networks can extract the specific features of presented patterns of speech signal with reduced number of topological structures.
2. The achieved results by MLP and LVQ neural networks present increase when the order of two-dimensional DCT time matrix was increased. It was verified that the neural network can extract particularity patterns of speech signal, due to the increase in number of parameters. By means of results, it also made sure of efficiency of the bi-dimensional parametrization through mel-cepstral coefficients and TCD on modeling of local and global variations of speech signal, providing appropriate and reduced patterns to neural network, where these patterns contain essential elements for good perform of classifier.

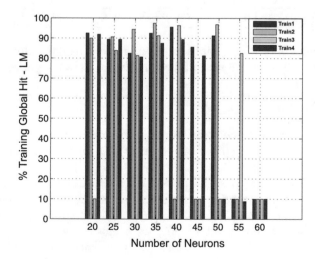

Fig. 5. MLP C_4^{jm}: Global hit after training stages for LM algorithm with single hidden layer

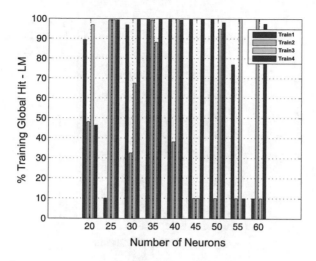

Fig. 6. MLP C_9^{jm}: Global hit after training stages for LM algorithm with single hidden layer

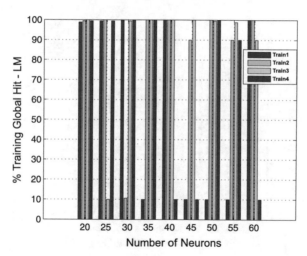

Fig. 7. MLP C_{16}^{jm}: Global hit after training stages for LM algorithm with single hidden layer

3. The initialization of weights influence the achieved results by LVQ and MLP neural networks. The determination of a right weight set to initialize the training process allow that the network has a fast convergence and not be lead up to local minimum greater than other existing minimums in the error surface. Thus, it is possible to achieve satisfactory results in relation to training time and generalization.

4. It was established that LVQ neural networks can be satisfactory used in pattern recognition problems, in special, the encoding of speech signal with

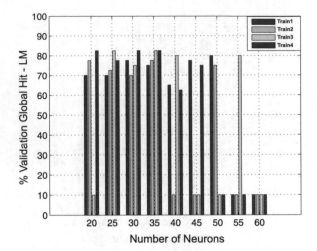

Fig. 8. MLP C_4^{jm}: Global hit after validation stages for LM algorithm with single hidden layer

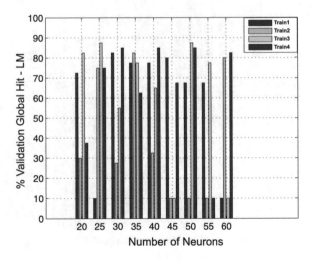

Fig. 9. MLP C_9^{jm}: Global hit after validation stages for LM algorithm with single hidden layer

a reduced parameter proposed in this work. This get obvious by closer performance than MLP network, that has many applications in patterns classifications and it is common used in speech recognition researches.

5. The obtained results by two configurations may be improved through application of optimization techniques in initialization of weigh, aiming to do convergence time reduce and increase the generalization of the neural networks.

6. It is summarized in Table 5 the achieved results by MLP and LVQ networks at the end the testing. The results show that the number of neurons used by

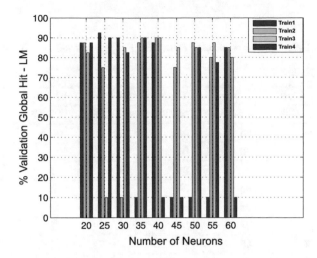

Fig. 10. MLP C_{16}^{jm}: Global hit after validation stages for LM algorithm with single hidden layer

two configurations for each pattern C_N^{jm} were similar and they used a reduced number of process units from n specified set.

6 Analysis with Other Classifier Methods

Finished performance analysis of MLP and LVQ configurations during learning and testing stages and, consequently, adequate topologies for proposed application identified, the achieved outcomes with adopted procedures were related to two others classifiers, represented by Support Vector Machine (SVM) and Gaussian Mixture Models (GMM). SVM and GMM algorithms used identical training and testing sets applied to LVQ and MLP networks, Ω_{NL}^{Tr} and Ω_{NL}^{T}. This way, the proposed methodology for speech recognition systems was corroborate.

For SVM training, it was made 100 iterations of elaborated algorithm and it was utilized as models the 10 machines that shown the best results in recognition. In this work, SVM classifier was trained with second order polynomial kernel function and Radial Basis Function with $\sigma = 0.03$. Final results from comparing classifiers are exhibited in Tables 6 and 7.

Table 4. Selection of best test results MLP C_4^{jm} trained by LM algorithm

	C_4^{jm}		C_9^{jm}		C_{16}^{jm}	
	1 layer	2 layer	1 layer	2 layers	1 layer	2 layer
	30-T4	45-T1	55-T3	50-T3	20-T1	50-T1
Loc_F1	94	88	90	92	95	92
Loc_F2	95	90	94	93	91	89
Loc_F3	83	82	93	91	97	97
Loc_F4	87	82	95	95	93	94
Loc_F5	91	88	91	96	90	95
Loc_F6	95	94	89	93	97	93
Loc_F7	67	65	76	85	92	93
Loc_F8	91	85	91	92	92	93
Loc_F9	75	73	82	77	82	79
Loc_F10	84	77	78	66	73	84
Loc_M1	72	77	74	83	84	86
Loc_M2	89	81	82	80	91	90
Loc_M3	86	86	89	85	90	91
Loc_M4	82	78	88	89	85	93
Loc_M5	72	71	65	64	59	69
Loc_M6	67	66	83	78	89	89
Loc_M7	73	73	69	77	74	88
Loc_M8	80	76	69	72	71	83
Loc_M9	73	70	88	84	82	82
Loc_M10	87	93	82	87	93	92
Mean	82.15	79.75	83.4	83.95	86	88.6

Table 5. Synthesis of the results for MLP and LVQ configuration

	C_4^{jm}		C_9^{jm}		C_{16}^{jm}	
	Number of neurons	% Test hit	Number of neurons	% Test hit	Number of neurons	% Test hit
LVQ	40	75.2	20	79.15	25	85.8
MLP	30	82.15	25	83.4	20	86

Table 6. Final results of classifiers for female locution recognition

		C_4^{jm}	C_9^{jm}	C_{16}^{jm}
Loc_F1	LVQ	91	88	94
	MLP	94	90	95
	SVM-Poli	68	62	65
	SVM-RBF	74	76	78
	GMM	92	84	88
Loc_F2	LVQ	90	86	95
	MLP	95	94	91
	SVM-Poli	65	65	66
	SVM-RBF	80	80	80
	GMM	94	89	82
Loc_F3	LVQ	80	81	90
	MLP	83	93	97
	SVM-Poli	60	60	77
	SVM-RBF	78	78	80
	GMM	88	88	95
Loc_F4	LVQ	84	82	96
	MLP	91	91	90
	SVM-Poli	66	72	72
	SVM-RBF	78	72	82
	GMM	70	72	83
Loc_F5	LVQ	79	90	99
	MLP	95	89	97
	SVM-Poli	66	72	72
	SVM-RBF	78	72	82
	GMM	70	72	83

Table 7. Final results of classifiers for male locution recognition

		Order of matrix = 2	Order of matrix = 3	Order of matrix = 4
Loc_M1	LVQ	77	81	87
	MLP	89	82	91
	SVM-Poli	70	66	73
	SVM-RBF	76	80	80
	GMM	57	64	72
Loc_M2	LVQ	77	80	85
	MLP	86	89	90
	SVM-Poli	67	63	71
	SVM-RBF	76	63	81
	GMM	80	87	91
Loc_M3	LVQ	77	80	88
	MLP	82	88	85
	SVM-Poli	62	63	70
	SVM-RBF	78	80	78
	GMM	52	67	77
Loc_M4	LVQ	77	64	74
	MLP	72	65	59
	SVM-Poli	68	63	69
	SVM-RBF	76	80	80
	GMM	66	70	71
Loc_M5	LVQ	79	77	90
	MLP	87	82	93
	SVM-Poli	66	66	74
	SVM-RBF	76	80	82
	GMM	72	74	86

References

1. Petry, F.E.: Speech recognition: a current perspective: in spite of limitations, areas of application are growing, and voice communication with computers may well be commonplace by the 21st century. IEEE Potentials **2**, 18–20 (1983). https://doi.org/10.1109/MP.1983.6499579
2. Husnjak, S., Perakovic, D., Jovovic, I.: Possibilities of using speech recognition systems of smart terminal devices in traffic environment. Proc. Eng. **69**, 778–787 (2014)
3. Špale, J., Schweize, C.: Speech control of measurement devices. IFAC-PapersOnLine **49**, 13–18 (2016). https://doi.org/10.1016/j.ifacol.2016.12.003
4. Breen, A., et al.: Voice in the user interface. In: Interactive Displays: Natural Human-Interface Technologies. https://doi.org/10.1002/9781118706237.ch3 (2014)
5. Bisio, I., et al.: Gender-driven emotion recognition through speech signals for ambient intelligence applications. IEEE Trans. Emerg. Top. Comput. **1**, 244–257 (2013). https://doi.org/10.1109/TETC.2013.2274797
6. Weng, F., et al.: Conversational in-vehicle dialog systems: the past, present, and future. IEEE Signal Process. Mag. **33**, 49–60 (2016). https://doi.org/10.1109/MSP.2016.2599201
7. Yang, Y., Li, L.: The design and implementation of a smart e-receptionist. IEEE Potentials **32**, 22–27 (2013). https://doi.org/10.1109/MPOT.2012.2213851
8. Singh, T., Yadav, N.: Voice recognition based advance patient's room automation. IJRET: Int. J. Res. Eng. Technol. **4**, 308–310 (2015)
9. Silva, W.L.S.: Intelligent genetic fuzzy inference system for speech recognition: an approach from low order feature based on discrete cosine transform. J. Control Autom. Electr. Syst. **25**, 689–698 (2014)
10. Bellegarda, J.R., Monz, C.: State of the art in statistical methods for language and speech processing. Comput. Speech Lang. **35**, 163–184 (2016). https://doi.org/10.1016/j.csl.2015.07.001
11. Youcef, B.C.: Speech recognition system based on OLLO French corpus by using MFCCs. In: Lecture Notes in Electrical Engineering. https://doi.org/10.1007/978-3-319-48929-2_25 (2017)
12. Sarma, M., Sarma, K.K.: Acoustic modeling of speech signal using artificial neural network: a review of techniques and current trends. In: Intelligent Applications for Heterogeneous System Modeling and Design. https://doi.org/10.4018/978-1-4666-8493-5.ch012 (2015)
13. Lee, C.H., Siniscalchi, S.M.: An information-extraction approach to speech processing: analysis, detection, verification, and recognition. Proc. IEEE **101**, 1089–1115 (2013). https://doi.org/10.1109/JPROC.2013.2238591
14. O'Shaughnessy, D.: Acoustic analysis for automatic speech recognition. Proc. IEEE **101**, 1038–1053 (2013). https://doi.org/10.1109/JPROC.2013.2251592
15. Silva, W.: Sistema de Inferência Genética-Nebuloso para Reconhecimento de Voz (System of fuzzy- genetic inference for speech recognition) Thesis, Federal University of Maranhão (2015)
16. Bridle, J.S.: Neural networks or hidden markov models for automatic speech recognition: is there a choice? In: Pietro, L., De Mori, R. (eds.) Speech Recognition and Understanding: Recent Advances, Trends and Applications, vol. 75, pp. 225–236. Springer, Heidelberg (1992)
17. McCrocklin, S.M.: Pronunciation learner autonomy: the potential of automatic speech recognition. System **57**, 25–42 (2016). https://doi.org/10.1016/j.system.2015.12.013

18. Picheny, M.: Trends and advances in speech recognition. IBM J. Res. Dev. **55**, 1–18 (2011). https://doi.org/10.1147/JRD.2011.2163277
19. Haton, J.P.: Neural networks for automatic speech recognition: a review. In: Chollet, G., et al. (eds.) Speech Processing, Recognition and Artificial Neural Networks: Proceedings of the 3rd International School on Neural Nets "Eduardo R. Caianiello", 1999, pp. 259–280. Springer, London (1999)
20. Nightingale, C., Myers, D.J., Linggard, R.: Introduction neural networks for vision, speech and natural language. In: Nightingale, C., Myers, D.J., Linggard, R. (eds.) Neural Networks for Vision, Speech and Natural Language, vol. 1, pp. 1–4. Springer, Netherlands (1992)
21. Siniscalchi, S.M., Svendsen, T., Lee, C.-H.: An artificial neural network approach to automatic speech processing. Neurocomputing **140**, 326–338 (2014). https://doi.org/10.1016/j.neucom.2014.03.005
22. Lippmann, R.P.: Review of neural networks for speech recognition. Neural Comput. **1**, 1–38 (1989). https://doi.org/10.1162/neco.1989.1.1.1
23. Hu, Y.H., Hwang, J.N. (eds.): Handbook of Neural Networks for Speech Processing. CRC Press, Washington DC (2014)
24. Kim, M.W., Ryu, J.W., Kim, E.J.: Speech recognition with multi-modal features based on neural networks. In: King, I., et al. (eds.) Neural Information Processing: 13th International Conference, ICONIP 2006, Hong Kong, China, October, 2006. Proceedings, Part II. Lecture Notes in Computer Science (Lecture Notes in Neural Information Processing), vol. 4233, pp. 489–498. Springer, Heidelberg (2006)
25. Veselý, K., Burget, L., Grézl, F.: Parallel training of neural networks for speech recognition. In: Sojka, P., et al. (eds.) Text, Speech and Dialogue: 13th International Conference, TSD 2010, Brno, Czech Republic, Sept 2010. Lecture Notes in Computer Science (Lecture Notes in Text, Speech and Dialogue), vol. 6231. Springer, Heidelberg (2010)
26. Yam, J.Y.F., Chow, T.W.S.: A weight initialization method for improving training speed in feedforward neural network. Neural Comput. **30**, 219–232 (2000)

Knowledge-Based Expert System Using a Set of Rules to Assist a Tele-operated Mobile Robot

David Adrian Sanders[1]([✉]), Alexander Gegov[1], and David Ndzi[2]

[1] Faculty of Technology, Portsmouth University, Portsmouth, UK
`tele-op-research@serg.org.uk`
[2] School of Engineering and Computing, University of the West of Scotland, Hamilton, UK

Abstract. This paper firstly reviews five artificial intelligence tools that might be useful in helping tele-operators to drive mobile robots: knowledge-based systems (including rule based systems and case-based reasoning), automatic knowledge acquisition, fuzzy logic, neural networks and genetic algorithms. Rule-based systems were selected to provide real time support to tele-operators with their steering because the systems allow tele-operators to be included in the driving as much as possible and to reach their target destination, while helping when needed to avoid an obstacle. A bearing to an end-point is added as an input with an obstacle avoidance sensor system and the usual inputs from a joystick. A recommended direction is combined with the angle and position of a joystick and the rule-based scheme generates a recommended angle to rotate the mobile robot. That recommended angle is then blended with the user input to assist tele-operators with steering their robots in the direction of their destinations.

1 Introduction

Five artificial intelligence tools are reviewed: knowledge-based systems (including rule based systems and case-based reasoning), fuzzy logic, automatic knowledge acquisition, neural networks and genetic algorithms. Each artificial intelligence tool is outlined and briefly reviewed. A Knowledge-based expert system using a set of rules is selected to help tele-operators to drive their mobile robots.

Applications of these tools have become more widespread and more complex mobile robot applications may require greater use of hybrid tools that combine the strengths of two or more of the tools. The tools and methods have minimal computation complexity and can be implemented on single robots or systems with low-capability microcontrollers. The appropriate deployment of the new AI tools will contribute to the creation of more efficient and effective mobile robot and tele-operated systems.

A rule-based system that describes knowledge in terms of IF…THEN…ELSE is selected. The moving machine obtains information about its local surrounding environment from sensors while moving towards a more global destination. Assistance is made available to help tele-operators avoid obstructions.

© Springer International Publishing AG 2018
Y. Bi et al. (eds.), *Intelligent Systems and Applications*,
Studies in Computational Intelligence 751,
https://doi.org/10.1007/978-3-319-69266-1_18

Systems presented in this paper help tele-operators drive when they cannot see (possibly because of smoke or because the view is obscured) or when the robot is in a remote setting away from the driver.

Tele-operated structures are often open-loop. Operatives communicate their desired direction and speed using a joystick. The robot then tends to move in the desired speed and direction. Tele-operator demands are processed and blended with inputs from the ultrasonics along with a more global destination end point to help the operatives to drive their robots. Local and global planning are mixed inside a knowledge-based expert system using a set of rules to help the operatives to steer their robots. Local information from ultrasonics [1] is blended with a global path.

Navigation for tele-operated mobile robots is discussed within academic literature [1–4]. Usually they have used a local algorithm and aimed to circumvent obstructions [5] and suggest movements based on local sensors [4].

Some work has planned initial paths for mobile robots and then modified them locally [1]. In this work, a local planner produces drive to motors attached to the driving wheels depending on input received from: transducers using ultrasonics, the joystick, and the global targets. The robots respond swiftly to the desires of the operatives and to unanticipated obstructions but has a tendency to move towards the goal objective on every occasion.

Huq defined a fuzzy blending of schemas that depended on context [6] and that eliminated a few restrictions that had become apparent in previous approaches. It used navigation based on goal orientation as well as avoiding obstacles within the robot path. Fuzzy logic has been blended with Genetic algorithms to solve mapping and location problems [7]. That automatically looked for a suitable local plan. Bennewitz and Burgard described a method to create random real time routes within undefined environments without using vision [1, 8], while tracking trajectories [9]. Hwang and Chang described techniques to avoid obstacles that used a fuzzy decentralized sliding-mode of control [10]. The potential field method was improved by Song and Chen by resolving some of the local minima problems [5] and Nguyen defined an obstacle avoidance system that used Bayesian Neural Networks [11].

A technique that improves a minimum-cost route is presented in this paper. A joystick mainly controls the speed but some simple AI systems also provides input [12–15]. The AI methods use rules that were perception based and similar to those described by Parhi and Singh, who used them for an independent self-directed mobile robot [1] and by Sanders et al. [16] who considered a tele-operated mobile robot.

Algorithms trade path length against distance to an obstacle(s). Rules generate a suggested steering angle and that steering angle is merged with the contribution provided by a joystick to create the signals to drive the mobile robot motors.

The system and the techniques were successfully proven using simulation and then the sensors and microcontrollers were mounted on a mobile robot (Fig. 1).

Many sensors can be used for obstacle avoidance, for example: structured light or laser [17]; ultrasonics [18]; or infra-red [19]. The more comprehensive methods sometimes perform poorly indoors [20] but simpler and more local sensors can successfully determine position, for example: gyros, odometers, tilt, and ultrasonic [21, 22]. Images of the space ahead of the robot can be converted into a digital format and can be useful when the view ahead of the robot is unobstructed but vision systems can

Fig. 1 Bobcat II mobile robot base avoiding an obstacle while being driven along a corridor

need more processing and they can be more complicated [23]. They are getting cheaper and computing power is quickly increasing [24]. The most accurate source of knowledge about the surroundings and situation comes from the human tele-operator but diminished visibility, separation and imperfect environmental information can reduce the ability of a human teleoperator [25].

Ultrasonics were selected for detecting ranges because it was inexpensive, uncomplicated, straightforward and rugged [26].

The paper continues with a review of the five artificial intelligence tools that were considered for this work followed by a description of the input from the sensors and joystick. Then the kinematics of the mobile robot base are described before discussing control and the artificial intelligence rule based tool selected. Then the testing and the results are described and the paper finishes with some discussion and conclusions.

2 Review of Some Artificial Intelligence Tools

Artificial Intelligence (AI) can improve teleoperation of mobile robots. AI has produced some useful tools for teleoperation that automatically solve problems normally requiring human brainpower. Five such tools are reviewed in this Section: fuzzy logic, knowledge-based systems, inductive learning, neural networks and genetic algorithms.

New advances are allowing seamless interactions between computers and people and the introduction of AI into teleoperation promises to make it more flexible, efficient and reliable. Tele operated mobile robots are exceeding human performance and as they merge with humans more intimately and we combine computer capacity with brain power to analyse, deliberate and make decisions, then we might be on the verge of a new assistive robot age.

2.1 Knowledge-Based Systems

Knowledge-based systems (sometimes called expert systems) are computer programs representing knowledge about solving problems. These systems typically have two principal parts, knowledge-bases and inference-mechanisms. Knowledge-bases hold knowledge about a domain that can be stated as arrangements of 'IF–THEN' rules, frames, factual statements, procedures, objects and cases.

Inference mechanisms manipulate stored knowledge to generate solutions. Knowledge manipulation methods include using constraints and inheritance (in object-oriented expert systems a frame-based expert systems), recovery and reworking of case examples (in a case-based system) and applying inference rules (within a rule-based system), corresponding to control procedures (forward or backward chaining) and search strategies (breadth or depth first).

Rule-Based Systems describe knowledge in terms of IF…THEN…ELSE. Decisions can be made using specific knowledge. They represent knowledge and decisions in ways that are understandable to human beings. Because of the rigid rule-base structure they can be poorer at handling uncertainty and imprecision. Typical rule-based systems have four fundamental components:

- the rules;
- an inference engine (or a semantic reasoner), that surmises information or acts depending on the interaction between the rules and the input(s);
- short-term memory;
- and user interfaces or alternative devices to input and output signals.

Case-Based Reasoning adapts solutions from earlier problems and applies them to existing problems. Solutions are stored in a database. The solutions can represent human experience. When a new problem is encountered, systems compare it with previous problems and select a problem that is most like the new problem. It then acts using the previous solution and records whether the action was successful or a failure. Case-Based Reasoning is effective at representative knowledge in a way that is easy and well-defined for humans, but they can also learn from previous examples by creating extra new solutions.

Case-based reasoning has been formalized as a process with four steps:

i. Retrieve: Recover cases from short-term memory that are applicable to solving a target problem. Cases include a problem, its solution, and, often, comments concerning the way that a solution was originated.

ii. Reuse: Map a solution from a previous case onto the target problem. The solution may need to be adapted automatically to fit a new situation.

iii. Revise: After mapping a previous solution onto a target situation, test the solution and revise it if necessary.

iv. Retain: Once successfully adapted then store the resultant occurrence as a new case within short term memory.

CBR is frequently described as an expansion of Rule-Based Systems. Both CBR and Rule-Based Systems are useful for denoting knowledge clearly but CBR systems can also learn from the past by automatically creating new cases.

A lot of expert systems are created using 'shells'; ready-made programs that are expert systems (including inferencing and knowledge storage but lacking domain knowledge). Sophisticated expert systems can be created using 'development environments'. Development environments are more flexible than shells. They provide ways for operators to employ their own inferencing and ways of representing knowledge.

Expert systems are probably the most mature methods from amongst the five tools considered here and lots of development tools and commercial shells are available. The building of a system can be relatively simple once domain knowledge has been extracted. Because they are relatively easy to develop, many applications have been created, for example for automatic robot programming and sequence planning.

2.2 Fuzzy Logic

A rule-based expert system cannot handle a situation not explicitly included within their knowledge base (that is, situations not fitting within the 'IF' statements within the rules). Rule-based systems cannot generate solutions when the encounter an unusual situation. They are consequently considered to be shallow systems which can fail in a 'brittle' fashion, rather than gradually, as a human expert would.

Fuzzy logic reflects the qualitative and inexact nature of human reasoning. They can help an expert system to be more robust. Exact values for variables are exchanged for linguistic descriptions, represented by fuzzy sets. Based on this representation, the inferencing takes place. For example, an assembly speed of 35 thingamabobs per minute could be replaced by 'normal' as a linguistic description of the variable 'assembly speed. A fuzzy set defining the term 'normal assembly speed' might be:

normal assembly speed = 0.0/below 15 thingamabobs per minute +0.5/15–25 thingamabobs per minute +1.0/25–35 thingamabobs per minute +0.5/35–45 thingamabobs per minute +0.0/above 45 thingamabobs per minute.

The values 0.0, 0.5 and 1.0 are the degrees or grades of membership of the production ranges below 15 thingamabobs per minute (or above 45 thingamabobs per minute.), 15–25 thingamabobs per minute (35–45 thingamabobs per minute), and 25–35 thingamabobs per minute to the given fuzzy set. A grade of membership equal to 1 indicates full membership and a null grade of membership corresponds to total non-membership.

Knowledge within expert systems using fuzzy logic can be expressed as qualitative statements, (or fuzzy rules), such as 'If apartment is at normal temperature, then set warmness inputs to normal'.

Reasoning procedures known as compositional rules of inference enable conclusions to be drawn by generalisation (interpolation or extrapolation) from qualitative information within a knowledge base. For example, when the normal assembly speed is perceived as 'slightly below normal', a controlling fuzzy expert system may well determine that inputs should be increased to 'slightly above normal'. Even though that conclusion may not have been covered by any fuzzy rule within the system.

Fuzzy Expert Systems use fuzzy logic to manage uncertainty produced by inadequate or partly corrupted data. Fuzzy logic uses a mathematical theory of fuzzy sets to mimic human logic. Humans easily deal with ambiguity when making decisions but computers still find it challenging.

Fuzzy logic has been used in mobile robotics, especially for control when domain knowledge has been imprecise. Fuzzy Logic is useful when there is imprecision. For instance, for object recognition and scene interpretation. Fuzzy expert systems are suitable for ambiguous and imprecise situations. They cannot learn because system values cannot be changed.

2.3   Automatic Knowledge Acquisition

Learning programs often need a set of examples to use but it can be time consuming and difficult to get domain knowledge into a knowledge base. That can create a bottleneck during the construction of an expert system. Automatic knowledge acquisition techniques were created to deal with that.

An example of an approach is 'divide-and-conquer'. Here attributes are selected according to a strategy that divides an example set into several subsets. A decision tree is then built to classify examples. The decision tree represents knowledge that is generalised from a set of specific examples. This can then be used to handle situations not covered by the example set.

Another example is a 'covering approach'. An inductive learning program endeavours to locate groupings of attributes that are uniquely shared by examples within classes and then form rules with the IF part as combinations of those attributes and the THEN part as the classes.

Another example is the use of logic programming in place of propositional logic to depict examples and characterise new concepts. That uses a more potent predicate logic to characterise training examples and background knowledge and to convey new concepts. That allows results from induction to be defined as unspecific first-order clauses with variables.

There are many learning programs such as:

- ID3 (a divide-and-conquer program),
- FOIL (an ILP system adopting generalisation/specialisation methods),
- AQ program (which follows a covering approach),
- and GOLEM, (ILP system based on inverse resolution).

Most of these sorts of programs generate crisp decision rules but some algorithms have been created that also produce fuzzy rules.

Automatic learning has been tricky to use with tele-operated mobile robots because they require a set of examples in a rigid format and few mobile robot problems are described easily within rigid sets of examples. Automatic learning is generally more suitable for problems with discrete or symbolic attribute values rather than those with continuous-values. A recent application of inductive learning is in the control of a laser cutting robot.

2.4   Neural Networks

Neural networks can capture domain knowledge from examples. However, they do not archive the acquired knowledge in an explicit form such as in rules or decision trees. They can readily handle both discrete and continuous data. They also have a generalisation capability (as for fuzzy systems).

Neural network models distribute computation between several simpler units called neurons. Neurons are interconnected and operate in parallel so that, neural networks can be called parallel-distributed-processing systems.

The most popular neural network is the multi-layer perceptron, which is a feedforward network: all signals flow in a single direction from the input to the output of the network. Feedforward networks can perform static mapping between an input space and an output space: the output at a given instant is a function only of the input at that instant. Recurrent networks, where the outputs of some neurons are fed back to the same neurons or to neurons in layers before them, are said to have a dynamic memory: the output of such networks at a given instant reflects the current input as well as previous inputs and outputs.

Implicit 'knowledge' is built into a neural network during training. Some networks can be trained by presenting them with typical input patterns and the corresponding expected output patterns. Errors between the actual and expected outputs are used to modify weights on connections between neurons. This is "supervised training". In a multi-layer perceptron, the back-propagation algorithm for supervised training is often adopted to propagate the error from the output neurons and compute the weight modifications for the neurons in the hidden layers.

Some neural networks are trained in an unsupervised mode, where only the input patterns are provided during training and the networks learn automatically to cluster them in groups with similar features.

Artificial Neural Networks typically have inputs and outputs, with processing within hidden layers in between. Inputs are independent variables and outputs are dependent. ANNs are flexible mathematical functions with configurable internal parameters. To accurately represent complicated relationships, these parameters are adjusted through a learning algorithm. Once trained then ANNs can accept new inputs and attempt to predict accurate outputs. To produce an output, the network simply performs function evaluation. The only assumption is that there exists some continuous functional relationship between input and output data. Like expert systems, they have found a wide spectrum of applications in almost all areas of robotics, addressing problems ranging from modelling, prediction, control, pattern recognition and optimisation.

2.5   Genetic Algorithms

A genetic algorithm is a stochastic optimisation procedure inspired by natural evolution. A genetic algorithm can yield a global optimum solution within a complex multi-modal search space without specific knowledge about a problem.

Potential solutions to a problem must be represented as strings of numbers known as chromosomes and there must be a means of determining the goodness of each chromosome. A genetic algorithm operates on a group or population of chromosomes at a time, iteratively applying genetically based operators such as cross-over and mutation to produce fitter populations containing better solution chromosomes. The algorithm normally starts by creating an initial population of chromosomes using a random number generator. It then evaluates each chromosome. The goodness values of the chromosomes are used in the selection of chromosomes for subsequent operations. After the cross-over and mutation operations, a new population is obtained and the cycle is repeated with the evaluation of that population.

Genetic algorithms have found applications in tele-operation problems involving complex combinatorial or multi-parameter optimisation. Some recent examples of those applications are in Robot Path Planning.

2.6   Combining Systems

The purpose of a hybrid system is to combine the desirable elements of different AI techniques within a single system. The different AI methods each have their own strengths and weaknesses. Some effort has been made in combining different methods to produce hybrid techniques with more strengths and fewer weaknesses. An example is a Neuro-Fuzzy system which seeks to combine the uncertainty handling of Fuzzy Systems with the learning strength of Artificial Neural Networks.

The nodes of a Fuzzy Network are fuzzy rule bases and the connections between nodes are interactions in the form of outputs from nodes that are fed as inputs to the same or other nodes. A fuzzy network is a hybrid tool combining fuzzy systems and neural networks due to its underlying grid structure with horizontal levels and vertical layers. This tool can be suitable for modelling the environment because separate areas can be described as modular fuzzy rule bases interacting in sequential/parallel fashion and feed forward/feedback context. The main advantages from the application of this hybrid modelling tool are better accuracy due to the single fuzzification-inference-defuzzification and higher transparency due to the modular approach used. These advantages can be crucial because of uncertainties in the data and the interconnected structure of the environment.

3   Selection of the Rule Based Expert System

A knowledge-based expert system was selected to assist in teleoperation using an inference mechanism because they are relatively simple and good at representing knowledge and decisions in a way that is understandable to humans.

A knowledge base was created that contained domain knowledge as a combination of 'IF–THEN' rules. An inference mechanism manipulated the knowledge to produce solutions to driving problems.

The rule-based system used IF…THEN…ELSE to make decisions. It had four basic components: a list of rules, an inference engine, temporary working memory and a joystick user interface.

4 Charting the Environment in Front of the Robot

The sensors that were used were similar to those described in [27, 28]. Ultrasonic sensors were mounted above the driving wheels on the front of the mobile robot. The time taken for an ultrasonic pulse to be reflected back to a sensor represented the distance to an obstacle. The robot is presented and explained in [29].

An imaginary potential field was placed around detected objects within the software [5, 21]. As the ranges to obstacles altered then the sensor system modified pulse lengths. The range-finder progressively elongated pulses if obstacles were not being sensed so as to build-up the range until an obstacle was eventually detected. That technique provided an earlier warning of upcoming problems.

Histogramic In-Motion Mapping was used to filter out false readings. Volumes in front of the mobile robot were separated into left and right matrixes, with NEARBY, INTERMEDIARY and DISTANT compartments in each matrix. There was also a matrix that represented the volume in the center where the ultrasonic volumes intersected. That matrix represented the case when obstacles are detected by both sets of transducers. If an obstacle was perceived in front of the mobile robot then it was categorised as NEARBY, INTERMEDIARY or DISTANT. Transducers were mounted on the chassis of the mobile robot in such a way that the ultrasonic envelopes intersected and covered the environment in front.

If something was sensed then a quantity correlated with a cell was increased by a comparatively big amount, e.g.: ten to fifteen. Remaining cells reduced in value by a lesser amount, e.g.: five, downwards towards 0. The result was a histogrammic representation of obstacles in front of the robot. A cell quickly increased in value if an obstacle entered it. Random misreads simply incremented for a solitary misread before the cell then reduced to zero again. When obstacles appeared in other cells then those cells rapidly increased in value. If the obstacle moved away from the initial cell, then the value of the initial cell decreased back to 0. A consistent and dependable range was obtained within 0.4 s.

5 The User Input

A Penny & Giles joystick was used that contained two potentiometers. Two A/D converters determined joystick location.

Data from the joystick was represented in Cartesian coordinates but were converted to polar coordinates.

$$|J|\angle\theta.$$

Where $|J|$ is a representation of the distance that the joystick has been shifted away from its central position. $|J|\angle\theta$ represented the velocity that an operative desired. $\angle\theta$ was the angle that represented the desired bearing from the robot position.

The confidence of the operative in their decision was assumed to be represented by how long the joystick remained in that position.

$$|J| = \sqrt{((JA * JA) + (JB * JB))} \tag{1}$$

where JA and JB represented the position of the joystick in Cartesian co-ordinates.

|J| and θ established joystick position and from that, the chosen velocity (bearing and speed) could be calculated. The position and the level of confidence were logged within a matrix so that each cell within the matrix consisted of two values:

- "*Confidence*" specified the amount of time that a joystick had remained still.
- "*Magnitude*" quantified the desired speed.

Jstickin was used as an input to the rule base and gave a level of confidence about the intentions of the operatives.

Histogrammic representation carried out pseudo-integration. If operatives held their joysticks still, then the value of cells associated with that position grew. Other cells reduced in value. The cell that had the largest amount within it signified the position of the joystick.

The function *JstickArray* identified the position occupied by the joystick and *AngConf* increased for that position and the value of the empty cells reduced. Cell values reduced quickly but incremented more slowly.

JstickArray cells rose to their highest value in roughly 0.4 s but decreased to 0 in roughly 100–200 ms.

A weight was set to direct the rate of increasing in value and a separate weight was set to direct the rate of decreasing in value. These weights were determined through experimentation and testing. The two weights can be established and adjusted depending on the tasking and the abilities of the human operative.

6 Mobile Robot Kinematics

Kinematics are represented in Fig. 1 and are explained here. There were a pair of larger front wheels and a pair of smaller trailing casters at the back. Turning the driving wheels moved the robot. Turning them separately turned the robot so that the direction changed. If r represents the radius of the driving wheels, then 2r represents the diameter (Fig. 2). Exploiting the notation used within [1], then W represents the distance between the driving wheels. The centre of gravity of the machine is C and P is placed at the junction of a line drawn through the centre of the machine and a line through the wheel axis. Distance between P and C is d.

Figure 3 shows the kinematics of the machine.

An assumption was made that there was not any slip between the floor and the wheels.

$$Vel_{tot} = 0.5 * (vel_{right} + vel_{left}) \tag{2}$$

$$\omega = 1/W * (vel_{right} - vel_{left}) \tag{3}$$

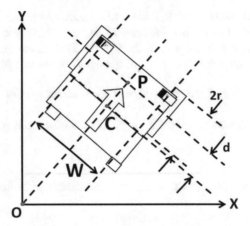

Fig. 2 Mobile robot geometry

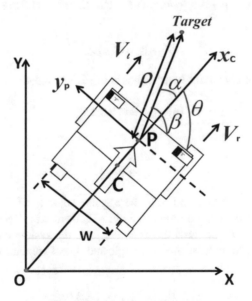

Fig. 3 Mobile robot kinematics

$$vel_{right} = r\omega_{right} \quad and \quad vel_{left} = r\omega_{left} \tag{4}$$

where v was the linear velocity ω was the angular velocity. Position in global coordinates was [O X Y]. In vector notation that was:

$$q = [x \, y \, \theta]^{T} \tag{5}$$

where x and y were the global coordinates of P as shown in Fig. 2. θ was the orientation of [P x y]. They were the local coordinates as shown in Fig. 3, from the horizontal axis. These defined configuration (5). The machine was assumed to be rigid. Wheels were assumed not to slip. That meant that the machine could only move in a direction that was normal to the axis of the wheels. That meant that the velocity where the wheel contacted the floor and was orthogonal to the wheel plane was 0.

$$(dy/dt)\cos\theta - (dx/dt)\sin\theta - d\theta/dt = \text{zero} \tag{6}$$

Kinematics restrictions are not time dependent, so they can be considered as

$$A^T(q)\ dq/dt = \text{zero} \tag{7}$$

where $A(q)$ represents the input matrix that is associated with the constraints. Then

$$C^T A(q) = 0 \tag{8}$$

where $C(q)$ is a full-rank matrix of the set of linearly independent vector fields covering the null space of $A^T(q)$. v_{tot} is a function of vector time found from Eqs. (7) and (8) for time t.

$$dq/dt = C(q)v_{tot} \tag{9}$$

For the machine, the constraint matrix shown in (6) becomes

$$A^T(q) = [-\sin\theta \ \cos\theta - d] \tag{10}$$

and

$$v_{tot} = [v\ \omega]^T \tag{11}$$

where v is the linear velocity and ω is the angular velocity of P (along the machine axis). So, kinematics (9) can be represented in a dq/dt matrix. Because the machine is considered to only move forwards then $v = -v_{tot}$ and the system can be simplified and represented by a simpler matrix. Steering angle and wheel velocity were generated by a controller such that:

$$\text{SAngle} = (v_{left} - v_{right})/W,$$

SAngle was used to drive the machine to follow a desired path.

7 Control and Rules

The controller calculated ω and v to move the machine from a configuration, e.g.: ρ_0, α_0, β_0, to a new position and orientation. Considering linear control [30]

$$v = K_\rho \rho \tag{12}$$

$$\omega = K_\alpha \alpha + K_\beta \beta \tag{13}$$

A matrix can depict this closed-loop system to drive the machine to

$$(\rho, \alpha, \beta) = (0, 0, 0).$$

where (0, 0, 0) represented a target destination.

Control was successfully simulated and then the controller was tested on the robot. Joystick inputs and the output of the ultrasonic sensors were merged using a rule set that aimed to avoid objects. The first set of rules that were tested were a combination of 4 inputs (Fig. 4). They were: 1. Joystick steering angle; 2. Range to obstacles that both sensors detected; 3. Range to obstacles on the left and 4. Range to obstacles on the right.

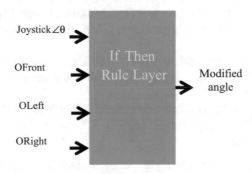

Fig. 4 The first set of rules tested

The steering angle used by the controller was modified by the input from the ultrasonic sensors. The input form the ultrasonic sensors represented the environment ahead. The resultant movement efficient and safe. If $\angle\theta$ was to the right then the robot tended to rotate clockwise. If $\angle\theta$ was to the left, then it turned anticlockwise.

The first system performed satisfactorily if the operatives could see their mobile robots but not if the operatives could not see the them. To overcome this limitation, the systems were enhanced and upgraded. The initial set of rules were revised to include a target destination. That target destination could assist the operatives if they could not see what was happening (Fig. 5). The new revised system now included a target point as well as the original environmental information (about the volume ahead) and the steering angle provided by the joystick. That addition considerably enlarged the number of rules.

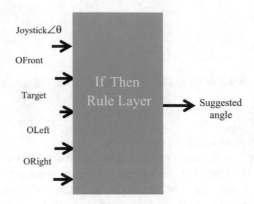

Fig. 5 The augmented system with a Target added

Some of the rules are shown here in their updated form:

CASE 1—the destination and an obstacle are to the left:

Rule 1: *If* Jstick = 0° *and* OLeft = INTERMEDIARY *and* ORight ≤ DISTANT *and* OFront ≤ DISTANT *and* Target Angle = 70°, *then* recommended adjustment to the steering angle = 0°

Rule 2: *If* Jstick = 0° *and* OLeft = INTERMEDIARY *and* ORight ≤ DISTANT *and* OFront ≤ DISTANT *and* Target Angle = 60°, *then* recommended adjustment to the steering angle = −10°

Rule 3: *If* Jstick = 0° *and* OLeft = INTERMEDIARY *and* ORight ≤ DISTANT *and* OFront ≤ DISTANT *and* Target Angle = 50°, *then* recommended adjustment to the steering angle = −25°

CASE 2—destination is to the right and an obstacle is to the right:

Rule 4: *If* Jstick = 0° *and* OLeft ≤ DISTANT *and* ORight = INTERMEDIARY *and* OFront ≤ DISTANT *and* Target Angle = 75°, *then* recommended adjustment to the steering angle = 15°

Rule 5: If Jstick = 0° *and* OLeft = ≤ DISTANT and ORight = INTERMEDIARY *and* OFront ≤ DISTANT *and* Target Angle = 60°, *then* recommended adjustment to the steering angle = 30°

Rule 6: *If* Jstick = 0° *and* OLeft = ≤ DISTANT *and* ORight = INTERMEDIARY *and* OFront ≤ DISTANT *and* Target Angle = 30°, *then* recommended adjustment to the steering angle = 25°

CASE 3—destination is to the right and an obstacle is ahead:

Rule 7: *If* Jstick = 0° *and* OLeft = NEARBY *and* ORight = NEARBY *and* OFront ≤ DISTANT *and* Target Angle = 20°, *then* recommended adjustment to the steering angle = 15°

Rule 8: *If* Jstick = 0° *and* OLeft = NEARBY *and* ORight = NEARBY *and* OFront ≤ DISTANT *and* Target Angle = 25°, *then* recommended adjustment to the steering angle = 20°

Rule 9: *If* Jstick = 0° *and* OLeft = NEARBY *and* ORight = NEARBY *and* OFront ≤ DISTANT *and* Target Angle = 30°, *then* recommended adjustment to the steering angle = 25°

The new system worked as satisfactorily as the previous system with the new rule set but it worked especially well when the operatives could not see the mobile robot.

The mobile robot path is in Fig. 6 with the additional rules being applied. The red dashed lines and arrow heads are the angles to the destination. The objects around the robot are shown by the blue boxes and the approach directions are shown by blue solid lines.

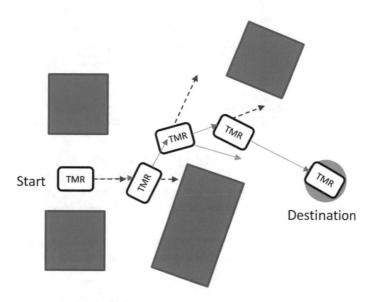

Fig. 6 Mobile robot being driven through objects (blue boxes) using the updated set of rules with calculated directions (red dashed line) and approach directions (blue solid line)

8 Experimentation and Results

A typical simulation of the system is shown in Fig. 7.

After the algorithms had successfully been tested in simulation, then the hardware and software were mounted onto the mobile robot base. Standard test routes at Portsmouth University were used for the tests.

Obstacles were avoided. When and object was detected by the sensors that was relatively near to the robot, then the mobile robot steered away from any potential impact. The operator could overrule the system by moving their joystick if, for example, an operative wanted to move the mobile robot closer to something.

The system began to take effect when sensors detected an object as DISTANT or nearer. If the ultrasonic transducers sensed something in front of the robot while it was moving towards a destination, then the robot rotated to pass along the side of the

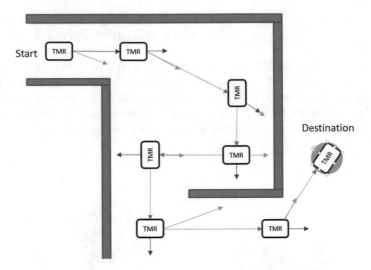

Fig. 7 Typical simulation of a robot path with the system using the revised rules and exhibiting the behaviour of the robot as it avoids local minima (for example the corners of the inner walls)

obstacle. When nothing was ahead, and the joystick was held forward, then the mobile robot headed in the direction of the destination. That reduced time taken by a significant amount when the operator could not see the mobile robot. The modified rules quickly changed the bearing and course so that the mobile robot moved in the direction of the destination.

Results from a real time experiment and from a simulation of the mobile robot are presented in Figs. 8 and 6. To show the way in which the system was validated.

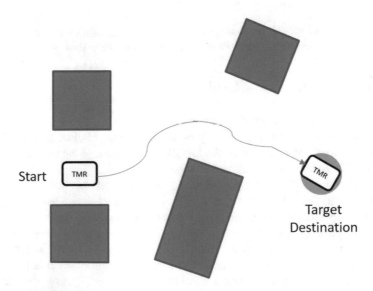

Fig. 8 Results from applying the revised rules in a real-time experiment

The performance of the system was compared with the rests form using the system described in [1]. This system tended to complete tasks faster than the earlier system. Figure 9 compares the time taken by the two systems. In each case, the robot was steered along a standard route.

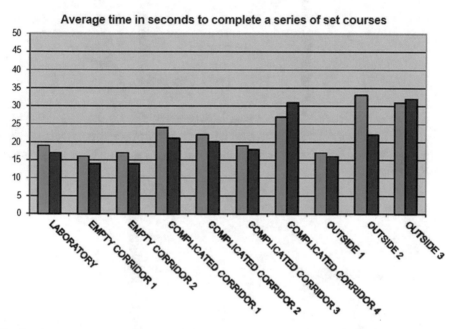

Fig. 9 Comparing the new system against a mobile robot without any sensors to assist the operator. Average time taken to complete a series of set courses is shown. Right hand bars show the time taken when the ultrasonic sensors are assisting the operator and left hand bars show the time taken when the sensor system is switched off

The new systems completed courses more quickly in most cases. Figure 10 shows two anomalies. The more complicated routes required the mobile robot to turn more often. The simpler routes did not need a sensor system as they could easily be achieved (and more quickly) by a tele-operator without any sensors to assist them.

Including destination as an additional input to the rule base made driving less effective in some of the easier sections of routes when the operative could see the robot and the surroundings. The tele-operative didn't need sensors to assist them. Two routes are shown as examples in Figs. 11 and 12.

The rules tended to attract the mobile robot toward the destination, as shown in Fig. 10. The first rule set was only influenced in direction by the joystick input. Although the path taken was usually less efficient when the tele-operator could not see the mobile robot, the route was always completed. The difference in the routes taken is presented in Fig. 12.

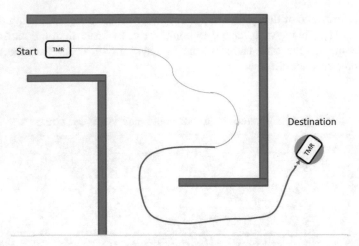

Fig. 10 Mobile robot path when the updated rule set was applied and the tele-operative can't see the mobile robot

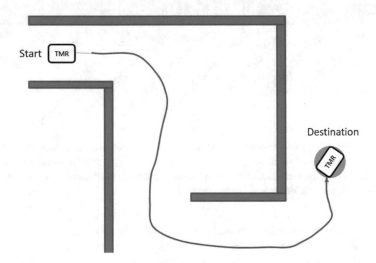

Fig. 11 Mobile robot path when the updated rule set was applied and the tele-operative can see the mobile robot

When an experienced tele-operator can see the mobile robot, and can drive well then the operative can override the rule set and take a more efficient route.

The robots efficiently reached destinations.

The rule based methods gave a quicker response in most cases and reduced computation time compared with other recent approaches. Figure 11 shows a real-time path.

The robot successfully avoided both moving and static objects. When sensors detected close objects, then the robot turned to avoid collision.

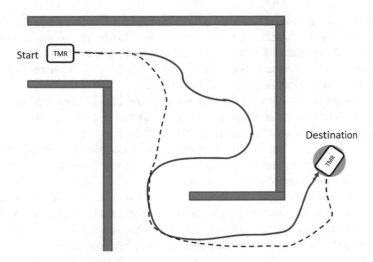

Fig. 12 The different paths taken by a mobile robot when using the revised rules. The dotted blue line shows the route taken when the tele-operator can see the robot and the solid red line shows the route taken when the tele-operator cannot see the mobile robot and the robot must rely only on the feedback from the ultrasonic sensors

Avoiding collisions was a high priority and that would primarily override other behaviour but if the operator held the joystick in the same position (roughly) then that input from the joystick would be integrated over time and the wishes if the operative would override the system.

When an obstacle was detected ahead while the robot was moving toward a destination then the robot exhibited wall-following behaviour; the robot rotated to align with an object and then moved parallel to the side of the object.

If the sensors were not detecting any objects, then the robot moved in a direction that was an average of the angle directed by the joystick position and the angle to the destination. If the joystick requested a direction that was aligned with the direction to the destination, then the robot moved towards the destination.

The rule-based system described here performed well when results were compared with those from other recent systems.

9 Discussion

Artificial intelligence has produced several powerful tools. This paper has reviewed some of those tools: knowledge-based systems, fuzzy logic, automatic learning, neural networks, ambient intelligence and genetic algorithms

The rule-base system selected was less good at handling uncertainty and is poor at handling imprecision because of the rigid structure. Case-Based Reasoning systems are often considered to be an extension of Rule-Based Systems. They are good at representing knowledge in a way that is clear to humans, but they also have the ability to learn from past examples by generating additional new cases.

Case-Based Reasoning could have been used because that can adapt solutions from previous problems to current problems. Solutions could be stored within a database. When a problem occurred that a system had not experienced, it could compare with previous cases and select one that was closest to the current problem. It could then update the database depending upon the outcome.

Without statistically relevant data for backing and implicit generalization, there is no guarantee that any generalization would be correct. However, all inductive reasoning where data is scarce is inherently based on anecdotal evidence.

The use of AI brings us to a point in history when our human biology can appear too slow and over-complicated. To overcome this, we are beginning to mix sensor systems and some powerful new technologies to overcome those weaknesses, and the longer we use that technology, the more we are getting out of it. We use less energy, space, and time, but get more and more assembly output for less cost. The AI exceeded human performance in several tasks. As computers merge with us more intimately and we combine our brain power with computer capacity, then teleoperation should become easier and more efficient. AI can reduce mistakes and increase efficiency. Time taken therefore reduces.

10   Conclusions

Applications of the AI tools discussed in this paper have become more widespread due to the power and affordability of present-day computers. Many new mobile robot applications may emerge and greater use may be made of hybrid tools that combine the strengths of two or more of the tools reviewed here. The tools have minimal computation complexity and can be implemented on single robots or systems with low-capability microcontrollers.

The rule-based systems were robust and safe. They were simple and efficient in helping with driving. The rule based techniques were employed effectively. The robot quickly detected obstacles ahead and assisted operatives with their tasks.

Laboratory testing was compared with simulated paths and the rules were validated. Systems compared favourably with other contemporary structures described in the literature and that also validated the methods and systems.

Ongoing research is exploring the integration and combining of diverse AI techniques to extract the best from each technique [31–40].

References

1. Parhi, D.R., Singh, M.K.: Rule-based hybrid neural network for navigation of a mobile robot. Proc. IMechE Part B: J. Eng. Manuf. **224**, 11103–1117 (2009)
2. Nguyen, A.V., Nguyen, L.B., Su, S., Nguyen, H.T.: Shared control strategies for human–machine interface in an intelligent robot. In: 35th Annual International Conference of IEEE-Engineering-in-Medicine-and-Biology-Society (EMBC), Osaka, JAPAN. IEEE Engineering in Medicine and Biology Society Conference Proceedings, pp. 3638–3641 (2013)

3. Parhi, Z.R., Pradhan, S.K., Panda, A.K., Behra, R.K.: The stable and precise motion control for multiple mobile robots. Appl. Soft Comput. **9**(2), 477–487 (2009)
4. Sanders, D.A., Ndzi, D., Chester, S., Malik, M.: Adjustment of tele-operator learning when provided with different levels of sensor support while driving mobile robots. In: Proceedings of SAI Intelligent Systems Conference 2016, London, UK (in press)
5. Song, K.T., Chen, C.C.: Application of asymmetric mapping for mobile robot navigation using ultrasonic sensors. J. Intell. Robot. Syst. **17**(3), 243–264 (1996)
6. Huq, R., Mann, G.K.I., Gosine, R.G.: Mobile robot navigation using motor schema and fuzzy context dependent behaviour modulation. Appl. Soft Comput. **8**(1), 422–436 (2008)
7. Begum, M., Mann, G.K.I., Gosine, R.G.: Integrated fuzzy logic and genetic algorithmic approach for simultaneous localization and mapping of mobile robots. Appl. Soft Comput. **8** (1), 150–165 (2008)
8. Bennewitz, M., Burgard, W.: A probabilistic method for planning collision-free trajectories of multiple mobile robots. In: Proceedings of 14th European Conference on AI (ECAI), Berlin, Germany, 20–25 Aug 2000, pp. 9–15
9. Gueaieb, W., Miah, M.S.: An intelligent mobile robot navigation technique using RFID technology. IEEE Trans. Instrum. Meas. **57**(9), 1908–1917 (2008)
10. Hwang, C.L., Chang, N.W.: Fuzzy decentralized sliding-mode control of a car-like mobile robot in distributed sensor-network spaces. IEEE Trans. Fuzzy Syst. **16**(1), 97–109 (2008)
11. Nguyen, A.V., Nguyen, L.B., Su, S., Nguyen, H.T.: The Advancement of an obstacle avoidance Bayesian neural network for an intelligent robot. In: 35th International Conference of IEEE-Engineering-in-Medicine-and-Biology-Society (EMBC), Osaka, JAPAN. IEEE Eng in Medicine and Biology Society Conference Proceeding, pp. 3642–3645 (2013)
12. Sanders, D.A.: Using a self-reliance factor for a disabled driver to decide on the share of combined-control between a powered wheelchair and an ultrasonic sensor system. IEEE Trans. Neural Syst. Rehabil. Eng. (in press)
13. Sanders, D.A., Stott, I., Graham-Jones, J., Gegov, A., Tewkesbury, G.E.: Expert system to interpret hand tremor and provide joystick position signals for tele-operated mobile robots with ultrasonic sensor systems. Ind. Robot **38**(6), 585–598 (2011)
14. Robinson, D.C., Sanders, D.A., Mazharsolook, E.: Ambient intelligence for optimal manufacturing and energy efficiency. Assem. Autom. **35**(3), 234–248 (2011)
15. Sanders, D.A., Tewkesbury, G., Gegov, A., et al.: Fast transformations to provide simple geometric models of moving objects. In: Proceedings of Intelligent Robotics and Application Conference, Part I, vol. 9244, pp. 604–615 (2015)
16. Sanders, D.A., Sanders, H., Ndzi, D., Gegov, A.: Rule-based system to assist a tele-operator with driving a mobile robot. In: IEEE Proceedings of SAI Intelligent Systems Conference, pp. 987–994 (2016)
17. Larsson, J., Broxvall, M., Saffiotti, A.: Laser-based corridor detection for reactive navigation. Ind. Robot: Int. J. **35**(1), 69–79 (2008)
18. Sanders, D.A., Graham-Jones, J., Gegov, A.: Improving ability of tele- operators to complete progressively more difficult mobile robot paths using simple expert systems and ultrasonic sensors. Ind. Robot.: Int. J. **37**(5), 431–440 (2010)
19. Lee, S.: Use of infrared light reflecting landmarks for localization. Ind. Robot: Int. J. **36**(2), 138–145 (2009)
20. Milanes, V., Naranjo, J., Gonzalez, C., et al.: Autonomous vehicle based in cooperative GPS and inertial systems. Robotica **26**, 627–633 (2008)
21. Sanders, D.A., Stott, I.: A new prototype intelligent mobility system to assist tele-operated mobile robot users. Ind. Robot. **26**(6), 466–475 (2009)

22. Chang, Y.C., Yamamoto, Y.: On-line path planning strategy integrated with collision and dead-lock avoidance schemes for wheeled mobile robot in indoor environments. Ind. Robot: Int. J. **35**(5), 421–434 (2008)
23. Sanders, D.A.: Progress in machine intelligence. Ind. Robot **35**(6), 485–487 (2008)
24. Sanders, D.A., Lambert, G., Pevy, L.: Pre-locating corners in images in order to improve the extraction of Fourier descriptors and subsequent recognition of shipbuilding parts. Proc. Inst. Mech. Eng. Part B-J. Eng. Manuf. **223**(9), 1217–1223 (2009)
25. Sanders, D.A.: Comparing speed to complete progressively more difficult mobile robot paths between human tele-operators and humans with sensor-systems to assist. Assem. Autom. **29** (3), 230–248 (2009)
26. Sanders, D.A.: Analysis of the effects of time delays on the tele-operation of a mobile robot in various modes of operation. Ind. Robot **36**(6), 570–584 (2009)
27. Sanders, D.A., Gegov, A.: Different levels of sensor support changes the learning behaviour of wheelchair drivers. Assist. Technol. (in press)
28. Sanders, D.A., Bausch, N. (Ed. Liu, H.): Improving steering of a powered wheelchair using an expert system to interpret hand tremor. In: Proceedings of Intelligent Robotics and Applications (ICIRA 2015), Part II, vol. 9245, pp. 460–471 (2015)
29. Sanders, D.A., Stott, I.J., Robinsosn, D.C., et al.: Analysis of successes and failures with a tele-operated mobile robot in various modes of operation. Robotica **30**, 973–988 (2012)
30. Sanders, D.A., Tewkesbury, G.E., Stott, I.J., et al.: Simple expert systems to improve an ultrasonic sensor-system for a tele-operated mobile-robot. Sensor Rev. **31**(3), 246–260 (2011)
31. Sanders, D.A., Geov, A.: AI tools for use in assembly automation and some examples of recent applications. Assem. Autom. **33**(2), 184–194 (2013)
32. Sanders, D.A., Tewkesbury, G.E., Ndzi, D., et al.: Improving automatic robotic welding in shipbuilding through the introduction of a corner-finding algorithm to help recognise shipbuilding parts. J. Mar. Sci. Technol. **17**(2), 231–238 (2012)
33. Sanders, D., Lambert, G., Graham-Jones, J., et al.: A robotic welding system using image processing techniques and a CAD model to provide information to a multi-intelligent decision module. Assem. Autom. **30**(4), 323–332 (2010)
34. Sanders, D.A., Tewkesbury, G.E.: A pointer device for TFT display screens that determines position by detecting colours on the display using a colour sensor and an artificial neural network. Displays **30**(2), 84–96 (2009)
35. Bergasa-Suso, J., Sanders, D.A., Tewkesbury, G.E.: Intelligent browser-based systems to assist internet users. IEEE Trans. Educ. **48**(4), 580–585 (2005)
36. Sanders, D.A., Tan, Y., Rogers, I., et al.: An expert system for automatic design-for-assembly. Assem. Autom. **29**(4), 378–388 (2009)
37. Sanders, D.A., Lambert, G., Pevy, L.: Pre-locating corners in images in order to improve the extraction of Fourier descriptors and subsequent recognition of shipbuilding parts. Proc. IMechE Part B—J. Eng. Manuf. **223**(9), 1217–1223 (2009)
38. Sanders, D.A.: Controlling the direction of walkie type forklifts and pallet jacks on sloping ground. Assem. Autom. **28**(4), 317–324 (2008)
39. Sanders, D.A.: Progress in machine intelligence. Ind. Robot: Int. J. **35**(6), 485–487 (2008)
40. Geov, A., Gobalakrishnan, N., Sanders, D.A.: Rule base compression in fuzzy systems by filtration of non-monotonic rules. J. Intell. Fuzzy Syst. **27**(4), 2029–2043 (2014)

Fuzzy Waypoint Guidance Controller for Underactuated Catamaran Wave Adaptive Modular Vessel

Jyotsna Pandey[1(✉)] and Kazuhiko Hasegawa[2]

[1] Graduate School of Engineering, Osaka University, 2-1 Yamadaoka, Suita,
Osaka, Japan
jyotsanapandey30@gmail.com
[2] Division of Global Architecture, Department of Naval Architecture & Ocean
Engineering, 2-1 Yamadaoka, Suita, Osaka, Japan
hase@naoe.eng.osaka-u.ac.jp

Abstract. The development of a GPS based position control system for Wave Adaptive Modular Vessel (WAM-V) able to navigate between waypoints is discussed in this chapter. A fuzzy reasoned double loop controller is proposed for navigation path planning of WAM-V. For outer loop fuzzy controller is used to feed the desired heading to the inner loop. In the inner loop, a PID feedback controller is used to correct the desired course generated by the fuzzy reasoned algorithm. The control system provides the required feedback signals to track the desired heading which is obtained from the fuzzy algorithm. After PID generates the appropriate command, the thrust isallocated to the port side and starboard side thrusters along with the command from lookup table. Using the proposed controller, several experiments are conducted at Osaka University free running pond facility. The WAM-V is equipped without rudder, thus it is driven by a combination of different thrusts to control both speed and heading. Several experimental results with different sets of waypoint validate the proposed algorithm. The obtained results affirmed that the proposed fuzzy waypoint guidance control algorithm is powerful to realize the navigation path planning. The waypoint navigation experimental results show that the fuzzy guided waypoint controller scheme is simple, intelligent and robust. The goal of this research is to present a solution to the waypoint control problem for the underactuated catamaran vessel (WAM-V), which is achieved successfully.

Keywords: Navigation system · Waypoint guidance · Fuzzy logic
Marine robotics · Underactuated catamaran vessel

1 Introduction

Currently oceanography or ocean environmental sensing (meteorological survey) is carried out using satellites, buoys, research vessels or ships. However, remote surveillance of oceanography data using satellites and airplanes is restricted due to cloud cover, temporal/geographical coverage as well as spatial resolution.

© Springer International Publishing AG 2018
Y. Bi et al. (eds.), *Intelligent Systems and Applications*,
Studies in Computational Intelligence 751,
https://doi.org/10.1007/978-3-319-69266-1_19

Meanwhile, manned research vessels are expensive for ocean surveillance, whereas the use of moored buoy due to lack of controllability and self deployability is not so attractive option for spatial sampling purposes. Due to all of the above mentioned reasons Autonomous Surface Vehicles (ASV) and Autonomous Underwater Vehicles (AUV), due to their various capabilities for payload, communication and autonomy, have emerged as the best option for in situ measurement of oceanography data as well as a complimentary observing system (port protection, mine countermeasures, and surveillance missions). It is foreseen that there will be a future market for autonomous marine vehicles capable of doing different marine operations without the assistance of a human pilot.

In 90s technology have been grown up and many ASV's have been developed for different missions around the world. In 2009 Marine Advanced Research Inc. has developed a new generation surface vessel called Wave Adaptive Modular Vessel (WAM-V). WAM-V technology contains a huge amount of potential in various marine applications. An autonomously navigated WAM-V can conveniently replace different dangerous coastal tasks and provide disaster assistance. WAM-V technology provides unparalleled support in diverse rescue missions. Autonomous ship navigation can be divided into two major areas of research, collision avoidance and path following is a very challenging problem in terms of navigation, controller design and guidance algorithm, one should have a robust controller, which takes into account of external disturbances. Path tracking uses positional information, typically obtained from GPS, IMU and vehicle odometry, to control the vessel speed and steering to follow a specified path. Fuzzy logic control is a practical alternative for a variety of challenging control and guidance problems. It shows an advantage over conventional autopilot because they are robust and guarantee the optimality of the system performance. An autopilot was developed for ship using fuzzy theory and the behaviour of a human controller is translated into a fuzzy mathematical model (Amerongen et al. 1977). Ship Auto-Navigation Fuzzy Expert System (SAFES) was developed by Hasegawa et al. (1986), Hasegawa (1990, 1993) for harbor manoeuvring, congested waterway navigation and collision avoidance. This navigation path planning algorithm using fuzzy theory, which is quite similar to the human control. The fuzzy control rules are constructed based on the human operator's experience. Several guidance algorithms such as pure pursuit (PP), line of sight (LOS) and constant bearing (CB) are widely studied in the literature for waypoint navigation and control applications (Fossen 1994; Fossen and Johansen 2006). An algorithm based on Dubin's path is also used as a path planning algorithm (Beard and Maclain 2012). International regulations for preventing Collisions at Sea (COLREGs) rules and regulations based algorithms are used in many researches for collision avoidance and path planning. One of the fuzzy based algorithm for collision avoidance is proposed under COLREGS guideline, which has ability to handle static and/or moving obstacles (Lee et al. 2004).

A double loop fuzzy autopilot is proposed for waypoint tracking of ships. The inner loop implements the ship course control and the outer loop implements the tracking control (Cheng and Yi 2006). The model predictive control (MPC) scheme using line of sight (LOS) algorithm for underactuated marine surface vessels is proposed and simulated (Oh and Sun 2010). The fuzzy switched PID controller for ship track keeping was proposed. In this controller, the fuzzy PD controller was used to improve the response and reduce overshoot time and the fuzzy PI controller was used to improve the accuracy (Jia et al. 2012). Performance analysis of PID and fuzzy controller was studied for ship tracking and it was concluded that the performance of fuzzy controller is better than PID, in terms of tracking efficiency and computational time (Sanjay et al. 2013).

To design a robust controller, the dynamics of the system and its manoeuvring performance should be known. Pandey and Hasegawa (2015, 2016a) studied the maneuvering performance of WAM-V in calm and deep water and calculated the manoeuvring parameters of WAM_V with the help of towing tank and free running experiments. The MMG type of mathematical model was derived for shallow draft, twin-hull and twin- propeller vessel WAM-V. The fuzzy guided waypoint algorithm was simulated using the MMG mathematical model for WAM-V (Pandey and Hasegawa 2016b). Ahmed and Hasegawa (2016) presented the simulation and experimental results of the same fuzzy waypoint guidance algorithm for Esso Osaka Ship and compared the experimental and simulation results.

Control design for high speed autonomous vehicles such as WAM-V is challenging due to uncertainty in dynamic models, significant sea disturbances, underactuated dynamics and overestimated/underestimated of hydrodynamic parameters. The purpose of this chapter is to present a solution to the waypoint navigation problem for a class of underactuated catamaran vessels. The remainder is organized as follows: in the next section configuration of WAM-V is described. In Sect. 3 mathematical model is presented. The description of waypoint algorithm and fuzzy guiding rules with the reasoning are illustrated in Sect. 4. The control structure and scheme is presented in Sect. 5 followed by the experimental results on Sect. 6. The conclusion is described in the last Sect. 7.

2 Configuration of WAM-V

A WAM-V is a shallow-draft high speed catamaran vessel, with tremendous controllability and maneuverability. WAM-V is designed to adapt to the shape of the water surface and equipped with springs, shock absorbers, and ball joints which gives enough agility to the vessel and damping stresses to the structure and payload as shown in Fig. 1. Two propellers attached to the aft part of each pontoons with special hinges that keep the propeller in the water all the times. High frequency waves are absorbed by the air filled pontoons. The 2:1 length-to-beam ratio, in addition to ball joint and suspension system, makes the WAM-V a stable platform. The propulsion system of the vehicle consists of a pair of Minn. Kota transom mount trolling motors installed on the port and starboard side. In the original state of the thrusters, the RPM of the motor and its orientation are controlled by a foot pedal. The thrusters are custom designed in order to control it autonomously instead of pedal control. This electric thrusters are designed for 12 V, powered by 12 V deep cycle lead acid battery. The vehicle features Furuno SC-30 satellite compass, which provides highly accurate attitude information for navigation. The SC-30 consist of compact GPS antenna with a built- in processor. The hardware components comprising the GNC are located at the center of the hull. The main computer and various peripheral devices, such as serial-to-USB interfacing hardware, voltage regulator, DC to AC converter and the wireless LAN hub, are housed in a sealed plastic fiber box.

The main particulars of WAM-V are listed in Table 1.

3   Mathematical Model

A successful control system design requires knowledge of the system to be controlled. The mathematical model for ship manoeuvrability is the most important issue for a ship manoeuvring simulation. Numerous methods were proposed to predict and reproduce

Fig. 1. Configuration of WAM-V

Table 1. Main particulars of WAM-V

Parameters	Measurements
Hull length	3.91 m
Overall vehicle height	1.27 m
Overall vehicle width	2.44 m
Payload	136 kg (Maximum)
Full load displacement	255 kg
Draft	0.165 m
Primary sensors	GPS, camera, LRF, INS, hydrophone-pinger

the ship manoeuvrability during the long history of ship dynamics research. Apparently by the end of 1970's, it appeared that there were many manoeuvring mathematical models was established. There was a well established MMG model proposed in the Japan Towing Tank Conference (JTTC) by Ogawa et al. (1977, 1978a, b). The model was proposed by a research group called Manoeuvring Modeling Group (MMG) (Yasukawa and Yoshimura 2015). The MMG model explicitly includes the individual open water characteristics of hull/propeller/rudder and their interaction.

For the dynamic mathematical modeling of the WAM-V, several assumptions were adopted, namely, the WAM-V is assumed to be rigid and have planar motion by neglecting heave, pitch and roll motion. For aquatic applications, WAM-V motion can be described by 3-degrees of freedom which lies in the plane parallel to the surface of the water, namely surge, sway and yaw. The z-axis is chosen so as to be perpendicular to the x-y plane (positive downward). Figure 2 shows the Earth-fixed coordinate and body-fixed coordinate system of WAM-V.

Coordinate systems are defined for the control system design by $O - X_0 Y_0$ (i.e., earth fixed coordinate system) and $G - XY$ (i.e., body fixed coordinate system). The origin of the body-fixed coordinate is assumed to coincide with the center of gravity of WAM-V. The manoeuvring equations are shown in Eqs. (1)–(3)

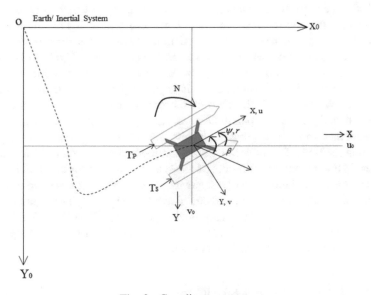

Fig. 2. Coordinate system

$$m(\dot{u} - vr) = X_{\dot{u}}\dot{u} + X_{vv}v^2 + X_u(U) + X_P \tag{1}$$

$$m(\dot{v} + ur) = Y_{\dot{v}}\dot{v} + Y_{\dot{r}}\dot{r} + Y_v v + (Y_r + X_{\dot{u}}u)r + Y_{vvv}v^3 \tag{2}$$

$$I_z\dot{r} = N_{\dot{v}}\dot{v} + N_{\dot{r}}\dot{r} + N_v v + N_r r + N_{vvv}v^3 + N_P \tag{3}$$

where, (u, v, r) is the vehicle's surge velocity, sway velocity and yaw rate respective, m is the vehicle's mass and I_z is the moment of inertia about the z-axis.

X_H, Y_H and N_H are the hydrodynamic forces acting on the hull. X_P, Y_P and N_P are the forces caused by propellers. As there is no rudder, so rudder forces are zero.

The propeller force is given in Eqs. (4) and (5)

$$X_P = (1 - t_P)(T_S + T_P) \tag{4}$$

$$N_P = (1 - d_{NP})(T_S - T_P)y_P \tag{5}$$

The distance between the propellers action point and the baseline is y_p Where t_P is the thrust deduction factor generated by the propeller in x-direction. d_{NP} is the propeller influence factor.

To simulate the manoeuvring motion, the manoeuvring derivatives in the equation should be determined. Using these all the equations the dynamic simulation of the vehicle is achieved. Faltinsen (2005) performed his research on catamaran vessel and found that determining the hydrodynamic coefficients with the use of purely theoretical methods to predict the ship maneuverability is still underdeveloped. Although Inoue et al. (1981) derived some semiempirical formulae to calculate the manoeuvring derivatives of conventional ships. Motora (1959, 1960) provided an empirical method for estimating the added masses and added moment of inertia from model tests with various conventional ships, which is known as Motora's chart. The WAM-V is neither similar to conventional ship nor like a usual catamaran vessel, so it doesn't satisfy the assumptions to use those formulae. So it is decided to calculate the manoeuvring derivatives of WAM-V with the help of the captive model experiments and system identification method. Certain parameters such as $Y_r, N_r, X_{\dot{u}}, Y_{\dot{r}}, N_{\dot{v}}$ and $N_{\dot{r}}$, which could not be determined from the captive model tests are estimated by means of the parameter identification methods. Parameter identification method can predict or tune the system parameters in the mathematical model of a dynamic system from measured data of free running experiments. Pandey and Hasegawa (2015) conducted a captive model test in the Osaka University towing tank and calculated some hydrodynamic parameters such as $X_{vv}, Y_v, Y_{vvv}, N_v, N_{vvv}$ and hull resistance $X_u(U)$ for WAM-V. The rest of the parameters such as $X_{\dot{u}}, Y_r, Y_{\dot{r}}, Y_{\dot{v}}, N_r, N_{\dot{r}}$ and $N_{\dot{v}}$ are calculated with the help of system identification technique as given in Pandey and Hasegawa (2016c).

4 Waypoint Guidance Algorithm

A fuzzy waypoint algorithm and fuzzy reasoning, which includes fuzzy rules are defined as follows.

4.1 Algorithm Description

The waypoint heading guidance algorithm is realized by fuzzy control methods used for Ship Auto-Navigation Fuzzy Expert System (SAFES). Waypoints are selected based on the interest of the user input. The fuzzy logic decides waypoint-heading commands targeting two waypoints at one time However, near the waypoint point, the fuzzy reasoning algorithm decides the appropriate course define by the next two waypoints as follows Eq. 6.

$$\psi_1 = \psi_1 + (\psi_2 - \psi_1) * CDH \tag{6}$$

where ψ_1 is order of course and ψ_1 is course of the shortest path to the next WP, ψ_2 is course of the shortest path to the second next WP. Correction for degree of heading (*CDH*) is reference degree to the second WP ($0 \leq CDH \leq 1$). In this algorithm, to judge the nearness of the waypoint, *TCPA* (time to closest point of approach) and *DCPA* (distance of the closest point of approach) are used for fuzzy reasoning. Figure 3a shows the course changing command near a course changing point (WP) and (b) shows the bearing relationship between the ship and waypoint. If *DCPA* is very big and *TCPA* is also very big, then *CDH* is very small. It means if the ship is far from second waypoint then the command course will consider only for the first waypoint. As the values of DCPA and TCPA are decreasing the ship is reaching closer to the first WP. The algorithm is described step by step in detail below.

At first the waypoints (WP) are initialized as $W_i = (X_i, Y_i)$, $W_{i+1} = (X_{i+1}, Y_{i+1})$ and current vehicle position (P) and vessel heading ψ is obtained from the sensor. The distance between the ship and nearest waypoint (*D*) is calculated by Eq. 7.

$$D = \sqrt{(X_0 - X_t)^2 + (Y_0 - Y_t)^2} \tag{7}$$

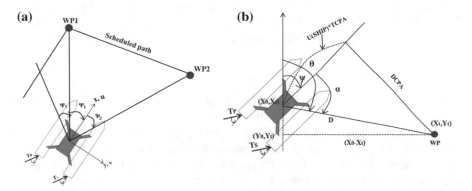

Fig. 3. **a** Course command near a course changing point. **b** Bearing relationship between ship and waypoint

Encountering angle of waypoint from the vertical axis (θ) and bearing angle of waypoint (α) from the ship is calculated with the help of Eqs. 8 and 9. Here, if the value of ψ, θ or α becomes negative, then 2π is added to make them positive.

$$\theta = a \tan 2 \frac{(Y_t - Y_0)}{(X_t - X_0)} \tag{8}$$

$$\alpha = \theta - \psi \tag{9}$$

Finally, $TCPA, DCPA$ are calculated with the help of Eqs. 10 and 11.

$$DCPA = D|D \sin \alpha| \tag{10}$$

$$TCPA = \frac{D \cos \alpha}{U_{ship}} \tag{11}$$

Another important point to be considered is the scale effect. There should be some difference on the nearness between a large ships and a small ships. Therefore, the following equations are used for non-dimensionalised $TCPA$ and $DCPA$. The nearness is then reasoned from $DCPA'$ and $TCPA'$ instead of $DCPA$ and $TCPA$ using the Eqs. 12 and 13 is calculated.

$$DCPA' = \frac{DCPA}{L} \tag{12}$$

$$TCPA' = \frac{TCPA}{\frac{L}{V}} \tag{13}$$

When $TCPA'$ becomes negative the waypoint is switched and next set of waypoints is targeted. The nearness of the waypoint is judged using fuzzy logic by calculation of TCPA', DCPA' and CDH, which is discussed in the next section. The waypoints are stored in a database and are used to generate the desired trajectory to follow.

4.2   Fuzzy Reasoning

Fuzzy logic controller mimics human reasoning and logic by developing sets of distributive membership functions (Driankov et al. 1996).

The fuzzy controller block diagram is shown in Fig. 4, where there are $TCPA'$ and $DCPA'$ as a crisp inputs. The minimum-maximum method is utilized to represent the premise and implication. The fuzzy output set is defuzzified using the center of gravity (COG) method. Finally, CDH is calculated as a crisp output.

The linguistic variables are defined as:

$$TCPA' = \{SA, SM, ME, ML, LA\}$$

$$DCPA' = \{SA, SM, ME, ML, LA\}$$

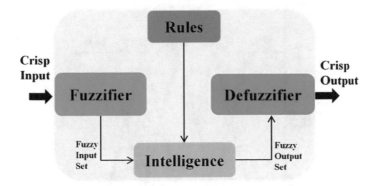

Fig. 4. Fuzzy controller architecture

$$CDH = \{SA, SM, ME, ML, LA\}$$

where SA = SMALL, SM = SMALL MEDIUM, ME = MEDIUM, ML = MEDIUM LARGE, LA = LARGE.

The complete fuzzy system involves some 25 fuzzy rules with 5 linguistic variables. The triangular membership functions of $TCPA'$, $DCPA'$ and CDH are shown in Fig. 5. The fuzzy control rules to reason CDH is shown in Table 2.

Defuzzyfication is the process of converting degree of membership of the output linguistic variable into numerical values. There are various methods such as center of area (COA), Center of Sums (COS) or Center of Maximum (COM) etc. In this research Center of Gravity (COG) method is used. Equation 14 shows the function equation and Fig. 6. shows the graphical representation of the COG function, which chooses the y-coordinate of the center of gravity of the area below the graph (y) μ. This defuzzification can be interpreted as a weighted mean, i.e., each value y is weighted with (y) μ and the integral in the denominator serves for normalization.

$$y' = \frac{\int_Y \mu(y)y\,dy}{\int_Y \mu(y)\,dy} \tag{14}$$

The characteristics of fuzzy control surface, that is a graphical representation of the function, CDH, $TCPA'$ and $DCPA'$ is depicted from Fig. 7.

The guidance algorithm is expected to perform specific types of manoeuvre, depending on the magnitude of the distance of the waypoint. Some of the important points of the algorithm are described as follows

1. The algorithm doesn't ensure the reachability of waypoints rather reaching the target is the main concern.
2. This algorithm targets two waypoints at a time and chooses an optimal path between two.
3. When the distance to the waypoint is larger it targets to steer towards the waypoint.
4. When the distance to the waypoint is smaller it starts targeting to the next set of waypoints.

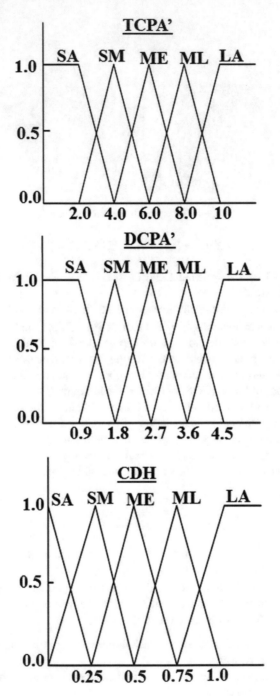

Fig. 5. Membership functions for course changing algorithm

Table 2. Fuzzy control rules

DCPA'	TCPA'				
	SA	SM	ME	ML	LA
SA	LA	ML	ME	SM	SA
SM	ML	ME	SM	SA	SA
ME	ME	SM	SA	SA	SA
ML	SM	SA	SA	SA	SA
LA	SA	SA	SA	SA	SA

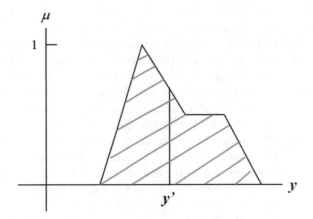

Fig. 6. COG function for defuzzyfication

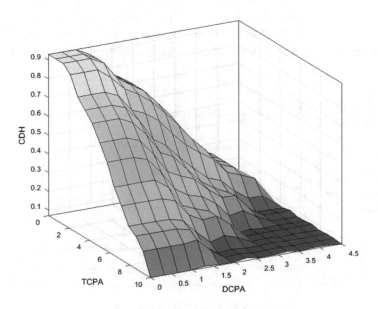

Fig. 7. Fuzzy control surface

5. Nearly zero distance to waypoint makes only small course corrections.

5 Control Design and Control Scheme

In this section the guidance, navigation and control (GNC) architecture are introduced for the waypoint guidance of WAM-V. In general, motion control systems are constructed as three interacting systems. These three systems are the guidance, navigation and control systems. The guidance system provides a reference model with a calculated desired heading, the navigation system acquires position and attitude of the vehicle, while the control system allocates thrust to the actuators to ensure that desired position and velocity are satisfied. The guidance system keeps track of the desired heading angle that the object shall follow. The desired reference is continuously computed based on the current position given by the navigation system and a target defined by the guidance algorithm. The desired heading is fed to the control system, so that the autopilot can follow the calculated result. Control system layout designed contains two loops as shown in the Fig. 8. For outer loop fuzzy controller is used to feed the desired heading to the inner loop. In the inner loop a PID feedback controller is used to correct the desired course generated by the fuzzy reasoned algorithm. The control system provides the necessary feedback signal to track the desired heading comes from the fuzzy algorithm. After PID generate the appropriate command the thrust is allocated to the port side and starboard side thrusters along with the command from lookup table. This system processes information to infer the state of the WAM-V plant and to generate an appropriate command for the actuators so as to reduce the reference heading and actual heading.

In this control system, waypoints are input. The PID control system provides the necessary feedback signal to track the desired heading comes from the guidance algorithm (τ_{com}). Equation 15 shows governing equation of PID controller. Where ψ_1 is desired heading calculated with the help of fuzzy reasoning. ψ and $\dot{\psi}$ are the vessel's

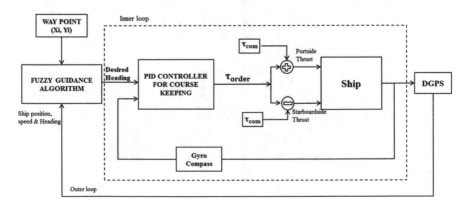

Fig. 8. Control system layout

current heading and yaw rate correspondingly. Where a proportional gain (K_p) Integral gain (K_i) and derivative gain (K_d) are the regulator design parameters. These parameters are tuned in order to avoid the overshoot and steady state errors.

$$\tau_{order} = K_p(\psi_1 - \psi) - K_d\left(\dot{\psi}_1 - \dot{\psi}\right) + K_i \int_0^t (\psi_1 - \psi) \tag{15}$$

The control law has a concise form and easy to implement in practice due to a smaller computational burden with only few online parameters being tuned. Thrust allocation is an important part of the system. It allocates the desired command to the actuators using the command information from the controller. Under actuated system is a system in which there is a less actuator than needed to satisfy the control objectives. The problem of control allocation for ships and underwater systems are briefly described in the literature. Generally, over actuated system is chosen as it improves the safety of control system in case of actuator fails (Fossen and Johansen 2006) but the same time under actuated system are preferred in order to reduce the cost, maintenance and easily controllability.

In the manoevring experiments several combinations of propeller revolution are checked. During the experiment the GPS data was stored and analyzed offline to see the turning response of the WAM-V, with the change in the thrust of port and starboard thruster. At various differentials thrust cases of turning circle tests the input supplied $(T_P$ and $T_S)$ yaw rate and vessel speed data are collected. This data is plotted in the form of a surface plot graph. These graphs show the functional relationship between a designated dependent variable. Figure 9 shows the data table obtained from the free running experiment test.

The WAM-V is controlled by differential thrust for surge, sway and yaw motion control. The desired heading can be generated by differential thrust combinations. The WAM-V can be modelled as two inputs $(T_P$ and $T_S)$ and the single output system as shown in Fig. 10.

In order to linearize the model at an operating point, it is assumed that the vehicle is running at a constant speed 0.8 m/s. This corresponds both the thrusters running at 10 N. τ_{com} represents the common mode thrust and τ_{order} is differential mode thrust coming as a PID output. In order to maintain the velocity of the vessel τ_{com} is generated from a lookup table. The differential mode thrust varies depending on the direction of manoeuver. The final command supplied to port side and starboard side thrusters are given by Eqs. 16 and 17.

$$T_{port} = \tau_{com} + \tau_{order} \tag{16}$$

$$T_{starboard} = \tau_{com} - \tau_{order} \tag{17}$$

(a)

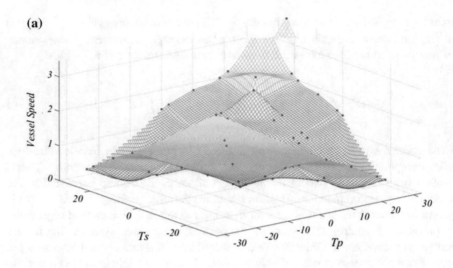

(b)

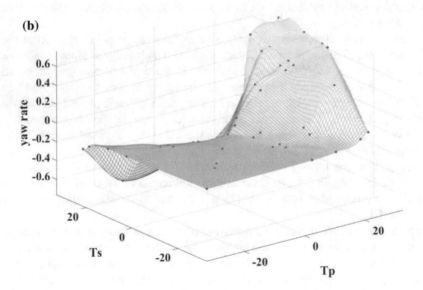

Fig. 9. **a** Surface plot between yaw rate and Tp, Ts. **b** Surface plot between vessel speed and Tp, Ts

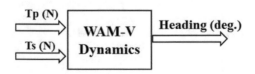

Fig. 10. Block diagram representation of two inputs WAM-V

6 Experimental Results and Discussion

Using the proposed waypoint guidance controller in above section, several experiments were conducted at the Osaka University pond facility to validate the robustness and efficiency of autonomously navigated WAM-V through waypoints. To begin with the experiment, initially random waypoints belonging within the domain of the latitudinal and longitudinal boundary of the pond were selected. During the waypoint selection process, the physical navigational constraint of the WAM-V was always taken care off. Table 3 delineates the set of waypoint selected to demonstrate and validate the robustness of the proposed waypoint algorithm. The set of the waypoint given in Table 3 has been converted to Cartesian coordinated, where the origin, i.e., (0, 0) coordinate represented the launching point.

The results of each experiments is briefly discussed in the section below. To maintain the generality of the experimental result same steps was followed with each set of experiments.

1. First Experiment

For this experiment first row of waypoints corresponding to Table 3 was selected. Figure 11a shows x-y plot of the controlled trajectory followed by the WAM-V in order to navigate through the selected set of waypoints. Figure 11b shows the time history of the allocated thrust. The thrust is allocated to both the sides of the thrusters according to the desired heading. Port side thrust (Tp) is higher that starboard side due

Table 3. Experimental set of waypoints

Description	Waypoints (x, y)		
	I	II	III
Figure 11	5.8, −11.8	6.0, −13.9	–
Figure 12	2.7, −1.2	7.8, −3.4	11.9, −8.7
Figure 13	10.2, −1.0	22.0, −6.7	34.5, −5.5

to the turning direction. From the time history graph of the allocated thrust it can be clearly seen that the controller was successful in guiding the WAM-V in a desired turning direction. In Fig. 11c time history of heading error graph, $TCPA'$, $DCPA'$ and CDH are shown.

The $TCPA'$ graph shows that when the WAM-V is reaching to the waypoints it is gradually decreasing and tending towards zero. The $DCPA'$ plot shows the distance approaching towards the waypoint, which is decreasing with time. The value of CDH is calculated with the fuzzy reasoning. The heading error is positive just for a small time and then it became negative and slowing tending towards zero. This test demonstrates how the action taken by the vessel in order to reach the waypoint. This experiment proves the effectiveness of the waypoint algorithm.

2. Second Experiment

Figure 12a shows the x-y plot of the controlled trajectory followed by the WAM-V in order to navigate through the selected set of waypoints. From the time history of the

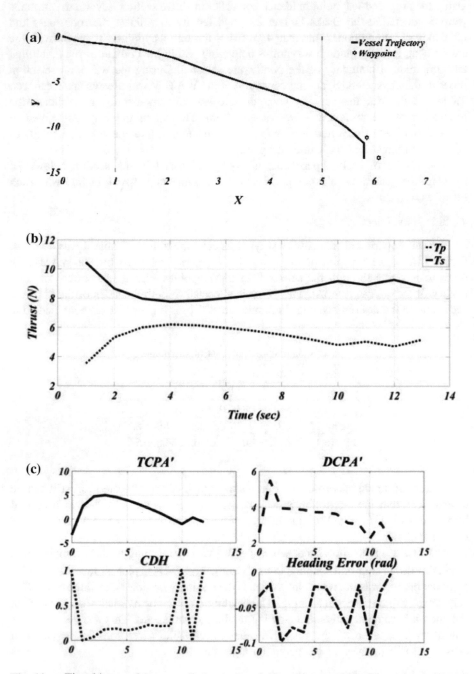

Fig. 11. **a** Time history of the controlled trajectory. **b** Time history of the allocated port side and starboard side thrust. **c** Time history of *TCPA'*, *DCPA'*, *CDH* and heading error

trajectory it can be concluded that the controller and the proposed way point algorithm are robust enough to guide the WAM-V through the desired set of waypoint. Figure 12b shows the time history of the allocated thrust. From the time history graph of the allocated thrust it can be clearly seen that the controller was successful in guiding the WAM-V in the desired direction. The thrust behaviour shows that the T_P is higher than T_S in order to follow the desired course. The same has been reflected in Fig. 12c which shows the time history of heading error graph and the time history of $TCPA'$, $DCPA'$ and CDH has also been shown. $TCPA'$ become negative three times, which shows the reachability of every waypoint. $DCPA'$ shows the distance of the waypoint from the vessel. CDH is calculated according to the fuzzy rules defined and shows the degree of closeness to the waypoint. The heading error remained slightly negative and gradually tends to zero.

3. Third Experiment

Figure 13 shows the result of the third experiment, where 3 waypoints are chosen in a zigzag manner. Figure 13a shows the x-y plot of the controlled trajectory followed by the WAM-V in order to navigate through the selected set of waypoints. The vessel trajectory shows good agreement in terms of the reachability of the waypoint. Figure 13b shows the time history of the allocated thrust. The thrust graph shows the variation of the thrust between port side and starboard side, according to the reference course. From the time history graph of the allocated thrust it can be clearly seen that the controller was successful in guiding the WAM-V in a zigzag pattern and as the external disturbances such as wind tries to deviate WAM-V from the controlled path the controller acts instantly to bring back WAM-V to zero error path. The same has been reflected in Fig. 13c which shows the time history of heading error graph and the time history of $TCPA'$, $DCPA'$ and CDH has also been shown. The $TCPA'$ plot became negative whenever the waypoint passed and switch to the next waypoint. The $DCPA'$ plot shows the time history of the distance covered in order to chase each waypoint. CDH calculated with the fuzzy reasoning. The heading error is also in a negative side. This algorithm is not very accurate in terms of the reachability of waypoint but it is very robust for navigational path planning. In some of the real applications, it is not necessary to reach the waypoint but to navigate through the points are very important and this algorithm is very feasible.

7 Conclusion

In this research WAM-V is introduced with its navigational and control performance. A fuzzy reasoned double loop controller is proposed for navigation path planning of WAM-V. The outer loop generates the desired course to the waypoint with fuzzy reasoning. The inner loop takes necessary action for course correction. The fuzzy controller is designed based on the human operator's manipulating experience. Based on the value of the $DCPA'$ and $TCPA'$, the nearness of the next waypoint is measured and the reference degree to the second next waypoint is calculated by fuzzy controller. Therefore, based on the nearness of two consecutive waypoints, fuzzy controller

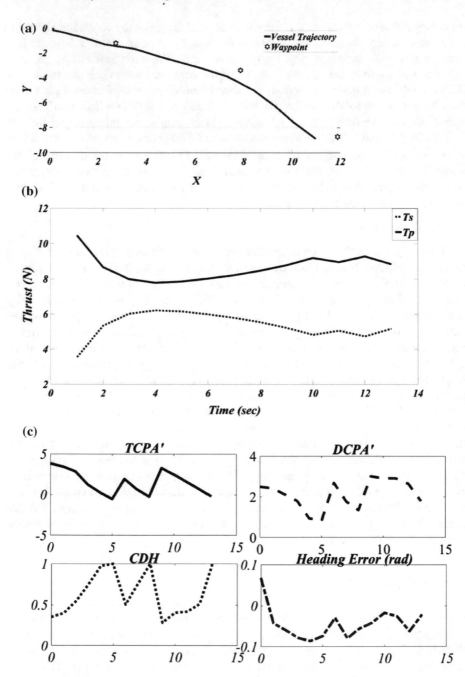

Fig. 12. a Time history of the controlled trajectory. **b** Time history of the allocated port side and starboard side thrust. **c** Time history of *TCPA'*, *DCPA'*, *CDH* and heading error

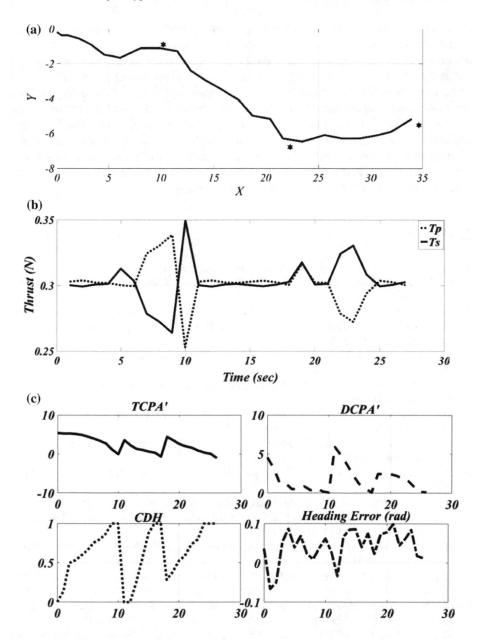

Fig. 13. **a** Time history of the controlled trajectory. **b** Time history of the allocated port side and starboard side thrust. **c** Time history of $TCPA'$, $DCPA'$, CDH and heading error

gradually modifies the desired course. Switching of the waypoint is done when $TCPA'$ becomes negative. This procedure continues for the rest of path navigation. The thrust allocation problem is solved with the help of lookup table. The lookup table is formed based on the free running experimental results conducted in order to study the manoeuvrability of WAM-V. Based on the proposed control scheme experiments are conducted at the Osaka University pond facility. The experimental results of waypoint navigation show that the fuzzy reasoned waypoint algorithm followed by feedback controller gives satisfactory performance. The thrust is allocated according to the desired heading. The waypoint navigation experimental results show that the fuzzy guided waypoint controller scheme is simple, intelligent and robust. The goal of this research is to present a solution to the waypoint control problem for the underactuated catamaran vessel (WAM-V), which is achieved successfully. There are many applications possible using this algorithm for path planning.

References

Ahmed, Y.A., Hasegawa, K.: Fuzzy reasoned waypoint controller for automatic ship guidance. In: 10th International Federation of Automatic Control—CAMS, vol. 49, issue 23, pp. 604–609 (2016)

Amerongen, J.V., Naute Lenke, H.R., Veen der Van, J.C.T.: An autopilot for ships designed by with fuzzy sets. Proc. IFAC Conference on Digital Computer Applications to Process Control, The Hague, pp. 479–487 (1977)

Beard, R.W., Maclain T.W.: Small Unmanned Aircraft Theory and Practice. Princeton University Press (2012)

Cheng, J., Yi, J.: A new fuzzy Autopilot for way-point tracking control of ships, pp. 451–456. IEEE—ICFS, Canada (2006)

COLREGs—convention on the international regulations for preventing collisions at sea. International Maritime Organization (IMO) (1972)

Driankov, D., Hellendoorn, H., Reinfrank, M.: An Introduction to Fuzzy Control, 2nd edn. Springer-Verlag, Berlin (1996)

Faltinsen, O.M.: Hydrodynamics of High-Speed Marine Vehicles. Cambridge University Press (2005)

Fossen, T.I.: Guidance and Control of Ocean Vehicles. Wiley, New York (1994)

Fossen, T.I., Johansen, T.A.: A survey of Control Allocation method for Ships and Underwater Vehicles. In: MED 06 14th, Italy (2006)

Hasegawa, K. et al.: Ship auto-navigation fuzzy expert system (SAFES). J. Soc. Naval Archit. Jpn. 445–452 (1986)

Hasegawa, K.: Automatic navigator-included simulation for narrow and congested waterways. In: Proceedings of Ninth Ship Control Systems Symposium, vol. 2, pp. 110–134 (1990)

Hasegawa, K.: Knowledge-based automatic navigation system for harbour manoeuvring. In: Proceedings of 10th Ship Control System Symposium, vol. 2, pp. 67–90 (1993)

Inoue, S., Hirano, M., Kijima, K., Takashima, J.: A Practical Calculation Method of Ship Maneuvering Motion, vol. 28, no. 325. ISP (1981)

Jia, B., et al.: Design and Stability Analysis of Fuzzy Switched PID controller for Ship Track Keeping. J. Transp. Technol. 2, 334–338 (2012)

Lee, S.M., et al.: A Fuzzy Logic for Autonomous Navigation of Marine Vehicles Satisfying COLREG Guidelines. Int. J. Control Autom. Syst. 2(2), 171–181 (2004)

Maritime Advanced Research Inc. Wave Adaptive Multipurpose Vessel. http://www.wam-v.com/

Motora, S.: On the measurement of added mass and added moment of inertia for ship motions (Part 1, 2 and 3). J. SNAJ **105**, **106** (1959, 1960). (In Japanese)

Oh, S.R., Sun, J.: Path following of underactuated marine surface vessels using line-of-sight based model predictive control. Ocean Eng. 289–295 (2010). Elsevier

Ogawa, A., Koyama, T., Kijima, K.: MMG report on the mathematical model of ship manoeuvring. Bull. Soc. Naval Archit. Jpn. 575–28 (1977). (in Japanese)

Ogawa, A., Kasai, H.: On the mathematical model of manoeuvring motion of ships. Int. Shipbuild. Prog. **25**(292), 306–319 (1978a)

Ogawa, A., Kasai, H.: On the mathematical model of manoeuvring motion of ships. Int. Shipbuild. Prog. **25**(292), 306–319 (1978b)

Pandey, J., Hasegawa, K.: Study on Manoeuverability and control of an autonomous wave adaptive modular vessel (WAM-V) for ocean observation. In: IAIN World Congress. IEEE (2015)

Pandey, J., Hasegawa, K.: Study on turning manoeuvre of catamaran surface vessel with a combined experimental and simulation method. In: 10th IFAC Conference on Control Applications in Marine System CAMS, vol. 49, issue 23, pp. 446–451 (2016a)

Pandey, J., Hasegawa, K.: Path following of underactuated catamaran surface vessel (WAM-V) using fuzzy waypoint guidance algorithm. In: IEEE Intellisys, pp. 995–1000 (2016b)

Pandey, J., Hasegawa, K.: Manoeuvring mathematical model of catamaran wave adaptive modular vessel (WAM-V) using the system identification technique. In: Proceedings of 7th PAAMES and Advance Maritime Engineering Conference, 2016 13–14 October, Hong Kong, pp. 1–6 (2016c)

Sanjay, N., et al.: Performance Analysis of Ship tracking using PID/Fuzzy controller. IJCTT **6**, 1858–1861 (2013)

Yasukawa, H., Yoshimura, Y.: Introduction of MMG standard method for ship manoeuvring predictions. JASNAOE **20**, 37–52 (2015)

Trust and Resource Oriented Communication Scheme in Mobile Ad Hoc Networks

Burhan Ul Islam Khan[1](✉), Rashidah F. Olanrewaju[1],
Roohie Naaz Mir[2], S. H. Yusoff[1], and Mistura L. Sanni[3]

[1] Department of ECE, Kulliyyah of Engineering, International Islamic Malaysia,
Kualalumpur, Malaysia
burhan.iium@gmail.com
[2] Department of CSE, National Institute of Technology, Hazratbal, Srinagar,
Kashmir, India
[3] Faculty of Technology, Department of CSE, Obafemi Awolowo University,
Ile-Ife, Osun State, Nigeria

Abstract. Attaining high security in mobile ad hoc networks (or MANETs) is the utmost concern in the present era of wireless ad-hoc communication and efforts are continuously being made by the researchers, in order to provide a feasible solution for the same. Impervious security has to be ensured in MANETs because of their involvement in transacting highly sensitive information. However, it has been discerned that during the communication process, the decentralized and dynamic nature of MANETs impedes the security of mobile nodes. This study is an attempt to enhance the security in routing techniques of MANETs by overhauling the existing security system after its critical evaluation. It has been observed that the cryptographic techniques in use prove to be inefficient or fail in some of the current scenarios. Therefore, a non-cryptographic method has been put forward that strengthens the process of authenticating nodes in MANETs by taking into account two factors viz. trust and resource, unlike the conventional ones. On analyzing the performance with respect to throughput, packet delivery ratio, end-to-end delay and computational time, the proposed system proves to be better than the previous standard secure routing scheme.

Keywords: Mobile ad-hoc network · Secure routing
Node authentication · Trust · Resource

1 Introduction

Mobile Ad hoc Networks or MANETs are networks composed of self-organizing nodes and are characterized by an architecture that is highly decentralized [1]. MANETs can be set-up effortlessly thus providing an economical solution for communication for a range of applications [2]. It finds broad applications in tactical networks, emergency services, commercial and civilian environment, enterprise/home networking, etc. [3]. This shows that MANETs are not restricted in their scope, i.e., they have been adopted by civil societies for safety as well as comfort besides their applications in provisional rescue operations [4]. Typically, communication in ad hoc networks [5] can be exemplified by

© Springer International Publishing AG 2018
Y. Bi et al. (eds.), *Intelligent Systems and Applications*,
Studies in Computational Intelligence 751,
https://doi.org/10.1007/978-3-319-69266-1_20

a standard connection via media like Wi-Fi, cellular networks or satellite transmission. Only some ad hoc networks are limited to a small area of wireless communication devices such as a collection of cell-phones or laptops while the majority of them are linked to the Internet. However, the dynamicity and highly mobile nature of MANETs makes them insecure as a result of which high vigilance is required during data transmission in MANETs [6].

Mobile ad hoc networks being self-configuring are independent of a central manager such as a fixed access point or base station. This implies that each node in a mobile ad hoc network acts as a router such that the mobile nodes that are remote can easily establish communication with other nodes in the network. In such a communication scenario, routing functionality mainly depends on the Medium Access Control layer of Open System Interconnection (OSI) model, and thus a framework that reactively or proactively sets up paths through multiple hops is recommended. As a result of this, mobile ad hoc networks are threatened with certain security issues much like those in conventional wireless communication like exposed or hidden terminals [7]. Other issues that may arise include interruption by attacker [8], problems in data delivery, route path, etc. In order to evade such problems, various approaches have been put forward in the past such as True Flooding Algorithm (TFA), Routing efficiency [9], Route Discovery algorithm, Route Request/Reply method, multi-swarm optimization, etc. All the approaches mentioned above have proven that the detection of malicious nodes in an ad hoc network is quite possible so that a secure path can be established for communication by the mobile nodes [10].

The issues and challenges in MANETs have been studied by numerous researchers for more than ten years till now; the popular ones being issues related to Quality of Service (QoS), energy, security, routing, etc. [11, 12]. However, the most prominent and challenging issues have been the provision of security in routing the data packets during communication. Among the major components of any security technique that include availability, integrity, non-repudiation, authenticity, confidentiality, authenticity is considered as the primary security issue [13] since the nodes in an ad-hoc network may be compromised if not authenticated properly [14]. Moreover, the prevention of malicious node entry into ad hoc networks is of utmost importance else they may destroy the entire network. A number of attacks implicated on mobile ad hoc networks have been analyzed but there has not been a suitable solution to ward-off these attacks. So, in a way, the security issue of MANETs is still unsolved [15, 16]. Although numerous techniques exist to tackle the security issues in ad hoc networks [17], only some methods were found to be highly resistant to potential adversaries in the decentralized architecture of MANET.

Another point of concern that arises due to the characteristics of ad hoc networks is the quantification of uncertainty or probability which plays an essential role in decision making and evaluating the performance of MANETs. For example, during node communication in MANETs, there may be significant degradation in performance as a result of packet dropping thereby increasing uncertainty in MANET communication. This paves way for the introduction of routing mechanisms based on trust factor which can neutralize the attacks on MANETs like packet-dropping attacks. Several routing protocols have been put forward in the past that employ uncertainty for calculating trust within an ad hoc network. Therefore, there is a close association between uncertainty and trust [18]. Furthermore, the necessity of developing and designing energy efficient and secure routing mechanisms in MANETs can be ascertained from the wide range of applicability of MANETs in the transaction of sensitive information in strategic communication networks, particularly in combat and surveillance operations [19, 20]. It is, therefore, advocated that secure and efficient routing mechanisms should be designed that can defend against the prevalent attacks on ad hoc networks such as data hijacking, interception, jamming, etc. [2]. Various approaches have been employed in the past for securing the routing process in MANETs like security mechanisms based on game theory [21, 22], pricing [23, 24], cryptography [25, 26], trust [27, 28], etc.

The remaining paper is organized as follows: Section 2 throws light on the related work that is followed by the description of problem statement in Sect. 3. Section 4 gives a brief description of the contribution of proposed system. Sect. 5 contains the algorithm implementation of the proposed trust and resource based system followed by the discussion of results in Sect. 6. Finally, the concluding remarks of the paper are given in Sect. 7.

2   Related Work

The absence of a central trusted authority in a mobile ad hoc network makes it open to a wide range of security attacks. For establishing a secure communication link among the mobile nodes in a network, various routing protocols have been put forward. However, those protocols are vulnerable to malicious node attacks, and thus there is a dire need to prevent and detect these attacks in time to avoid the breakdown of the entire network. This section has discussed recent studies that have been carried out for securing communication in ad hoc networks. The work presented in [6] focuses on the existing security problems, routing attacks and the possible solutions for mitigating attacks on routing policies that are based on node cooperation. The discussion in [29] has critically analyzed and significantly illustrated the techniques employed for addressing the malicious node behavior in ad hoc networks. Our next work [30] introduces a novel framework that utilizes game theory for investigating the malicious behavior as well as proposed a scheme to potentially defy against the malicious activities of selfish and malicious nodes.

Authors in [31] have enhanced the conventional OLSR (Optimized Link State Routing) for preventing Denial-of-Service (DoS) attack in mobile ad hoc networks. The outcome of the study has shown about 45% improvement in data delivery. In [32], a statistical regression-based method has been used for ensuring better trust management over the routing protocol. A conventional key agreement protocol has been adopted in [33] for securing group communication in mobile ad hoc networks. Authors have developed a lightweight cryptographic protocol in [34] by adopting a networking coding based scheme for securing the communication among the nodes.

The self-organizing, decentralized and open nature of mobile ad hoc networks makes it quite vulnerable to attacks. Therefore, security as the chief concern in MANET, authors in [35] have taken into consideration two attacks, namely DDoS and Vampire attacks. A small network comprised up of about thirty nodes has been considered that are separated into three clusters with each cluster having 10 nodes. After taking into consideration several parameters and conducting simulations, the results show effective mitigation against both the attacks by the usage of trust-based policy. In order to authenticate the nodes entering a mobile ad hoc network, another novel approach has been discussed in [14] that implements cryptographic algorithm and ones complement to ensure secure routing. The proposed mechanism has the ability to be embedded in a wide range of routing algorithms for providing security and enhancing the network efficiency. The mitigation of attacks on the nodes in the network has also been shown by authors. The analysis of the performance of this approach has been done in terms of control overhead, and it has been found to have a minimal computation overhead.

Identity-based encryption for securing communication between the transmitting and receiving nodes in a mobile ad hoc network has been utilized in [36]. Adoption of received signal strength for the determination of link quality is studied by authors in [37]. Utilization of trust management schemes over the opportunistic network was seen in the work carried out by [38]. A study on trust prediction was also conducted by [39] making use of fuzzy logic. The routing protocol is more of a unicast type and performs secure routing over shortest path. Similar study towards trust-based routing was also carried out by [40]. Developed over Petri net, the presented technique employs fuzzy logic using trust reasoning system. The authors in [41] have developed a secure routing protocol for incorporating the features of unlinkability and unobservability in the communication.

The study conducted in [42] shows the usage of probability and game theory for effective representation of the variety of node actions like cooperation, declination, attacks besides node reporting thereby forming a strategic profile of the nodes in a MANET. A game specification has been designed for formulating a mathematical model that makes decisions using game theory after taking into consideration the strategies adopted by malicious, selfish and regular nodes. In [43], a line has been drawn to differentiate the kinds of nodes in a mobile ad hoc network on the basis of their characteristic behavior, and a review of the contributions of past works pertaining to node misbehavior has also been presented. The related work in which game theory is employed for addressing the node misbehavior in MANETs has been highlighted by authors. A novel approach namely Elephant Swarm Optimization (ElePSO) was developed for MANETs in [7]. This approach formulated the behavioral features of a swarm of elephants like sensitiveness to environment, inimitable memorization and rapid reaction for identifying established links and nodes, optimizing slot assignment and adaptive scheduling as per traffic in a mobile ad hoc network. A hybrid cross-layer scheduling was put forward by us that minimized network contention and provided collision avoidance with the least overheads by making use of multiple layers of OSI.

In [4] an analytical model for secure routing has been presented, referred to as Dual Threshold-based Authentication for Secure Routing (or DTASR) that ensures resistance to attackers by employing authentication based on trust as well as resource thresholding. A secure mitigation mechanism has been put forward that defies unauthorized access in MANETs. On comparison with previous routing schemes, this system has shown an improved performance in communication. An Uncertainty Analysis Framework (UAF) for secure routing based on trust has been discussed in

[18] which calculates values like Belief-Disbelief-Uncertainty (BDU) in a mobile ad hoc network. This framework models network uncertainty after its integration with AODV trust variants that make use of direct, indirect and global trust; and analyzes the effect of various trust models on BDU values of MANET through several test conditions. The experimental evaluation reveals that routing strategies based on trust show minimum 5% gain in network belief and around 3% increase in packet delivery ratio.

The work presented in [2] discusses Markov chain based energy efficient and secure multi-path routing technique for MANETs. This routing mechanism tries to compute the various routes between a pair of source and destination and then forwards the packet along energy efficient route chosen stochastically. This also helps in secure data transmission along the routes since the path chosen for forwarding data is random. Thus, it becomes difficult for the attacker to hijack, jam or intercept the data packets since it will not be practically feasible to listen to all the routes leading to the destination from the source. When the performance of the proposed technique was analyzed, its numerical results revealed enhanced performance gain with respect to delay, energy consumption, security and throughput [2]. The authors in [44] have presented a routing framework that addresses the issues related to energy drainage and routing overhead in the mobile nodes. This study shows distinction than the existing simulation mechanisms in the involvement of auxiliary nodes for reducing the routing overhead during communication in MANETs. The results obtained from this study reveal that the routing overhead has been considerably reduced besides enhancing energy efficiency in comparison to the conventional DSDV and AODV routing protocols.

For enhancing the security in MANETs and minimizing the vulnerabilities of mobile nodes, the authors have developed a certificate authority distribution and threshold revocation method based on trust. Trust values—direct and indirect—are used for computation of the trust value, and the secret key is distributed to every node in the network by a certificate authority. This follows the computation of a threshold revocation method based on trust and thus the elimination of misbehaving nodes [45]. True flooding approach has been employed in [10] for developing a framework named Flooding Factor based Framework for Trust Management (F3TM) for mobile ad hoc networks. In this approach, the identification of attacker nodes is done on the basis of trust values. Authors have developed Route Discovery Algorithm for discovering a secure and efficient route to forward data by making use of Experimental Grey Wolf algorithm to validate the nodes in the network. Then, the identified route is optimized using Enhanced Multi-Swarm Optimization. The framework has been simulated on NS2 for comparing and assessing its performance with other standard frameworks viz. PRIME and CORMAN by taking into account metrics like packet delivery ratio, throughput, delay, and overhead. The simulation results validate better security reliance on F3TM in comparison to the standard frameworks.

Thus, it can be observed that numerous secure routing techniques have been developed for ensuring improved communication at the expense of cost-effective and simple security mechanisms in mobile ad hoc networks. However, none of the same has been able to emerge as a standard security solution for MANETs. Until this day, none of the secure routing protocols proposed for MANETs have been standardized by Internet Engineering Task Force (IETF).

3 Problem Statement

On conducting a review of the existing research works, several issues were found to be present in these approaches. The major problems that were detected in the previous works include the following: (i) Employing complex cryptographic approach, (ii) inadequate balance between security and communication performances, (iii) Most of the routing solutions presented so far are somewhat specific to the adversaries, and (iv) no benchmarking of the prevalent routing approaches. This gives us the definition of the problem statement of the proposed system as, "The development of a non-cryptographic routing algorithm for MANETs that can provide the maximum level of resistance to security attacks is computationally challenging and would thus affect the communication performance."

4 Proposed System

The proposed study takes into consideration analytical modeling approach which lays emphasis on non-cryptographic methods to ensure security during the communication process in MANETs. Therefore, the proposed system contributes to the development of a secure and simple routing policy capable of authenticating nodes in a network without the reliance on cryptographic techniques. Furthermore, the authenticity of the inbound request or communication is identified by trust and resource-based threshold. Lastly, this system ensures to provide a light-weight algorithm maintaining an excellent balance between security and performance of the communication in a MANET. The development and design of the proposed system is made up of two core components.

4.1 Design Principle of Resource-Based Thresholding Scheme

The main goal of the thresholding scheme is the selection of the best path where algorithm of the lined up authentication technique can be implemented. However, for availing better performance of the lined up authentication system, the node involved in communication must have enough bandwidth, cut-off memory, least residual energy and better mobility. Consequently, in the process of route discovery, the proposed system enables S (source) to calculate probabilities $(\alpha_1, \alpha_2, \alpha_3, \ldots)$ from its neighboring nodes $(i_1, i_2, i_3, \ldots)$. These probabilities shall be later compared with T_α, the threshold value. The nodes whose probabilities exceed T_α are retained in the next phase whereas those with probabilities lesser than T_α are discarded (Fig. 1).

The probability factor $(\alpha_1, \alpha_2, \alpha_3 \ldots)$ was calculated by the evaluation of shortest distance (d_1, d_2, d_3) plus the time (t_1, t_2, t_3) of intermediary hops $(i_1, i_2, i_3 \ldots)$ with regard to the total distance and time of the link proceeding towards the destination node D. For avoiding infinite loop generation, the calculation of probability factor is continued only for some fixed time duration t. The objective function of the proposed system, f(x) can be mathematically represented by,

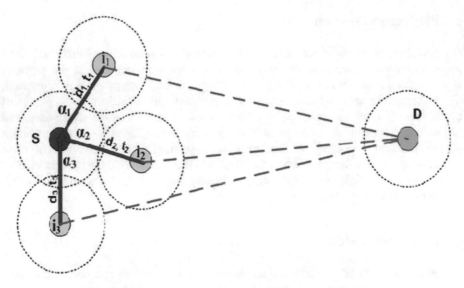

Fig. 1. Pictorial representation of communication from S to D

$$f(x) = \sum_{n=1}^{N} \alpha_n [\forall \alpha_n \subseteq \arg_{\min}(d_n t_n)] \tag{1}$$

In accordance with (1), the objective function f(x) initially tries to compute values of the probability factors, α followed by the sorting of probabilities on the basis of the lowest probability values for distance as well as time of hops. Employing a broadcast method, the proposed system assumes a fair distribution of the routing information that results in the knowledge of probability values of time and distance in memory. When fresh routing mechanism is employed, the previous knowledge of distance and time in memory is updated to new values and in this way, the routing information is updated precisely and without any memory overhead. In the preliminary phase, the thresholding value T_α is initialized, that is calculated based on the initial time and distance among the nodes. When the broadcast mechanism is used, the nodes receive updates of the past threshold that gets revised then on the basis of the requirements of node involved in communication. The result of such a mechanism is the generation of the best value of the objective function, i.e., those routes that have the highest probability of quickly arriving at the destination. The proposed system then performs the process of lined up authentication only for those nodes that comply with the objective function. This algorithm has been further discussed in Sect. 5.

4.2 Design Principle of Trust-Based Authentication Scheme

This phase focuses on authenticating only those nodes that comply with the objective function as represented in (1). If we consider that some value of probability, say α_3 was calculated to be lesser than T_α hence it was discarded. So, no routing would be performed with hop i_1. The source node will opt for routing through hops i_2 and i_3 as can be seen in Fig. 2. Now, let us consider that probability factors α_1 and α_2 have value greater than T_α. Only in this scenario, the source node S shall opt for routing. Nevertheless, the hops have to be authenticated before routing can be performed through them. The proposed system then selects next set of threshold T_β and probability factor β_1 on hop i_2 and β_2 on hop i_3. The design of this threshold probability relies on the state-based trust incorporating three distinct states of trust viz. suggested trust, Capacity of Transmitting (CoT) and undeviating trust. Therefore, hops with higher values of β shall be taken into account for routing data packets securely from source to destination. On the other hand, in case there were two β values greater than threshold value T_β, then the hop with least travel time, shortest distance and higher residual energy is selected by the system.

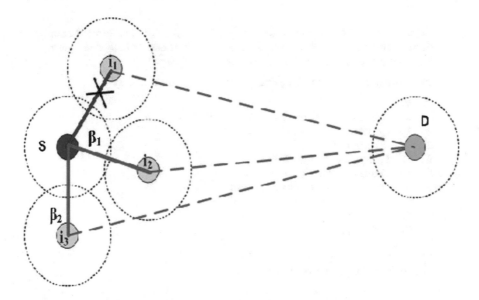

Fig. 2. Pictorial representation of in-line authentication system

As a result, routing is performed by the proposed system on comparison of resource probability factor (α_1, α_2, α_3,...) with threshold value T_α followed by a comparison of hops chosen with trust-based probability factor (β_1, β_2, β_3,...) with threshold value T_β. The discussion of the proposed algorithm implementation for both the schemes mentioned above has been presented in the following section that shall bring further insights into the potential factor of the algorithm design for ensuring secure routing in mobile ad hoc networks.

5 Algorithm Implementation

This section presents a brief discussion on the analytical design of the proposed system with the help of Algorithm-1 which represents resource-based thresholding scheme whereas Algorithm-2 represents trust-based or thresholding scheme. The simulation parameters shall consist of about 40–100 nodes making use of Random Waypoint Model with a mobility of 0–30 m/s. The nodes are assumed to have the highest transmission range of about 300 m in order with 802.11 MAC protocol. The simulation study has been carried out in Matlab assuming simulation time of 1000 s. The core algorithm has been described below:

Algorithm-1: Algorithm for Resource-Based Thresholding Scheme

Input: N (Total Number of nodes), d_i (Transmission distance of neighbor node i), d_S (Transmission distance of Source S), d_{is} (distance between i and S)

Output: Route with higher resource probability

START:
1. Initialize number of nodes (N)
2. Apply Random mobility model
3. S → beacon → in
4. If d_{si} < (d_i + d_s)
5. S calculates α (i_n)
6. For round = 0 to max(round)
7. T_α = g_{best}(d, t)
8. If α(i_n) < T_α
9. Discard d_{si}
10. Or else
11. Select min(d_{si})
12. Apply Objective Function

$$f(x) = \sum_{n=1}^{N} \alpha_n [\forall \alpha_n \subseteq arg_{min}(d_n t_n)]$$

13. f(x) → Apply Algorithm-2
END

Algorithm-1 employs the random mobility model on the input of simulation parameters. Source S generates a beacon that is received by intermediary hops (i_n, n = 1, 2,..., N) as indicated in Line-3. If nodes S, as well as i, fall within each other's ranges of transmission (Line-4), then S begins to compute resource probability values of intermediary nodes (i_n) and returns all those values to S (Line-5). For all the simulation iterations (Line-6), the resource-based threshold value T_α is calculated by the global best result of time and distance for every node leading from i to D (Line-7). The individual probability factor α shall then be calculated for each intermediary hop (Line-8). If the probability factor of intermediary hop α(i_n) is found lesser than the threshold value

T_α, then that route (d_{Si}) is discarded or else selected (Line 9 and line 11). Finally, the best route is found by applying the objective function on the selected routes. If more than two routes have values greater than T_α, then the algorithm chooses only those routes that have a lesser value of t from the input arguments to the objective function.

Algorithm-2 takes the objective function input from algorithm-1, and it essentially calculates the most secure route making use of trust-based thresholding scheme. Then, cumulative trust β is calculated from the empirical relationship represented in Line-5 of Algorithm-2, where it can be observed that trust is calculated from capacity of transmitting (CoT), suggested trust (δ_{sug}) and undeviated trust (δ_{ut}). The simulation study also takes into consideration some constraint factors for computing trust e.g., p, q and r, where p, q, r $\in$ [0, 1] and p + q + r = 1 due to the adoption of probability theory.

Algorithm-2: Algorithm for Trust-Based Thresholding Scheme

Input: objective function (f(x)), p (probability), R (regular nodes), M (malicious nodes), E_{res} (residual energy), bw (bandwidth)

Output: Route with higher trust probability

START:
1. Select routes compliant of f(x)

2. Evaluate undeviated trust

$$\delta_{ut} = \begin{cases} \dfrac{p(R)-p(M)}{p(R)+p(R,M)} & p(R) > p(M) \\ 0 & or-else \end{cases}$$

3. Evaluate suggested trust

$$\delta_{sug} = \begin{cases} p(R)-p(M), & p(R) > p(M) \\ 0 & or-else \end{cases}$$

4. CoT = {E_{res}(i), bw(i), mobility state (i)}
5. Evaluate total trust β

$$\beta = \delta_{ut}(i,j).p + \delta_{sug}(i,j).q + CoT(j).r$$

6. Initialize T_β = 0.05
7. If β < T_β
8. Discard d_{Si}
9. Or else
10. Select d_{Si}
11. Forward data packet
END

The foremost essential component is the calculation of undefeated trust as can be seen in line-2 of algorithm-2. It is assumed that if an intermediate node contributes in the transmission of packets then it is termed as regular or else malicious node. A deeper look into Line-2 reveals that calculation of undefeated trust can be done only after computing regular node probability p(R), malicious node probability p(M), and the probability of unknown mobile node p(R, M). Correspondingly, Line-3 represents trust factor that is offered by other neighboring nodes for nodes S and i. The last variable Capacity of Transmitting (CoT) is calculated by the evaluation of mobility states of intermediate hops, residual energy, and bandwidth. The algorithm then begins with the computation of individual trust-based probability factor β of intermediate hops and selects only those hops that have values greater than trust threshold value T_β. When more than two routes have probability value greater than T_β, the system then chooses the probability factor randomly and selects the new route for routing data packets.

6   Result Discussion

This section deliberates upon the results that have been fetched after evaluation of the proposed system for the purpose of benchmarking and comparison. The proposed framework is compared with the state of art mobile ad hoc network routing protocol, namely Secured AODV [46] conceived by Cerri and Ghioni, for effective analysis. The performance metric selected for the purpose of analysis include throughput, packet delivery ratio, computational time and end-to-end delay.

6.1   Analysis of Computational Time

Computational time in the testbed has been formulated as the total time expended in executing the presented algorithms completely. The outcome as shown in Fig. 3 reveals the increase in computational time till 500th iteration in the proposed system, where the curve begins to exhibit small increment in computational time past 500th iteration.

The main reason for this outcome is the usage of recursive encryption technique in SAODV for catering up the encryption requisite over dynamic topology. As a result, SAODV performs several encryption rounds for a control message till an acknowledgement is received from its surrounding nodes. On the other hand, our proposed system employs an entirely different approach since there is no involvement of cryptography. By simply computing trust, the system proposed identifies an illegitimate request/access by any user with the help of two separate thresholding criteria with no constraint to recheck the trust value. Therefore, the computational time of the proposed system is reduced by about 60% in comparison to the standard SAODV which shows its robustness through its ability to deal with the security requirement under the dynamic mobile ad hoc network architecture.

6.2   Analysis of Packet Delivery Ratio

Computationally, Packet Delivery Ratio (PDR) has been formulated as the amount of data (packets) received by the destination over the total amount of data being forwarded

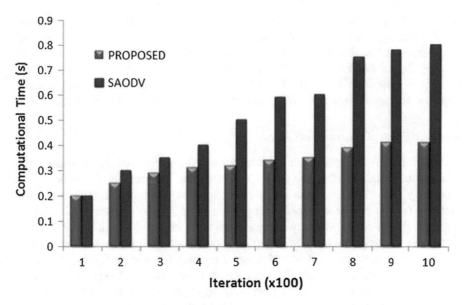

Fig. 3. Graphical demonstration of computational time analysis

by the sender node. Normally, complex cryptographic methods expend considerable time due to the sequence of encryption steps which leads to a reduction in the packet delivery ratio. The graphical outcome shown in Fig. 4 exhibits better packet delivery ratio in the proposed technique as compared to the existing SAODV with the increasing number of iterations. The underlying reason is that SAODV makes use of reactive routing protocol which is particularly concerned in routing stale information within the domains of the network. Because of this stale information, the algorithm is unable to single out an illegitimate query resulting in further encryption which reduces the

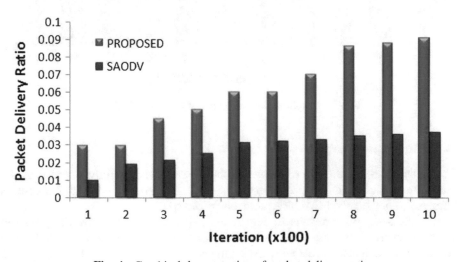

Fig. 4. Graphical demonstration of packet delivery ratio

clearance time considerably in order to fine-tune the SAODV encryption policies thereby leading to a decrease in packet delivery ratio. However, our technique utilizes lesser computational time for perceiving the intrusion owing to the dual threshold scheme (based on trust and resource) and thus clears the encryption stage rapidly yielding higher data transmission in unit time. The proposed approach shows an improvement over existing SAODV in packet delivery ratio by a percentage of 40–50.

6.3   Analysis of End-to-End Delay

End-to-End delay in the test bed has been formulated as the total time consumption for a datagram to arrive at the receiver node, taking into consideration the entire network. The delay factor in the proposed approach highly depends on the time consumed to perform encryption (or data processing) on each data packet forwarded. The result given in Fig. 5, when carefully analyzed, shows a higher delay in SAODV in comparison to the proposed system. The main explanation such a result is the encryption policy implemented in SAODV which has some major efficiency drawbacks due to its complex cryptographic nature. Consequently, this increases the time consumed by data to arrive at the receiver node thereby causing higher delay. The tendency of the current result is also in conformity with the computational time trend.

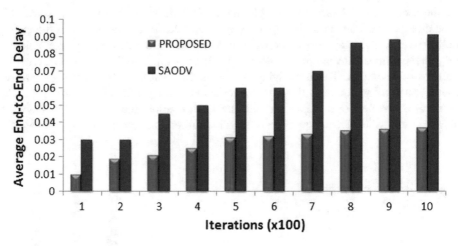

Fig. 5. Graphical demonstration of end-to-end delay

6.4   Throughput Analysis

Throughput measures the rate of delivering the messages successfully in unit time. As evident from Fig. 6, the proposed technique shows a rising trend in throughput as compared to the standard SAODV protocols in MANETs.

As indicated by the results in Fig. 3, the performance shown by the proposed system in terms of processing time is much lower in comparison to SAODV. The proposed approach also excels with respect to packet delivery ratio and throughput, as

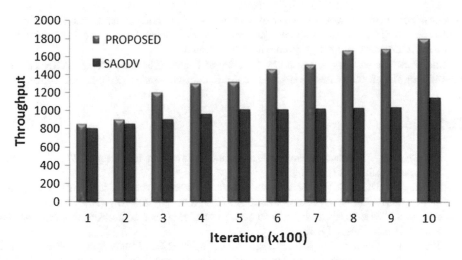

Fig. 6. Graphical outcome of throughput

shown in Figs. 4 and 6 respectively. Figure 5 highlights that our system shows better delay performance than SAODV; the underlying reason again being the dependence of SAODV on sophisticated cryptography due to which resource consumption becomes too massive. The higher delay encountered in SAODV is also due to spontaneous resists replying for receiver node. Further, the signature mechanism employed in the conventional SAODV leads to an increase in the computational time and thus thwarts the throughput.

7 Conclusion

For data transmission in MANETs, security emerges as the biggest challenge because of their inherent characteristic topology. The decentralization in mobile ad hoc networks makes it cumbersome to authenticate all nodes involved in communication during simulation effectively. Moreover, it has been observed from the previous works that most of the secure routing policies in MANETs are based on cryptographic algorithms. In this paper, a novel routing technique for mobile ad-hoc networks has been projected that implements a non-cryptographic algorithm. This technique can be utilized in providing an efficient as well as enhanced security solution to routing in MANET communication. The performance of the proposed system when evaluated with the standard SAODV reveals improvement in terms of computational time, packet delivery ratio, end-to-end delay, and throughput. This system employs probability theory and an analytical approach for determining the presence of reliable nodes in order to find the shortest route to establish communication. Furthermore, it also ensures that the shortest established path has the maximum trust factor that ascertains secure data transmission. This has repercussions in the form of rejecting intrusion by malicious nodes as the trust factor is determined at each step and broadcast to all nodes in the network.

Acknowledgements. I would like to express my sincere gratitude towards Prof. A. G. Lone; I could see myself becoming a researcher just because, coincidently he was teaching mathematics at the same Engineering College from where I graduated.

This work was partially supported by Ministry of Higher Education Malaysia (Kementerian Pendidikan Tinggi) under Research Initiative Grant Scheme number (RIGS15-150-0150).

References

1. Jin, M., Du, Z.: Management innovation and information technology. WIT Trans. Inf. Commun. Technol. **61**, 2124 (2014)
2. Sarkar, S., Datta, R.: A secure and energy-efficient stochastic multipath routing for self-organized mobile ad hoc networks. Ad Hoc Netw. **37**, 209–227 (2016)
3. Sarkar, S.: Wireless sensor and ad hoc networks under diversified network scenarios. Artech House, Boston, Mass. (2012)
4. Khan, B., Olanrewaju, R.F., Baba, A.M., Mir, R.N., Lone, S.A.: DTASR: dual threshold-based authentication for secure routing in mobile adhoc network. Int. J. Inf. Technol. Comput. Sci. (IJITCS) **22**(1), 1–9 (2015)
5. Vashist, P., Hema, K.: New multicast routing protocol in ad-hoc network. Int. J. Innovations Eng. Technol. (IJIET) **2**(2), 108–119 (2013)
6. Aluvala, S., Sekhar, K.R., Vodnala, D.: An empirical study of routing attacks in mobile ad-hoc networks. Proc. Comput. Sci. **92**, 554–561 (2016)
7. Khan, B., Olanrewaju, R.F., Ali, N.A., Shah, A.: ElePSO: energy aware elephant swarm optimization for mobile adhoc network. Pensee J. **76**(5), 88–103 (2014)
8. McNerney, P.J., Zhang, N.: A study on reservation-based adaptation for QoS in adversarial MANET environments. In: 8th International Wireless Communications and Mobile Computing Conference (IWCMC), pp. 677–682. IEEE (2012)
9. Lafta, H.A., Al-Salih, A.M.: Efficient routing protocol in the mobile ad-hoc network (MANET) by using genetic algorithm (GA). IOSR J. (IOSR J. Comput. Eng.) **1**(16), 47–54 (2014)
10. Ahmed, M.N., Abdullah, A.H., Chizari, H., Kaiwartya, O.: F3TM: flooding factor based trust management framework for secure data transmission in MANET. J. King Saud Univ.-Comput. Inf. Sci. (2016). https://doi.org/10.1016/j.jksuci.2016.03.004
11. Loo, L., Mauri, J., Ortiz, J.: Mobile Ad Hoc Networks: Current Status and Future Trends, p. 538. CRC Press, Inc. Boca Raton, FL, USA©2011 (2012)
12. Kennington, J., Olinick, E., Rajan, D.: Wireless Network Design. Springer, New York (2011)
13. Sadasivam, K., Yang, T.A.: Evaluation of certificate-based authentication in mobile ad hoc networks. In: IASTED International Conference on Networks and Communication Systems (NCS 2005), Krabi, Thailand (2005)
14. Aluvala, S., Sekhar, K.R., Vodnala, D.: A novel technique for node authentication in mobile ad hoc networks. Perspect. Sci. **8**, 680–682 (2016)
15. Amine. A,, Ait Mohamed, O., Benatallah, B.: Network Security Technologies: Design and Applications, p. 330. IGI Global (2013)
16. Pathan, A., Monowar, M., Fadlullah, Z.: Building Next-Generation Converged Networks: Theory and Practice. CRC Press, Boca Raton, FL (2013)
17. Pathan, A.: Security of self-organizing networks: MANET, WSN, WMN, VANET. Auerbach Pub, Boca Raton (2011)

18. Thorat, S.A., Kulkarni, P.J.: Uncertainty analysis framework for trust based routing in MANET. Peer-to-Peer Netw. Appl. 1–11 (2016)
19. Yavuz, A.A., Alagöz, F., Anarim, E.: A new multi-tier adaptive military MANET security protocol using hybrid cryptography and signcryption. Turk. J. Electr. Eng. Comput. Sci. 18 (1), 1–22 (2010)
20. O'rourke, C., Johnson, S.B.: Mobile ad hoc networking revamps military communications. Mil. Electron. Comput. (2011)
21. Sarkar, S., Datta, R.: A game theoretic model for stochastic routing in self-organized MANETs. In: IEEE Wireless Communications and Networking Conference (WCNC), pp. 1962–1967 (2013)
22. Ji, Z., Yu, W., Liu, K.R.: A game theoretical framework for dynamic pricing-based routing in self-organized MANETs. IEEE J. Sel. Areas Commun. 26(7), 1204–1217 (2008)
23. Anderegg, L., Eidenbenz, S.: Ad hoc-VCG: a truthful and cost-efficient routing protocol for mobile ad hoc networks with selfish agents. In: Proceedings of the 9th Annual International Conference on Mobile Computing and Networking, pp. 245–259. ACM (2003)
24. Shen, H., Li, Z.: A hierarchical account-aided reputation management system for MANETs. IEEE/ACM Trans. Netw. (TON) 23(1), 70–84 (2015)
25. Zapata, M.G.: Secure ad hoc on-demand distance vector routing. ACM SIGMOBILE Mob. Comput. Commun. Rev. 6(3), 106–107 (2002)
26. Sarkar, S., Kisku, B., Misra, S., Obaidat, M.S.: Chinese remainder theorem-based RSA-threshold cryptography in MANET using verifiable secret sharing scheme. In: IEEE International Conference on Wireless and Mobile Computing, Networking and Communications, pp. 258–262 (2009)
27. Buttyán, L., Hubaux, J.P.: Stimulating cooperation in self-organizing mobile ad hoc networks. Mob. Netw. Appl. 8(5), 579–592 (2003)
28. Sarkar, S., Datta, R.: AODV-based technique for quick and secure local recovery from link failures in MANETs. Int. J. Commun. Netw. Distrib. Syst. 11(1), 92–116 (2013)
29. Khan, B., Olanrewaju, R., Habaebi, M.: Malicious behaviour of node and its significant security techniques in MANET-a review. Aust. J. Basic Appl. Sci. 7(12), 286–293 (2013)
30. Khan, B., Olanrewaju, R., Mir, R., Adebayo, B.: Behaviour visualization for malicious-attacker node collusion in MANET based on probabilistic approach. Am. J. Comput. Sci. Eng. 2(3), 10–19 (2015)
31. Marimuthu, M., Krishnamurthi, I.: Enhanced OLSR for defense against DOS attack in ad hoc networks. J. Commun. Netw. 15(1), 31–37 (2013)
32. Venkataraman, R., Rama Rao, T., Pushpalatha, M.: Regression-based trust model for mobile ad hoc networks. IET Inf. Secur. 6(3), 131–140 (2012)
33. Lv, X., Li, H.: Secure group communication with both confidentiality and non-repudiation for mobile ad-hoc networks. IET Inf. Secur. 7(2), 61–66 (2013)
34. Zhang, P., Lin, C., Jiang, Y., Fan, Y., Shen, X.: A lightweight encryption scheme for network-coded mobile ad hoc networks. IEEE Trans. Parallel Distrib. Syst. 25(9), 2211–2221 (2014)
35. Dangare, N.N., Mangrulkar, R.S.: Design and implementation of trust based approach to mitigate various attacks in mobile ad hoc network. Proc. Comput. Sci. 78, 342–349 (2016)
36. Zhao, S., Aggarwal, A., Frost, R., Bai, X.: A survey of applications of identity-based cryptography in mobile ad-hoc networks. IEEE Commun. Surv. Tutorials 14(2), 380–400 (2012)
37. Dhanapal, J., Srivatsa, S.: Link quality-based cache replacement technique in mobile ad hoc network. IET Inf. Secur. 7(4), 277–282 (2013)
38. Xi, S., Liang, S., Jianfeng, M., Zhuo, M.: A trust management scheme based on behavior feedback for opportunistic networks. China Commun. 12(4), 117–129 (2015)

430   B. U. I. Khan et al.

39. Xia, H., Jia, Z., Li, X., Ju, L., Sha, E.: Trust prediction and trust-based source routing in mobile ad hoc networks. Ad Hoc Netw. **11**(7), 2096–2114 (2013)
40. Tan, S., Li, X., Dong, Q.: Trust based routing mechanism for securing OSLR-based MANET. Ad Hoc Netw. **30**, 84–98 (2015)
41. Wan, Z., Ren, K., Gu, M.: USOR: An unobservable secure on-demand routing protocol for mobile ad hoc networks. IEEE Trans. Wirel. Commun. **11**(5), 1922–1932 (2012)
42. Khan, B., Olanrewaju, R.F., Mir, R.N., Baba, A., Adebayo, B.W.: Strategic profiling for behaviour visualization of malicious node in Manets using game theory. J. Theoret. Appl. Inf. Technol. **77**(1), 25–43 (2015)
43. Khan, B., Olanrewaju, R.F., Anwar, F., Shah, A.: Manifestation and mitigation of node misbehaviour in adhoc networks. Wulfenia J. **21**(3), 462–470 (2014)
44. Joshi, S.S., Biradar, S.R.: Communication framework for jointly addressing issues of routing overhead and energy drainage in MANET. Proc. Comput. Sci. **89**, 57–63 (2016)
45. Rajkumar, B., Narsimha, G.: Trust based certificate revocation for secure routing in MANET. Proc. Comput. Sci. **92**, 431–441 (2016)
46. Phu, P., Yi, M., Kim, M.: Securing AODV routing protocol in mobile ad-hoc networks. In: Active and Programmable Networks, vol. 4388, pp. 182–187. Springer, Berlin (2009)

Hybrid Audio Steganography and Cryptography Method Based on High Least Significant Bit (LSB) Layers and One-Time Pad—A Novel Approach

Samah M.H. Alwahbani[(✉)] and Huwaida T.I. Elshoush

Faculty of Mathematical Science, University of Khartoum, Khartoum, Sudan
samahmahdi22@gmail.com, htelshoush@uofk.edu

Abstract. The paper proposes a novel chaos based audio steganography and cryptography method. It is a higher Least Significant Bit (LSB) layers algorithm in which the secret message is encrypted first by one-time pad algorithm. Two chaotic sequences of Piecewise Linear Chaotic Map (PWLCM) were used. In the encryption process, the key for one-time pad is generated by PWLCM chaotic map. In the steganography process, the second sequence of PWLCM is used to generate a random sequence. Then, indices of the ordered generated sequence were used to embed the encrypted message in randomly selected audio samples. The encrypted data were embedded on the higher layers other than the LSB using efficient bits adjustment algorithm, in order to increase the robustness against noise addition or MPEG compression. An analysis is discussed for the proposed scheme. For the steganography algorithm, the proposed scheme overcomes the main two problems for LSB coding, which are the low robustness of secret message extraction and destruction. For the former, the proposed method encrypts the secret message by perfect efficient algorithm which is the one-time pad. Regard of the second one, the secret message is hidden in higher layers which improve the robustness against signal processing manipulation. The main three steganography characteristics were tested and evaluated which are high capacity, perceptual transparency and robustness. Furthermore, the drawbacks of key generation and key distribution for one-time pad is sloved by using the chaotic maps. For the experimental results, waveform analysis and signal-to-noise Ratio are made, which show the high quality of the stego audio, and hence demonstrate the efficiency of the proposed scheme.

1 Introduction

Due to the rapid development of communication technologies, and with the emergence of Internet, large amount of data are distributed over open shared network. Therefore, data security is an important issue. There are two solutions to secure these communication, cryptography and steganography. Cryptography

© Springer International Publishing AG 2018
Y. Bi et al. (eds.), *Intelligent Systems and Applications*,
Studies in Computational Intelligence 751,
https://doi.org/10.1007/978-3-319-69266-1_21

scrambles the secret message in order to be unreadable by eavesdroppers while, steganography conceals the secret message into another object in order to be invisible by eavesdroppers [14].

On the other hand, Steganography and watermarking embed data transparently into carrier objects. Watermarking establish the identity of information to prevent unauthorized use. It embeds the information into digital file in such a way that its removal is hardly possible [37].

The main difference between steganography and watermarking, is that watermarking does not necessarily hide the fact of secret transmission of information from third parties. Besides preservation of the carrier signal quality, watermarking generally has the additional requirement of robustness against manipulations intended to remove the embedded information from the marked carrier object [37,40].

The following table (Table 1) shows some general comparisons between Cryptography, Steganography and Watermarking [20].

Table 1. Comparisons between cryptography, steganography and watermarking

	Cryptography	Steganography	Watermarking
Definition	The art and science of secret codes	The process of hiding digital information in a carrier signal	The art and science of hiding information
Secret message	It is scrambled and unreadable	It is hidden but not scrambled	It is invisible or perceptual visible depending on the requirements
Objective	It make the message unreadable for without knowing the algorithm and the key	It hide the existence of the message	It mark the digital files with copyright information and to avoid illegal copying
Security	No one can read the message without knowing the key	No one can percept the hidden message	Verify the authenticity or integrity of the digital files or to show the identity of its owners

Steganography is an art of secret communication. It is sub-discipline of information hiding that focuses on concealing the existence of messages [10]. The origin of word steganography comes from the Greek "steganos", which means covered or secret and "graphy" means writing or drawing. So, steganography

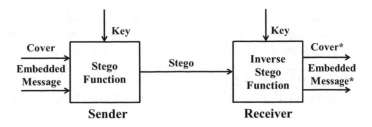

Fig. 1. A generic steganography system

means, literally, "Covered writing". It does not alter the structure of the secret message, but hides it inside a cover file so that it cannot be seen. The main goal of steganography is to communicate securely in a completely undetectable manner by preventing an unintended recipient from suspecting that the secret data exists [14]. The process in which a steganalyzer cracks the cover object to get the hidden data is often known as steganalysis [21].

Figure 1 shows a general model of the steganography. The following terms are used in the steganography:

- **Embedded Message**: The secret message to be embedded.
- **Cover**: The data in which the secret message will be embedded.
- **Stego**: A modified version of the cover that contains the secret message.
- **Key**: Secret data that is needed for the embedding and extracting processes.
- **Stego Function**: A steganographic function that takes the parameters: *Cover, Embedded Message* and *key* and produces stego as output.
- **Embedded Message***: The recovered secret message at the receiver side.
- **Cover***: The cover file at the receiver side.
- **Inverse Stego Function**: An Inverse steganographic function that has stego and key as parameters and produces Message* as output. The result of the extracting process should be identical to the input Message of the embedding process.

For the steganographic scheme, there are three main steganography characteristics should be satisfied [3,37]:

- **Robustness**: It should not be able to extract the secret message from the stego file without the knowledge of the secret key used in the extracting procedure.
- **Transparency**: The cover medium after being embedded with the secret data should be indiscernible from the original one.
- **High capacity**: The amount of information that a steganography scheme can successfully embed without introducing perceptual distortion should be as large as possible.

In addition to the these characteristics, the cover data should be able to survive when the stego medium has been manipulated such as by some lossy compression scheme.

This paper is organized as follows, Sect. 2 presents an overview of the current steganography techniques, while Sect. 3 discusses the various audio steganography techniques in some details. Some related work is discussed in Sect. 4. Then, a brief overview of chaotic map is presented in Sect. 5. Section 6 introduces One-time pad algorithm while Sect. 7 presents the proposed method, and Sect. 8 presents and analyzes the experimental results. At last, Sect. 9 concludes the paper.

2   Steganography Techniques

In the past, invisible ink and papers were used for secret communication. Recently, more practical media are used such as text, image, audio and video.

Due to the high redundancy of multimedia objects, they are suitable for data hiding. Generally, the steganography algorithms use the redundancy of the time, space or transform domain. For each domain, there are numerous techniques for hiding.

Based on the type of cover object, steganography methods are divided into five types which are introduced briefly in the following subsections.

2.1   Text Steganography

Hiding information in text involves several methods such as the insertion of spaces, resizing the text and change the style of text to hide the secret message [17]. However, text steganography is not widely used because the text has a very small amount of redundancy [19].

2.2   Audio Steganography

Audio steganography is the art and science of hiding digital data such as text, images and audio into audio files such as WAV and MP3 files. It based on the weakness of the Human Auditory System (HAS) in which the human ears do not notice the slight differences between the original and the stego audio file.

2.3   Image Steganography

An image file may store data in compressed, uncompressed format. Image steganography methods can be classified into two categories, spatial domain and frequency domain. Spatial domain methods involve direct manipulation of pixels in the image. Frequency domain techniques are based on modifying the Fourier transform of the image [5].

Compared with audio steganography, image steganography use the weakness of Human Visual System (HVS).

2.4   Video Steganography

Generally, video files are a collection of images and sounds, so most of the images and audio steganography techniques can be applied to video files. Therefore, video has high capacity to embed secret messages [9].

2.5   Protocol Steganography

In the layers of the OSI network model there exist covert channels that can be used by steganography. The secret messages can be hidden within messages and network control protocols used in network transmission. For example, secret data can be embeded in the optional fields of a UDP packet header [9].

3   Audio Steganography Techniques

There are many researchers have developed algorithms for audio steganography. These algorithms takes advantage of the perceptual properties of the HAS in order to add a secret message into a host signal transparently. The HAS system is more dynamic than Human Visual System (HVS). Therefore, increasing embedding capacity for audio is more difficult than that of the images.

In addition, several steganalysis techniques that can be applied against image steganography algorithms such as geometrical distortions and spatial scaling. These attacks cannot be implemented against audio steganography schemes. Therefore, audio steganography seems more secure than image steganography.

Several audio steganography techniques have been proposed and used. This section introduces them.

3.1   Least Significant Bit (LSB)

It is one of the most popular audio steganography methods. It replaces the least significant bit in some bytes of the cover file to hide a sequence of bytes containing the secret data. In this technique, a byte can be hidden in eight bytes of the cover file. Theoretically, there is a fifty percent propability that the replaced bit does not change, which helps to minimize quality distortion [14].

At the receiver side, he/she should access the sequence of samples that contains the secret message in order to extract it. The length of the secret message to be encoded should be smaller than or equal the total number of samples in the audio file. The important issue is how to choose the subset of samples that will contain the secret message and how the sender and receiver agree on the selection [16].

The simplest technique is to perform LSB substitution from the beginning of the audio and continue until the message has been completely embedded, leaving the remaining samples unchanged. Using this method, it is extremely easy for an attacker to recover the secret message. Also, the first part of the sound file will have different statistical properties than the second part of the sound file

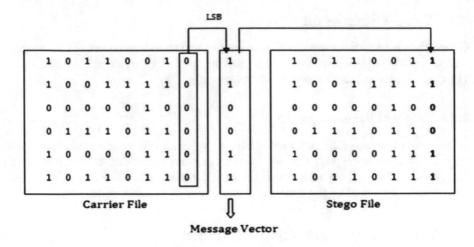

Fig. 2. Example of LSB coding

that was not modified. Such solution is to pad the secret message with random bits in order to make the message length is equal to the total number of samples [33].

Another approach is to use a pseudorandom numbers for selecting the audio samples randomly over the whole audio file. The sender and reciever agree on pseudorandom number generator and secret key, So that they can generate the same sequence of sample indices again. However, this pseudorandom number generator may lead to collision, in which a sample already modified with part of the message is modified again. This problem can be solved by keeping track of all the modified samples [33].

LSB coding considered the simplest audio steganography methods due to the fact that it does not require any complex operation or some special hardware which make it suitable for real time applications. The main advantage of the LSB coding method is a high data rate of the embedded data. For example, if only one LSB of the host audio sample is used and all samples are used for embedding with sampling rate 8,000 HZ, it gives a capacity of 8 kbps [8]. Figure 2 shows an example of LSB coding.

Subjective listening test conclude that in average, the maximum LSB depth that can be used for LSB based audio steganography method without causing noticeable audio quality degration is the 4th LSB layer when 16 bits are used for audio samples encoding [12].

However, LSB method has a significant security problem, the secret data are embedded in a very predictable way [5]. Furthermore, any random changes of the LSBs destroy the hidden message [24].

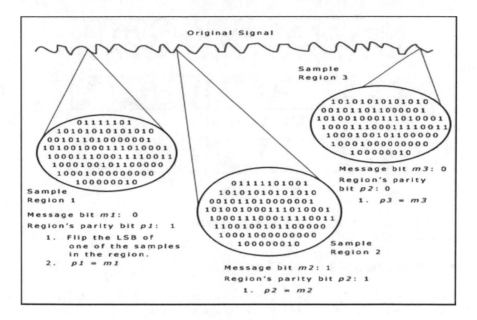

Fig. 3. Parity coding

3.2   Parity Coding

In this method, the signal is divided into groups or regions of samples, and it embeds each bit of the secret message in a sample region's parity bit. If the parity bit of a selected region does not match the secret bit to be encoded, the LSB of one of the samples in the region is inverted [14,18]. Figure 3 [11] shows the concept of this technique.

By using parity coding technique, the sender has more choices for encoding the secret bit. However, this method like LSB coding, it is not robust [22].

3.3   Echo Hiding

In echo hiding technique, the secret data are embedded into the audio signals as a short acoustic echo. Echo is a replication of sound received by the listener some time after the original sound. As the echo is audible, its amplitude must be decreased so that it becomes imperceptible. In order to hide data, bits whose values are 0 are represented by an echo delayed 1 ms; bits whose values are 1 are represented by an echo delayed 2 ms. The limitation of echo hiding technique is the low hiding capacity as it would be computationally intensive to insert echo for every bit to hide [5]. Figure 4 shows the concept of echo hiding.

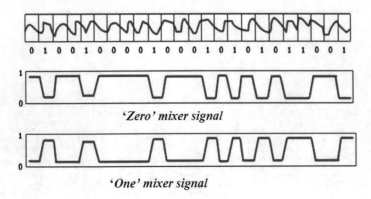

0 1 0 0 1 0 0 0 0 1 0 0 0 1 0 1 0 1 0 1 0 1 1 0 0 1

'Zero' mixer signal

'One' mixer signal

Fig. 4. Echo hiding

3.4 Silence Interval

This technique inserts a silence interval in the original audio signal to embed the secret data. The values that represent the length of the silence intervals are decreased by some value such as $0 < value < 2^n$ where n is the number of bits required to represent an element from the data to hide. The carrier audio file is then sent to the receiver having the new lengths for its silence intervals [5].

3.5 Phase Coding

In the phase coding technique, the phase of an initial audio segment is substituted with a reference phase that represents the hidden data. It uses Discrete Fourier Transform (DFT), which is a transformation algorithm for the audio signal. The phase coding technique encodes the secret message bits as phase shifts in the phase spectrum of a digital signal, achieving an inaudible encoding in terms of signal-to-noise ratio. Human Auditory System (HAS) cannot recognize the phase change in audio signal as easy as it can recognize noise in the signal, because the phase components of sound are not as noticeable as noise to the human ear [18,21]. The phase coding concept is shown in the Fig. 5.

At the receiver, he/she must know the segment length to extract the secret message. Then, he/she can use the DFT to get the phases and extract the information.

Phase coding is one of the complex and robust audio steganography techniques. On the other hand, it has a low data transmission rate because the secret message is hidden in the first signal segment only. This problem can be addressed by increasing the length of the signal segment. However, this would change phase relations between each frequency component of the segment more significantly which makes the encoding easier to detect [21].

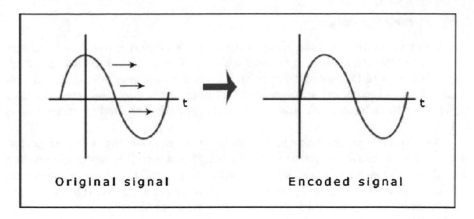

Fig. 5. Phase coding

Table 2. Comparisons between audio steganography techniques

Technique	Strengths	Weaknesses
LSB	• Simple • Low computational complexity • High capacity	• Easy to extract and destroy
Parity coding	• More robust than LSB • Low computational complexity • High capacity	• Easy to extract
Echo hiding	• Avoids problem with additive noise • Resilient to lossy data compression algorithms	• Low capacity
Phase coding	• Robust against signal processing manipulation	• Low capacity
Spread spectrum	• Increases transparency	• Occupies more bandwidth • Add noise for the cover audio
Silence intervals	• Resilient to lossy data compression algorithms	• Low capacity • Vulnerable to time scale modification

3.6   Spread Spectrum

The spread spectrum method spreads the secret information over the frequency spectrum of the audio file using a code which is independent of the actual signal. It is similar to LSB based system in which the secret message is spreaded randomly over the entire cover audio signal. It has a main drawback, that the final signal occupies a bandwidth more than what is actually required for transmission [14].

The Spread Spectrum method is capable of contributing a better performance in some areas compared to LSB coding, phase coding and parity coding techniques. It has a moderate data transmission rate and high level of robustness against removal techniques. However, the spread spectrum method has one main disadvantage that it can introduce noise into a sound file [14].

Table 2 presents a comparison between audio steganography techniques.

4   Related Work

In this paper the LSB based proposed methods for audio steganography are survayed. These methods have the advantages of simplicity and high capacity. There are many different LSB-based methods that were proposed for audio steganography. Some survey are presented at this section.

In the standard LSB coding method, the original host audio bit in the ith layer (i = 1, 2,...,16) is simply replaced with the bit from the embedded message bits which get high distortion. Hence, several reseraches were aimed to minimize the embedding distortion of the host audio using bits adjustment algorithms. In 2004, Cvejic and Seppanen [8] proposed a method that has high bit rate of LSB. It consists of two steps. In the first step, a secret message bit is embedded into the ith LSB layer of the host audio. Then, the impulse noise caused by the embedding is shaped in order to change its white noise properties. Also, Parthasarathy and Srivatsa [24,24] in 2009, Gadicha [12] in 2011 and [30] in 2012 emdedded the secret message at high LSB layers. Although the methods of [12,24] use secret key to select the group of samples for hiding in the host audio, they did not increase the LSB robustness significantly.

A method of multi-level steganography was introduced by Bandyopadhyay and Banik [4] in 2011. It has two layers, LSB and parity coding. At the first level, cover file C was embedded with the first secret message S_1. Assuming the stego file as C_1 which is cover file for the next level where secret message was denoted as S_2. Then, the final stego file created as C_{12}. So, C_{12} held both S_1 and S_2. The extraction of the hidden message when using Multi-Level steganography.

In 2012, Hmood et al. [13] hided images in the audio based on two LSB insertion method of the low part of the audio file. The secret image is converted from color image to gray image, encrypted and finally it was converted into binary representation. The original audio signal was broken up into two parts, low frequency and high frequency using Wavelet Transform, then the high frequency part was used to embed the encoded image. Signal to noise ratio was calculated to test the system performance.

Several methods were proposed that encrypt the secret message first before the steganography to enhance the security of LSB algorithms. Adhiya and Patil [3] introduced a hybrid method of audio steganography and cryptography. First, the message is encrypted by public key encryption algorithm. Then, the encrypted data is hidden in cover audio WAV file in randomly. Also, Padmashree and Venugopala in 2012 [23] proposed an audio steganography method in which the secret message is encrypted by the RSA method, then the ciphertext is embedded in the forth and fifth LSB bits. The experimental results show that there are less noise intrusion even after changing the forth and fifth LSB bit of the original audio.

For using symmetric algorithms, Zameer et al. [39] in 2011 presented LSB audio steganography method in which the data is encrypted first by AES algorithm. Also, to increase the hiding capacity, the data are compressed before the embedding. Two years later, Dengre [9] proposed LSB based audio steganography for AVI videos that encrypt the secret message using a group of symmetric algorithms including DES, Rijndael, RC2 and Triple DES algorithm. Then, it hides the encrypted data into LSB bit position of selected samples. Numerous tests were done using different videos sizes and different LSB layers.

In 2015, Yoon et al. [36] proposed a variant of honey encryption called visual honey encryption which employed an adaptive DTE in a Bayesian framework so that the proposed approach can be applied to more complex domains including images and videos. The authors applied this method to create a new steganography scheme which significantly improves the security level of traditional steganography.

Later after four months, in October 2015, Y. Castelan and B. Khodja [6] determined that it is best to inject covert data into the quiet portions of an MP3 file by configuring MP3Steg to use very high consecutive carrier frame values and very low consecutive non-carrier frame values (or a consecutive non-carrier frame value of 0). This will allow the user to take advantage of as much quiet MP3 file audio space as possible. When injecting into portions of an MP3 file that are not quiet or completely silent, it is best to configure MP3Steg to use low consecutive carrier frame values and high consecutive non-carrier frame values. This will make it so that any noticeable audio artifacts that are introduced are spread across the MP3 file's loud audio space as much as possible.

Ther are several researches has been done for audio steganography based on Genetic Algorithms. In 2009, Zamani et al. [37] proposed a genetic based LSB algorithm. The authors presented two solutions to LSB problems which are attacks that try to reveal the hidden message and distortions with high average power. For the first problem, the authors changed the other bits than LSBs in samples to increase the complexity for the steganalysis, and selected a group of samples—not all samples—for the change. To decrease the distortion, the message bits were embedded in higher layers and modify other bits to decrease the amount of the error.

In 2011 [38], more details and experiment results of the algorithm are presented by the author Zamani. The secret message is embedded into 2 bits of the cover object to achieve higher capacity. It shows that the genetic based algorithm get better PSNR (Peak Signal to Noise Ratio) than the original algorithm.

Singh et al. [34] and Santhi [31] in 2014 proposed a genetic based algorithm where the message is encrypted first by RSA algorithm to enhance the security.

In the literature, several methods increased the robustness of LSB coding by embedding the data in deeper LSB layers. However, hiding the data in higher LSB layer will increase the distortion, hence some methods introduced bits adjustments algorithm to decrease this distortion. In fact, the choices of hiding in the high LSB layers are not large enough for robustness against extraction. Other researches encrypted the message before hiding. But, most of the proposed methods use the standard LSB coding method which makes the message easy to be recovered by attackers. So, the steganography process will not be efficient. In addition, these encryption algorithms increase the computations of the overall system significantly, and so degrade the performance. An efficient solution for this problem is to select a group of samples for embedding in random positions in the host audio. However, no such solution. Therefore, This paper uses chaotic maps to generate pseudo-random numbers in order to be used for audio samples selections. In addition, the one-time pad encryption algorithm is used for message encryption before hiding. The following two sections present them.

5   Chaotic Map

In a scientific context, one general description of chaos is an unpredictable, non-periodic and random-like long-term evolution that results from deterministic nonlinear systems that exhibits sensitive dependence on initial conditions [15].

In the proposed scheme, the chaotic maps are used for the both processes, encryption and steganography. In the following, A Piecewise Linear Chaotic Map (PWLCM) is introduced briefly.

5.1   Piecewise Linear Chaotic Map

A Piecewise Linear Chaotic Map (PWLCM) is a chaotic map that is composed of multiple linear segments that is defined as follow:

$$y_n = F(y_{n-1}) = \begin{cases} y_{n-1} \times \frac{1}{p} & \text{if} \quad 0 < =y_{n-1} < \text{p} \\ (y_{n-1} - p) \times \frac{1}{0.5-p} & \text{if} \quad p < =y_{n-1} < 0.5 \\ F(1 - y_{n-1}) & \text{if} \quad 0.5 < =y_{n-1} < 1 \end{cases} \tag{1}$$

where $p \in (0,0.5]$ and $y_n \in [0,1]$ are the positive control parameter and the initial condition, respectively [2].

Compared with the logistic map, PWLCM has a wider range of control parameter choices. Also, the PWLCM has a better balance property and uniform invariant density function [2]. Hence, PWLCM is used for the two algorithms (encryption and steganography) of the proposed method.

6   One-Time Pad Algorithm

A one-time pad or Vernam's Cipher is a symmetric unbreakable cipher in which a secret random key is used only once to encrypt a message and then is discarded [29]. Each new message requires a new key of the same length as the new message. It produces random output that bears no statistical relationship to the plaintext. Because the ciphertext contains no information whatsoever about the plaintext, there is simply no way to break the code. It is the only cryptosystem that exhibits what is referred to as *perfect secrecy* [35].

For any plaintext of equal length to the ciphertext, there is a key that produces that plaintext. If an exhaustive search is done of all possible keys, many legible plaintexts are produced with no way to know which one is the intended plaintext. Therefore, the code is unbreakable.

The security of the one-time pad is entirely due to the randomness of the key. If the characters stream of the key is truly random, then the stream of characters that constitute the ciphertext will be truly random. Thus, there are no patterns or regularities that a cryptanalyst can use to attack the ciphertext.

The one-time pad offers complete security. However, in practice it has two fundamental drawbacks [35]:

• There is a practical problem of generating large quantities of random keys.
• Key distribution and protection. For every message to be sent, a new key of equal length is required. So, the key distribution is a significant problem.

In the proposed scheme, the chaotic maps solve these two drawbacks. The chaotic PWLCM map is used for the key generation which produce chaotic sequences with good randomness properties. Also, they have large key space to generate a huge number of random keys. For the key distribution problem, the sender and receiver needs only to exchange the initial conditions and system parameters for the chaotic map, and so there is no need to exchange the whole chaotic sequences.

7   Proposed Method

This research is an extended work for Alwahbani and Elshoush work [1]. In the [1], two chaotic maps were used, PWLCM and logistic map for message encryption and steganography, respectively. The secret message is encrypted by one-time pad algorithm using a key generated by PWLCM, and then it is embedded in LSB of randomly selected audio samples. These random numbers were generated by logistic map. Now in this paper, In addition of that, the encrypted data is embedded in high LSB layers, 2nd to 7th layers of the randomly selected audio samples. The embedding at high LSB layers increases the robustness against noise addition or MPEG compression. The proposed algorithm adjusts the other bits in order to minimize the distortion with efficient algorithm. Also, the chaotic maps are studied more, and the PWLCM is selected for using in the both, encryption and steganography, that instead of using logistic for steganography. PWLCM has larger key space and better balance property and uniform invariant density function.

The proposed method overcomes the LSB coding drawbacks:

- It has low robustness against the extraction of the secret message
- It has low robustness against the destroy of the secret message

For the first weakness, in the proposed method the message is encrypted by *perfect cipher*, one-time pad algorithm. Then, the encrypted message is hidden in randomly-selected audio samples of the higher layers of LSB. These random numbers for samples selection are generated by PWLCM chaotic map.

For the second weakness, the secret message is embedded in the higher layers of the samples' LSB using efficient embedding algorithm. The selection of the layer is critical because the random selection of the samples used for embedding introduces low power Additive White Gaussian Noise (AWGN). Obviously, changing the bits in higher layers than LSB introduce bigger errors. Therefore, a good adjustment algorithm for the other bits are used.

7.1 Message Encryption

The secret message is encrypted first by the one-time pad algorithm, then it will be embedded in the cover audio. The PWLCM chaotic map is used to generate a sequence of pseudo random numbers which is used as the key for the one-time pad. PWLCM has a good key space, chaotic properties and randomness behavior.

To generate the key for the one-time pad, both communicating parties should agree on the initial value and system parameter for PWLCM. The encryption algorithm is described as follow:

Algorithm 1   The Encryption Algorithm

    Input :

        M : plain message M of length L

        y_0 : Initial condition for the PWLCM

        p_1 : System parameter for the PWLCM

    Output :

        C : Encrypted message by one-time pad algorithm

Algorithm

1. Generate a sequence y_n of PWLCM with length L, using the initial condition y_0 and system parameter p_1 as a key.
2. Convert the chaotic sequence y_n to integer numbers as follows:

$$Y_i = (round(y_i \times 10^{14})) \, mod \, 2^8 \qquad i = 1, 2, \ldots L \qquad (2)$$

3. Encrypt the message M by one-time pad using the sequence Y_n as a key as follows:

$$C_i = (m_i + Y_i) \, mod \, 2^8 \qquad i = 1, 2, \ldots, L \qquad (3)$$

where C_i is the ciphertext, m_i is the secret message for embeding, Y_i is the quantized chaotic sequence and L is the message length.

7.2   Message Hiding

In the message hiding algorithm, instead of embedding the encrypted message bits on the audio samples sequentially, the encrypted message is embedded in cover audio sample using higher layer LSB algorithm. In addition, to increase the robustness of the LSB method, the samples indices, which will be used for the embedding process, are specified randomly by the using PWLCM sequence. Algorithm 2 presents the hiding algorithm.

Algorithm 2   *The Hiding Algorithm*

Input :

  c : cover audio

  C : cipher message of length L

  x_0 : Initial condition for the PWLCM

  p_2 : System parameter for the PWLCM

Output :

  s : stego audio

Algorithm

1. Generate another sequence of PWLCM x_n, of length L using x_0 and p_2 as initial condition and system parameter, respectively.
2. Sort the generated sequence x_n in ascending or descending order then put the indices of the unsorted sequence in vector which is called permutation vector P.

3. Use the permutation vector P to select the host audio samples for embedding the secret encrypted message C, in which each bit of the encrypted message is hidden into the **higher layer LSB** (2–7 depth) of the corresponding value of the permutation vector P, so the encrypted message is embedded in random positions of the host audio samples.

A main idea of the proposed method is the hiding of bits with minimal embedding distortion of the cover audio. The principle of the embedding algorithm is illustrated in the following Sect. 7.2.1.

On the receiver side, the message is extracted and then decrypted. For message extraction, the receiver uses the same initial condition and system parameter to generate the same sequence of PWLCM, and perform the same hiding algorithm to extract the encrypted message.

For the decryption, all the steps are the same, except the following step of the one-time pad:

$$m_i^* = (C_i - Y_i) \, mod \, 2^8 \qquad i = 1, 2, \ldots, L \qquad (4)$$

where m_i^* is the recovered secret message, C_i is the cipher text, Y_i is the quantized PWLCM sequence and L is the message length.

7.2.1   Embedding Bits at Higher Layer LSB This section presents the embedding higher layer LSB algorithm. It is based on Cvejic and Seppnen [8] algorithm. It embed the secret message at high LSB layers. The algorithm consider 2nd to 7th layers. Some experimental results are presented in Sect. 8 to compare the distortion after embedding at different layers.

The following algorithm aims to minimize the distortion after embedding the bits at high LSB layers. It consider the situations where the embedded bit and the corresponding bit at ith layer of the cover audio are different.

Algorithm 3   *The Higher Layer LSB Algorithm*

Input :

  c : cover audio sample

  n : no of bits per sample, 8 or 16

  b : bit to be embedded in cover c

  i : LSB layer for embedding $(1 < i < 8)$

Output :

  s : stego audio sample with bit b at the layer i of LSB (with efficient adjustment algorithm)

Algorithm

$s = c$

if $c > = 0$

  if $b = 0$

    if $c_{i-1} = 0$ **then** $s_{i-1}s_{i-2}\ldots s_0 = 11\ldots 1$

    else $s_{i-1}s_{i-2}\ldots s_0 = 00\ldots 0$

    if $c_{i+1} = 0$ **then** $s_{i+1} = 1$

    else if $c_{i+2} = 0$ **then** $s_{i+2} = 1$

    else if $c_{i+3} = 0$ **then** $s_{i+3} = 1$

    $\ldots$

    else if $c_n = 0$ **then** $s_n = 1$

  else $\%b = 1$

    if $c_{i-1} = 1$ **then** $s_{i-1}s_{i-2}\ldots s_0 = 00\ldots 0$

    else $s_{i-1}s_{i-2}\ldots s_0 = 11\ldots 1 and$

    if $c_{i+1} = 1$ **then** $s_{i+1} = 0$

    else if $c_{i+2} = 1$ **then** $s_{i+2} = 0$

    else if $c_{i+3} = 1$ **then** $s_{i+3} = 0$

    $\ldots$

    else if $c_n = 1$ **then** $s_n = 0$

else $\%c < 0$

  if $b = 0$

    if $c_{i-1} = 0$ **then** $s_{i-1}s_{i-2}\ldots s_0 = 11\ldots 1$

    else $s_{i-1}s_{i-2}\ldots s_0 = 00\ldots 0$

    if $c_{i+1} = 1$ **then** $s_{i+1} = 0$

    else if $c_{i+2} = 1$ **then** $s_{i+2} = 0$

    else if $c_{i+3} = 1$ **then** $s_{i+3} = 0$

    $\ldots$

    else if $c_n = 1$ **then** $s_n = 0$

  else $\%b = 1$

    if $c_{i-1} = 1$ **then** $s_{i-1}s_{i-2}\ldots s_0 = 00\ldots 0$

    else $s_{i-1}s_{i-2}\ldots s_0 = 11\ldots 1 and$

    if $c_{i+1} = 1$ **then** $s_{i+1} = 0$

    else if $c_{i+2} = 1$ **then** $s_{i+2} = 0$

    else if $c_{i+3} = 1$ **then** $s_{i+3} = 0$

    $\ldots$

    else if $c_n = 1$ **then** $s_n = 0$

The hybrid steganography-cryptography proposed method satisfies the good characteristics of steganography algorithms:

- **Perceptual transparency**: The secret message is hidden at high LSB layer algorithm using efficient flipping algorithm that minimizae the distortion (as illustrated in Sect. 8).
- **Capacity of hidden data**: In the proposed method, all the audio samples can be used for hiding as the standard algorithm.
- **Robustness**: The proposed algorithm is robust against message extraction and destruction. The secret message is encrypted by the one-time pad then is hidden in random positions of the cover audio at high LSB layers using efficient adjustment algorithm. To extract the secret message correctly, the chaotic sequences should be known, and the encryption-steganography algorithms have a large key space as will be discussed in Sect. 8. On the other hand, data hiding at high LSB layers defends against noise addition or MPEG compression.

8 Experimental Results

This section shows some experimental results to demonstrate the efficiency of the proposed algorithm.

For the testing, some WAV sound files were taken from the web page http://www.1speechsoft.com/voices.html. They are spoken sentence in English which is: *"This is an example of the AT & T natural voice speech engine; it is the most human sounding text to speech engine in the world"*.

8.1 Wave Form Analysis

A waveform of the cover audio and stego audio are visualized and compared. Figure 6 shows an example to embed a message: 10 bytes in a cover file "julia8" which is sample for testing, with sampling rate 8000 HZ, 8 bits for each sample, 65644 samples and 8.2055 s duration.

From Fig. 6 it is clear that there is a slight difference between the cover audio and the corresponding stego. The percentage differences between the cover and stego audio in Fig. 6 is 0.00060935%.

8.2 Signal to Noise Ratio

Signal-to-noise ratio (SNR) is a measurement of noise in a signal. It is a common measurement for the audio signal quality. It is described as follow [26]:

$$SNR(s(i), sn(i)) = 10 \times \log_{10} \frac{\sum_{i=1}^{i=L} s(i)^2}{\sum_{i=1}^{i=L} (s(i) - sn(i))^2} \tag{5}$$

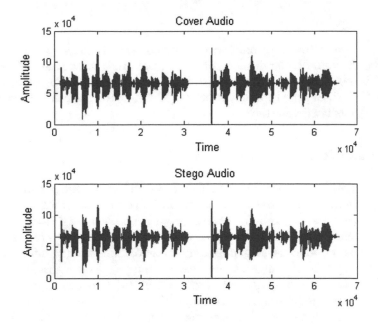

Fig. 6. Waveform of cover and stego audio for message of 10 bytes

where $s(i)$ and $sn(i)$ is the amplitude of the cover signal and the amplitude of the stego signal, respectively, and L is the number of samples. A high SNR indicates high precision of data, while a low SNR indicates noisy data.

SNR is used to measure the differences between the cover audio and the corresponding stego audio. Table 3 presents the SNR values for the audio file when different message sizes are embedded at different high LSB layers.

From Table 3, it is clear that the SNR are large even large messages are embedded in high LSB layers, which indicate good precision and quality of the stego file. On the other hand, there is an inverse relationship between the SNR on the hand and the LSB layers depth and messages sizes on the other hand.

Table 3. SNR for different message sizes and LSB layers

	Layer 2	Layer 3	Layer 4	Layer 5	Layer 6	Layer 7
10 bytes	39.8142	32.2441	32.2434	32.2421	32.2394	32.2349
100 bytes	19.8141	19.8138	19.8132	19.8121	19.8097	19.8051
1000 bytes	9.1675	9.1672	9.1667	9.1656	9.1635	9.1591

8.3   Key Space

Key space size is the total number of different keys that can be used for the algorithm. It should be large enough to resist the brute-force attack. A precision of 10^{-14} is used for the secret key. Both the encryption and hiding algorithm key space is $(10^{14})^2$. So, the total key space is $(10^{14})^4 = 10^{56}$, which is large enough to resist the exhaustive attack.

9   Conclusion

The paper presents a novel hybrid chaos-based audio steganography and cryptography algorithm. It is based on higher LSB layers method and one-time pad. A comprehensive study of audio steganography methods are described and discussed. Detailed analyses for these schemes and comparisons are made to be able to choose an efficient steganography method. LSB method offers the highest embedding rate and highest simplicity among all steganography methods, but it has low robustness against extraction and destroying which are solved in the proposed method. The embedding in the higher layer of LSB improves the robustness against signal processing manipulation, but it introduces large distortion. In the proposed method, an efficient flipping algortithm for the other bits are presented to overcome this issue.

A survey for chaotic maps is conducted to choose maps with good randomness properties and non-periodicity. This study uses Piecewise Linear Chaotic Map (PWLCM) for encryption and steganography. For the one-time pad encryption algorithm, a Chaotic sequence is generated and processed to be used as key for one-time pad. For message hiding, another sequence is used to generate a random sequence. Then, indices of ordered generated sequence are used to hide the encrypted data in random positions of the host audio samples. The one-time pad is used for encryption which is perfect cipher if a random key is used. However, it has two fundamental drawbacks; it is impractical to generate large quantities of random keys and the other problem is the key distribution.

The proposed method overcomes these problems by using the chaotic map (PWLCM). It produces random sequences with large key space which they are processed to be used as keys for the one-time pad. For the key distribution problem, the sender and receiver need only to exchange the initial conditions and system parameters for the chaotic maps, and so there is no need to exchange the whole long keys.

The experimental results demonstrate that the proposed innovative hybrid method is efficient. The key space of the proposed method is analyzed. The signal-to-noise ratio is tested which indicate that the stego audio has high quality, even if a large secret message is embedded in high layers such as 6th and 7th, and the degradation of SNR is considered small between embedding in some layer and the next one.

The proposed algorithm can be applied for other multimedia objects such as images and videos. Also, all the experimental results are carried out for text messages, whereas the proposed method deals with binary data. So, different file formats can be hidden on audio such as images, videos and even audio.

Acknowledgements. The authors would like to thank the University of Khartoum (UofK), Nile Center for Technology Research (NCTR) and Sudan's National Telecommunications Corporation (NTC) for their funding to this research paper.

References

1. Alwahbani, S.M., Elshoush, H.T.: Chaos-based Audio Steganography and Cryptography Using LSB Method and One-Time Pad, SAI Intelligent Systems Conference, London, UK, 21–22 September, pp. 1072–1078 (2016)
2. Awad, A., Saadane, A.: New chaotic permutation methods for image encryption. IAENG Int. J. Comput. Sci. **37**(4) (2010)
3. Adhiya, K.P., Patil, S.A.: Hiding text in audio using LSB based steganography. Inf. Knowl. Manag. IISTE **2**(3), 8–14 (2012)
4. Bandyopadhyay, S., Datta, B.: Higher LSB layer based audio steganography technique. Int. J. Electron. Commun. Technol. **2**(4) (2011)
5. Bassil, Y.: A two intermediates audio steganography technique. J. Emergi. Trends Comput. Inf. Sci. (CIS) **3**(11) (2012)
6. Castelan, Y., Khodja, B.: MP3 Steganography Techniques, RIIT'15, Chicago, IL, USA, 30 September–3 October 2015
7. Chattopadhyay, D., Mandal, M.K., Nandi, D.: Symmetric key chaotic image encryption using circle map. Indian J. Sci. Technol. (2011)
8. Cvejic, N., Seppanen, T.: Increasing robustness of LSB audio steganography using a novel embedding method. In: Proceedings of the International Conference on Information Technology: Coding and Computing (ITCC04), IEEE (2004)
9. Dengre, A.R., Gawande, A.D., Deshmukh, A.B.: Effect of audio steganography based on LSB insertion with image watermarking using AVI video. Int. J. Appl. Innov. Eng. Manag. (IJAIEM) **2**(6), 363–370 (2013)
10. Divya, S., Reddy, M.R.: Hiding text in audio using multiple LSB steganography and provide security using cryptography. Int. J. Sci. Technol. Res. **1**(6), 68–70 (2012)
11. Dutta, P., Bhattacharyya, D., Kim, T.H.: Data hiding in audio signal: a review. Int. J. Database Theory Appl. **2**(2), 1–8 (2009)
12. Gadicha, A.B.: Audio wave steganography. Int. J. Soft Comput. Eng. (IJSCE) **1**(5), 174–176 (2011)
13. Hmood, D.N., Khudhiar, K.A., Altaei, M.S.: A new steganographic method for embedded image in audio file. Int. J. Comput. Sci. Secur. (IJCSS) **6**(2), 135–141 (2012)
14. Jayaram, P., Ranganatha, H., Anupama, H.: Information hiding using audio steganography—a survey. Int. J. Multimedia Appl. (IJMA) **3**(3) (2011)
15. Hung, K.: A Study on Efficient Chaotic Image Encryption Schemes, Department of Electronic Engineering, City University of Hong Kong, Master thesis (2007)
16. Kumar, S., Barnali, B., Banik, G.: LSB modification and phase encoding technique of audio steganography revisited. Int. J. Adv. Res. Comput. Commun. Eng. **1**(4) (2012)
17. Mahajan, S., Singh, A.: A review of methods and approach for secure steganography. Int. J. Adv. Res. Comput. Sci. Softw. Eng. **2**(10), 67–70 (2012)

18. Malviya, S., Saxena, M., Khare, A.: Audio steganography by different methods. Int. J. Emerg. Technol. Adv. Eng. **2**(7), 371–375 (2012)
19. Mandal, P.C.: Modern steganographic technique: a survey. Int. J. Comput. Sci. Eng. Technol. (IJCSET) **3**(9), 444–448 (2012)
20. Meligy, A.M., Nasef, M.M., Eid, F.T.: A hybrid technique for enhancing the efficiency of audio steganography. Int. J. Image Gr. Signal Process. **8**(1), 36 (2016)
21. Nehru, G., Dhar, P.: A detailed look of audio steganography techniques using LSB and genetic algorithm approach. IJCSI Int. J. Comput. Sci. Issues **9**(2), 402–406 (2012)
22. Nosrati, M., Karimi, R., Hariri, M.: Audio steganography: a survey on recent approaches. World Appl. Program. **2**(3), 202–205 (2012)
23. Padmashree, G., Venugopala, P.S.: Audio stegnography and cryptography: using LSB algorithm at 4th and 5th LSB layers. Int. J. Eng. Innov. Technol. (IJEIT) **2**(4), 177–181 (2012)
24. Parthasarathy, C., Srivatsa, S.K.: Increased robustness of LSB audio steganography by reduced distortion LSB coding. J. Theor. Appl. Inf. Technol. **7**(1), 80–86 (2009)
25. Patil, B.A., Chakkarwar, V.A.: Review of an improved audio steganographic technique over LSB through random based approach. IOSR J. Comput. Eng. (IOSR-JCE) **9**(1), 30–34 (2013)
26. Prabu, A.V., Srinivasarao, S., Apparao, T., Rao, M.J., RaoK, B.: Audio encryption in handsets. Int. J. Comput. Appl. **40**(6), 40–45 (2012)
27. Priyanka, R.B., Vrushabh, R.K., Komal, K.P., Pingle, S.M., Sanghavi Mahesh, R.: Audio steganography using LSB, 1st international conference on recent trends in engineering & technology. Spec. Issue Int. J. Electron. Commun. Soft Comput. Sci. Eng. 90–92 (2012)
28. Rana, M., Rohit, T.: Genetic Algorithm in Audio Steganography (2014). arXiv:1407.2729
29. Rijmenants, D.: The complete guide to secure communications with the one time pad cipher. Cipher Mach. Cryptology (2012)
30. Roy, S., Singh, A.K., Parida, J., Sairam, A.S.: Audio steganography using LSB encoding technique with increased capacity and bit error rate optimization. In: CCSEIT-12, 26–28 October 2012
31. Santhi, V., Govindaraju, L.: Stego-audio using genetic algorithm approach. Res. J. Appl. Sci. Eng. Technol. **7**(22), 4806–4812 (2014)
32. Sheikhan, M., Asadollahi, K., Shahnazi, R.: Improvement of embedding capacity and quality of DWT-based audio steganography systems. World Appl. Sci. J. **13**(3), 507–516 (2011)
33. Singh, P.K., Aggrawal, R.K.: Enhancement of LSB based steganography for hiding image in audio. Int. J. Comput. Sci. Eng. **02**(05), 1652–1658 (2010)
34. Singh, G., Tiwari, K., Singh, S.: Audio steganography using RSA algorithm and genetic based substitution method to enhance security. Int. J. Sci. Eng. Res. **5**(5) (2014)
35. Stallings, W.: Cryptography and Network Security Principles and Practice, 5th edn. Prentice Hall, USA (2011)
36. Yoon, J.W., Kim, H., Jo, H., Lee, H., Lee, K.: Visual Honey Encryption: Application to Steganography, IH&MMSec15, Portland, Oregon, USA, 17–19 June 2015
37. Zamani, M., Manaf, A.A., Ahmad, R.B., Zeki, A.M., Abdullah, S.: A genetic-algorithm-based approach for audio steganography. World Acad. Sci. Eng. Technol. **30**, 355–358 (2009)

38. Zamani, M., Abdul Manaf, A.B., Zeidanloo, H.R., Chaeikar, S.S.: Genetic substitution-based audio steganography for high capacity applications. Int. J. Internet Technol. Secure. Trans. **3**(1), 97–110 (2011)
39. Zameer, F., Tarun, K.: Audio steganography using DES algorithm. In: Proceedings of the 5th National Conference, New Delhi, India (2011)
40. Zamani, M., Abdul Manaf, A.: Genetic algorithm for fragile audio watermarking. Telecommun. Syst. **59**, 291–304 (2015)

Generalised and Versatile Connected Health Solution on the Zynq SoC

Dina Ganem Abunahia[1], Hala Raafat Abou Al Ola[1], Tasnim Ahmad Ismail[1],
Abbes Amira[1(✉)], Amine Ait Si Ali[1,2], and Faycal Bensaali[1]

[1] College of Engineering, Qatar University, P. O. Box: 2713, Doha, Qatar
{da1105224,ha1104576,ti1207239,abbes.amira,f.bensaali}@qu.edu.qa,
amine.ali@northumbria.ac.uk
[2] Department of Computer and Information Sciences, University of Northumbria,
NE2 1XE, Newcastle, UK

Abstract. This chapter presents a generalized and versatile connected health solution for patient monitoring. It consists of a mobile system that can be used at home, an ambulance and a hospital. The system uses the Shimmer sensor device to collect three axes (x, y and z) accelerometer data as well as electrocardiogram signals. The accelerometer data is used to implement a fall detection system using the k-Nearest Neighbors classifier. The classification algorithm is implemented on various platforms including a PC and the Zynq system on chip platform where both programmable logic and processing system of the Zynq are explored. In addition, the electrocardiogram signals are used to extract vital information, the signals are also encrypted using the Advanced Encryption Standard and sent wirelessly using Wi-Fi for further processing. Implementation results have shown that the best overall accuracy reaches 90% for the fall detection while meeting real-time performances when implemented on the Zynq and while using only 48% of Look-up Tables and 22% of Flip-Flops available on chip.

1 Introduction

According to the UN World Health Organization statistics, the percentage of elderly will keep increasing to reach as much as twice the percentage of children in year 2050 (32% vs. 16%) [1]. As a result, the demand for medical attention is increasing. This is mainly because around one third of the elderly population over the age of 65 falls each year, and the risk of falls increases proportionately with age. Falls happen more frequently among elderly due to the following reasons: (1) Cerebrum degeneration or some diseases causing step imbalance; (2) Side effects of taking medications for people with chronic disease might cause dizziness and slow movements; (3) Health problems that might cause people to faint temporarily; and (4) Poor lighting environments, slippery floors and objects on the way of movement [13]. Other groups require special attention, such as Diabetics,

© Springer International Publishing AG 2018
Y. Bi et al. (eds.), *Intelligent Systems and Applications*,
Studies in Computational Intelligence 751,
https://doi.org/10.1007/978-3-319-69266-1_22

who are more likely to get low blood glucose causing fainting, spinal muscular atrophy patients who have movement inconsistency, neurological patients, like Parkinsons patients who suffer from imprecise movement, and cardiovascular patients who can have heart attacks and collapses. Hence, the necessity of developing integrated systems for healthcare has increased worldwide. This need is managed according to world-class standards to improve the health of the worlds population, and meet the needs of existing and future generations, and provide a healthy and lengthy life for all citizen to a greater extent. All health services will be accessible to the entire population. This work aims to develop a reconfigurable connected health platform using Zynq System on Chip (SoC) [16], in order to be used either indoors or in an ambulance environment to equip them with health monitoring technologies. The proposed wireless system deploys Bluetooth connectivity to establish communication between a prototyping board equiped with the Zynq SoC, and a Shimmer [3] wearable device that acquires both acceleration and Electrocardiogram (ECG) signals. The system is made of several stages: data acquisition using the Shimmer sensing device, data processing and analysis based on the Zynq SoC device, and the last stage is concerned with alerting and sending the encoded medical report to the doctor in case of a risk. The processing stage involves the implementation of k-Nearest Neighbors (KNN) classifier for fall detection using the Shimmer accelerometer data as well Advanced Encryption Standard (AES) and feature extraction using the ECG data. The structure of the remaining parts of the chapter is as follows: Sect. 2 describes the literature review and related work, Sect. 3 gives an overview of the proposed system, and Sect. 4 discusses the software and hardware implementation. Section 5 is concerned with the results and analysis. Section 6 concludes the chapter.

2   Literature Review

The area of connected health development systems for fall detection has been intensively searched. These systems can be implemented using different techniques, such as vision based falling detection. In this technique, some used special sensors as in [6], and others deploy surveillance systems as in [4] and [12]. In [2], an Omni camera was used to detect falling events. A new approach was proposed in [8] in which a MapCam (Omni camera) is used along with the personal information of each individual being captured on the camera which enhances the percentage of accuracy. Another technique is using ambience based devices which endeavor to fuse audio and visual data, besides sense through vibrating data. Image and video sensing can be achieved using multiple approaches: one method is by using signal strength measurements to track the estimated location of the user [7]. Another method is by extracting wavelet-based features from raw sensor and apply them to a TEO-based sound activity detector [14]. The last approach is uses wearable devices which are divided into two categories: motion based and posture based devices which are based on following [15]: Motion sensing method using an accelerometer and location sensing method using both accelerometer and gyroscope.This approach [11] uses sixMTw sensors, where each unit contains three tri-axial devices: accelerometer, magnetometer and gyroscope. Machine learning techniques based classifiers to distinguish between fall and daily life activities has been used. On the other hand, the system in [10] uses wearable camera and accelerometer for fall detection. It combines gradient local binary pattern features with edge orientation histograms in order to provide higher sensitivity. In [15] the proposed fall detection system uses a wearable device of single tri-axial accelerometer, and an algorithm that is based on thresholds of summing acceleration and rotation angle information. The summation acceleration is used as the first step to distinguish between high intensity movements from others. Moreover, there has been research efforts to accelerate some of the algorithms on hardware, such as implementation on heterogeneous computing platforms. For instance, a fall detection application on a heterogeneous computing platform, Zynq- 7000 SoC was deployed in [9]. The proposed solution in this system aims to use the power of the ARM cortex A9 processor of the Zynq platform together with the OpenCV libraries to achieve an efficient solution in terms of power consumption and execution time of the fall detection algorithm deployed in the system. The system design will be partitioned between the FPGA existing in the Zynq SoC, the Cortex A9 processor and the Graphical Processing Units (GPU) that will be also deployed for computer vision.

3   System Design

The designed system consists of three stages: data acquisition using the Shimmer sensor, data processing and analysis on the Zynq SoC prototyping board and finally alerting system. The system uses three hardware platforms: the Shimmer wireless healthcare sensing device, the NI myRIO Zynq SoC prototyping board, and a PmodBT2 UART Bluetooth module. The overall system is described in Fig. 1. The data acquisition phase uses an MMA7260qt 3-axis accelerometer to acquire acceleration as well as ECG Sensors to measure the heart muscle electrical activity. Both sensors are integrated into the Shimmer sensing device which is placed on the users chest. The ECG electrodes are connected to the right arm, left arm, right leg, and left leg from one side, and on the white, black, green, and red channels of the Shimmer from the other end respectively. A Bluetooth module is used to establish a connection between the Shimmer device and the Zynq SoC prototyping board, through this connection, the signals are transferred via Bluetooth to the Zynq SoC prototyping board using the integrated RN-42 Bluetooth module inside the Shimmer as a sender, and the Bluetooth module as a receiver. Data processing and analysis is performed on the Zynq SoC prototyping board using LabVIEW as a programming software environment. If an abnormality is detected, the alerting system is launched to check if the detection is a false alarm or if the user needs help. This system also helps to assess the state of the user after the fall. Finally, if a real fall is detected, an email is sent using Wi-Fi connection with an attached medical report to the health care providers such as hospitals or ambulances.

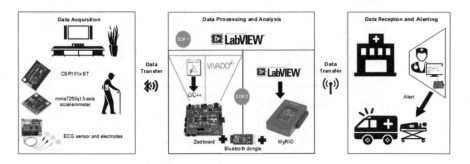

Fig. 1. System overview

4   System Implementation

The proposed solution consists of two main hardware platforms, the Shimmer sensing device and the Zynq SoC prototyping board which is myRIO. The Shimmer is used to acquire tri-axial acceleration and ECG signals from the user and then those signals are sent in real time via Bluetooth connection to the myRIO board which is located in the users side. This prototyping board has the implemented algorithm for fall detection and other algorithms used for ECG analysis and processing on the programmable logic. If the fall is detected, myRIO will send an alert via Wi-Fi to the doctors smart device. This email consists of a medical record about the information of the fall, as well as, ECG signals both encoded using AES. Consequently, the medical care can take action to reduce the risk of the fall or any related risks that might occur. The flowchart in Fig. 2 summarizes the different steps.

The system is implemented in multiple phases; the first phase involves software implementation, in this phase all the processing and testing is performed on a PC. The second phase involves the implementation of all algorithms on hardware that is capable of optimizing PC's performance. For that purpose, NI myRIO is chosen since it has the Zynq SoC which includes the ARM processor where simple code will be executed and an FPGA based programmable logic where computationaly intensive algorithms will be executed. The system setup is illustrated in Fig. 3.

4.1   Software Implementation

The software implementation where results are initially verified is important for building the system. The algorithms have been tested separately then they were combined to build the entire system. After the system is implemented and tested on a computer, computationally intensive parts are identified and implemented on the PL side of the Zynq SoC prototyping board for hardware acceleration while the remaining parts are implemented on the PS side.

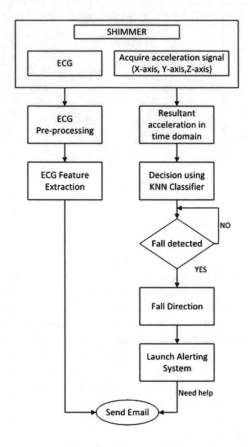

Fig. 2. Solution steps

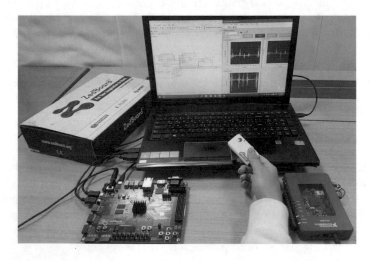

Fig. 3. System setup

4.1.1 Fall Detection

A fall detection database is constructed using the acquired tri-axial acceleration signals from the Shimmer platform. The database samples are collected to
construct a base and reference for the algorithm, through including different
scenarios to improve the accuracy of the system. As a result, the database consists of twelve subjects that perform various daily activities and falling scenarios.
Each of the twelve people performed multiple activities. The first type of activity which is "Activity of Daily Life (ADL)" involves four subtypes which are
"Jumping", "Jogging", "Picking something from the ground" and "Running".
This specific choice of ADL scenarios is made due to the fact that they produce
high acceleration resultant compared to other daily activities. Thus, they can act
as a separation line for distinguishing between falling and daily activities. This
decision was made after a series of trials. The seconfd type of activity which is
"Fall" involves four subtypes as well which are: "Back fall", "Left fall", "Right
fall" and "Front fall". For each person, three hundred samples are acquired for
each of the eight scenarios, then the resultant acceleration is calculated for each
tri-axial sample according to equation (1).

$$\text{Resultant acceleration} = \sqrt{Ax^2 + Ay^2 + Az^2} \qquad (1)$$

Among these three hundred resultant acceleration, only the three highest
values are taken and added to the database. This is applied for all of the subjects
which lead to: 12 people x 8 scenarios x 3 resultant acceleration extracted = 288
database samples. The Shimmer device was placed in the same position for
all subjects which is the chest. Furthermore, all experiments were done in the
same environment to achieve the requirements of a reliable system. The provided
system should distinguish between ADL and fall scenarios. Figure 4 represents
different experiments demonstrating different fall and ADL scenarios together
with the acquired acceleration signals.

The Shimmer platform can generate 3-axis acceleration using 3-axial
accelerometer. The acquired signals are transmitted via Bluetooth to the myRIO

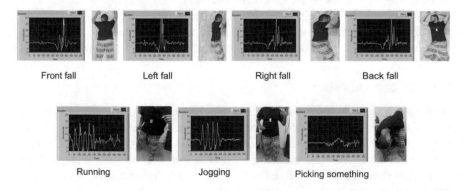

Fig. 4. Acceleration signals for ADL and various fall directions

for analysis in LabVIEW. Through LabVIEW application, the resultant of the three axis acceleration (Ax, Ay, Az) is calculated using Eq. (1) and then the resultant acceleration is sent to a sub vi that implements the KNN classification algorithm. The KNN classification is chosen to implement the fall detection algorithm because it requires less computation time and less data. However, it provides a high percentage of classification accuracy. The purpose of the KNN algorithm is to use a database in which the data points are separated into distinct pre-defined classes to predict the class of a new data point coming in. Predicting the class of the new data point can be achieved using one of the following distance functions: (2), (3) or (4).

$$\sqrt{\sum_{i=1}^{k}(x_i - y_i)^2} \tag{2}$$

$$\sum_{i=1}^{k}|x_i - y_i| \tag{3}$$

$$(\sum_{i=1}^{k}(|x_i - y_i|)^q)^{\frac{1}{q}} \tag{4}$$

The design of the proposed system uses the Euclidean distance for detecting the class of a new sample. The KNN classifier sub vi takes the sample resultant from time domain then calculates the distance between the new sample and all distance samples stored in the database and then all calculated values are stored in array X that is sorted ascendingly. After storing and sorting, the K smallest values of the distances array are selected and then the class of the sample is predicted depending on the class of the K selected samples. The simplified KNN algorithm pseudo code is shown in Fig. 5 where X are the training data stored in the database; Y presents class labels of X also stored in the same database and x is the new data point to be predicted.

In order to specify the fall direction of the user, several experiments were done to see the effect of the fall direction on of x- axis, y-axis, and z-axis values. Then a decision was made by choosing suitable values of x-axis and z-axis since the orientation of these two axes are affected with different directions: front, back, left and right. The process of detecting the direction of the fall is summarized in Fig. 6.

4.1.2   ECG Analysis

ECG features provide information about the heart rate, the condition of tissues within the heart, the conduction velocity as well as various abnormalities. It provides evidence for the diagnosis of cardiac diseases. The ECG analysis system contains several stages, starting with electrodes setup and wiring, to ECG acquiring, processing and analysis, and finally, results demonstration. Figure 7 shows an overview of the ECG analysis system.

Similar to the acceleration signals, Shimmer LabVIEW library has been used to acquire ECG signals. First of all, the sampling frequency is assigned to 51.2 Hz,

Algorithm 1: k-Nearest Neighbor
Classify(X, Y, x) // X: training data, Y: class labels of X, x: unknown sample
for i= 1 to m do
Compute distance d (Xi, x)
End for
Compute set I containing indices for the k smallest distances d (Xi, x)
Return majority label for {Yi where ε I}

Fig. 5. Algorithm 1

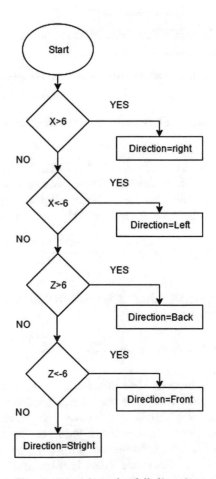

Fig. 6. Flowchart for fall direction

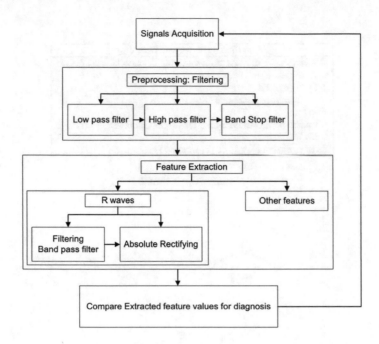

Fig. 7. ECG analysis system overview

which is suitable for the ambulatory ECG monitoring, i.e. monitor the heart while doing normal daily life activities. The Analog to Digital Converter (ADC) output of each ECG channel uses signed 24-bit digital format. The relationship between the ECG signal in mVolts and the ADC output is given by the Eq. (5).

$$\text{ECG Signal in mVolts} = \frac{((\text{ADC Output} - \text{ADC Offset}) \times \text{ADC Sensitivity})}{Gain} \tag{5}$$

Considering that the ADC Output is measured, and the ADC Sensitivity is described in equation (6):

$$\text{ADC Sensitivity} = \frac{V_{ref}}{\text{ADC Max}} = \frac{2420 \text{ mVolts}}{2^{23} - 1} \tag{6}$$

The values for the gain and ADC offset 5 must be inserted in Eq. (5). The gain is software configurable, whereas the nominal value of the ADC offset is 0. The channel inputs are connected, to calculate the offset of each channel. For example, for channel 1 (lead II), the mean value of the ADC output is calculated with the RA and LL inputs connected. This can be done if the uncalibrated data is saved into a file and the mean value is calculated for each channel. The following step after signals acquisition and extracting is filtering. A band stop filter is needed if the signal is experiencing interference from mains electricity. A 50 Hz frequency eliminator is required. Also, a high pass filter is necessary to eliminate low-frequency components of the signal. The cut-off frequency of

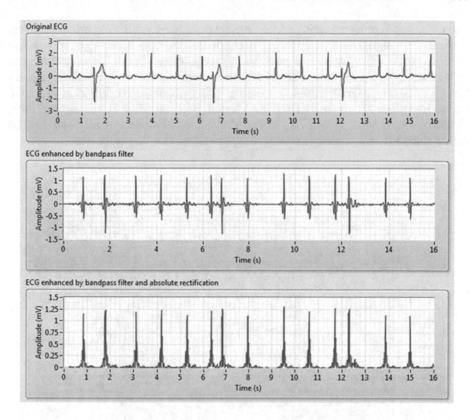

Fig. 8. ECG signal pre-processing for R waves detection

0.5 Hz is suitable for long term ECG monitoring, whereas a cut-off frequency of
0.05 Hz is recommended for ECG diagnostic. In the proposed work, the cut-off
frequency is a set as a choice according to the end user application. QRS detection algorithms are the foundation of ECG analysis, and can be used to estimate
the heart rate. The most widely used real-time QRS detection algorithm is The
Pan-Tompkins Algorithm. First, the signal is passed through a low pass and a
high pass filter to reduce the influence of the muscle noise, the baseline wander,
the T-wave interference, and the power line interference. This step is accomplished in the pre-processing phase. After filtering, the signal is differentiated
to provide the QRS slope information; then the signal is squared to make all
data points positive and to emphasize the higher frequencies. After squaring,
the signal is integrated using sliding window integration to obtain waveform
feature information. The size of the sliding window depends on the sampling frequency. The last step is to adjust a threshold that differentiates between peaks
of the signals, and the absence of peaks. ECG feature extraction firstly starts by
R-waves detection, then extracting other features. For normal ECG signals, they
can be easily detected. However, the process becomes harder for heart patients
where their ECG signals are abnormal. As a result, the ECG feature extraction

starts by signal enhancement, which contains two stages: filtering and rectifying. Initially, the signal is filtered using a bandpass filter, to allow signals within a specific range of frequencies to pass, and prevent others. Since R-waves of human ECG usually ranges between 10–25 Hz, these values were used as cutoff frequencies. Then, the filtered signal is rectified using absolute rectification. Figure 8 shows the processing result of an ECG signal, with negative R values, and large T-waves amplitudes. It can be noticed that after enhancement all beats can be easily detected, and heart rate variability analysis can be easily done.

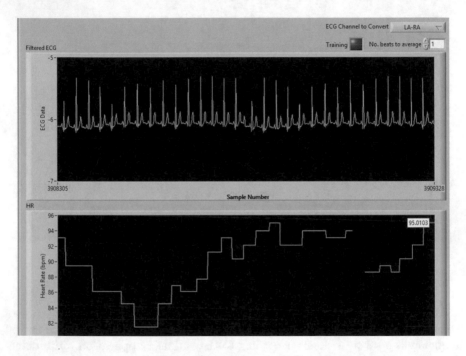

Fig. 9. Heart rate calculation

The heart rate (HR) in beats per minute (BPM) is calculated from the ECG using the R waves (part of the QRS complex defined above). The characteristics of the R peak are determined by training for an interval of a few seconds; these characters are used to calculate the heart rate. S1 is the sample number of the first R-wave detected, S2 is the sample number of the last R-wave detected, and NR represents the total number of R-waves between S1 and S2 inclusively. By knowing the sampling rate in Hz, which is Fs. These values can be substituted into the Eq. (7).

$$HR(BMP) = 60 \times \frac{F_s \times (N_R - 1)}{S_2 - S_1} \tag{7}$$

Figure 9 illustrates the heart rate calculation in the system. The ECG signal results from lead I (RA-LA) and the heart rate calculated which is 95.0103.

4.1.3   Alerting System

In the proposed solution, fall detection and minimizing false alarms are two major concerns. It is critical to confirm accurate detection and correct behavior of the system. Therefore, a complete alerting system is added to the functionality of the proposed system. Mainly to notify the caregivers about any abnormal activity with the user, and to minimize false alarms even further. Generally, falls can be categorized into three conditions:

Condition 1: Normal Condition of Fall

This is the condition when a fall is detected, but the elderly stays conscious. In this case, the user can turn off the alarm by pressing the "I am fine" button. This condition is crucial for eliminating false alarms, such that if the system detects a fall in ADL, the user can inform the system that it is not a fall.

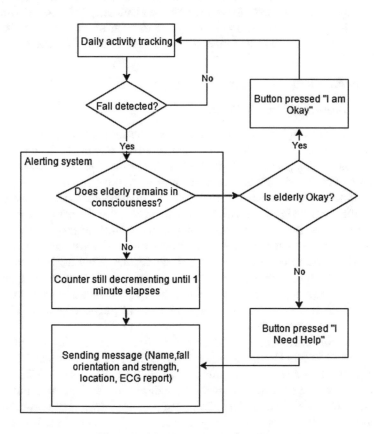

Fig. 10. Alerting system flowchart

Condition 2: Critical Condition of Fall

This condition results when a fall happens, and causes injuries, but elderly stays conscious. In this case, the user can press the button "I NEED HELP" to alert that the elderly needs help, and send the email.

Condition 3: Emergency Condition of Fall

This is the condition when the fall causes serious injuries, and the elderly is in fatal condition. As a result, an alert is sent to the emergency unit to provide emergency response and assistance as fast as it can.

For alerting system implementation, Data Dashboard for LabVIEW was used. This application runs on Android/IOS mobile phones and tablets. The system consists of two timers and two switches, and it follows the flowchart illustrated in Fig. 10. If a fall was detected, the first timer runs for 30 s to make sure that the user is conscious, if he/she did not respond within 30 s, an alert is sent immediately. If the user replied within the timeout, the second timer runs for 30 s to check if the user needs help. If the user chose "I am fine", it is more likely a false alarm that was stopped to reach the medical assistance.

4.2 Hardware Implementation

4.2.1 Implementation on Zedboard

The KNN software implementation gave a reasonably high detection rate within a short response time. The accuracy is expected to remain high with a considerable decrease in execution time if the fall detection algorithm is implemented on hardware. The Zynq SoC is chosen for hardware implementation since it contains both a processor system where software implementation is deployed and a programmable logic where hardware implementation and optimization are performed. The key element of the hardware implementation stage is to partition the system functionalities between hardware and software, and to decide on the appropriate interface between the two partitions. This decision is the most important step in hardware/software co-design approach since the system performances are directly dependent on it. Generally, software design on the Precessing System (PS) will be used to implement general purpose sequential processing tasks, an operating system, user applications and Graphical User Interfaces (GUIs), while computationally intensive data flow parts of the design are more suitably realized in the Programmable Logic (PL). For this project, the PS portion is responsible for sensor interfacing and data acquisition. Additionally, the ECG processing and encryption are implemented on the PS. Alternatively, the PL portion is responsible for KNN algorithm implementation to detect falls. The Advanced eXtensible Interface (AXI) interface handles the PS/PL communication to send acceleration signals from the PS to the PL for processing. However, the KNN algorithm is implemented on the ARM processor initially for testing and validating, after the design is verified, it is implemented on the PL. Vivado

Design Suite is used to implement the system on the Zedboard. Generally, the implementation on the Zedboard is partitioned into two sections: implementation on the PS, and the implementation on PL. PS implementation will be used to implement general purpose sequential processing tasks, an operating system, user applications and GUIs, while computationally intensive data flow parts of the design, and any software algorithms which exhibit significant parallelism can be strong candidates for implementation in PL. Figure 11 illustrates the design flow for a hardware implementation on the Zynq SoC. Through Vivado High Level Synthesis tool (HLS), an IP implementing the KNN algorithm has been designed and optimized. First a C-code is written to implement the KNN classifier for fall detection, it takes a form of a "predict" function. In addition, a C test bench is written such as it takes the main C function executing the "predict" function to self-check the results. Figure 12 shows the hardware block design of the overall design which includes the processing system, the AXI connections, and the KNN IP with the predict function.

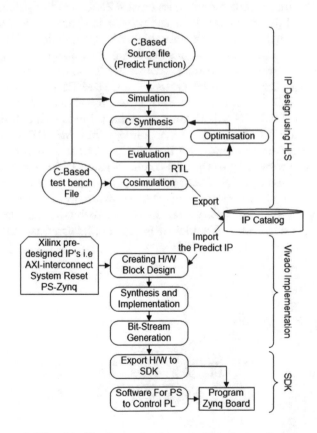

Fig. 11. Hardware implementation design flow

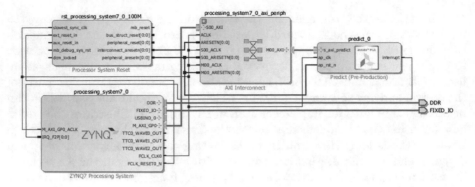

Fig. 12. Hardware block design

4.2.2 Implementation on MyRio

Even though the implementation results on the Zedboard are promising, implementing the remaining subsystems of the project is challenging. This is mainly because the Zedboard is only programmable with C/C++ and VHDL, while the software implementation of the proposed system is performed using LabVIEW. As a result and to simplify the task, NI myRIO Zynq SoC prototyping board is selected to implement the entire system. The ARM PS portion is responsible for Shimmer-myRIO communication and ECG processing, whereas the FPGA PL is responsible for KNN algorithm implementation. In this approach acceleration signals and ECG signals are acquired using LabVIEW program and transmitted to the host computer via Bluetooth. Then shared variables are used to send the values wirelessly to the myRio board using WiFi. Data was analyzed and processed on the Zynq SoC using ARM processor. As a result of the dual transmission before signals processing, data transmission consumed much time in case of weak Wifi signal causing fall detection to be delayed for about 30 s, and thus this approach was not sufficient for the application. In that matter, the

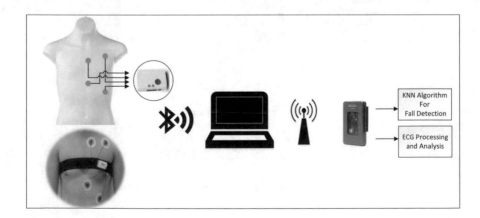

Fig. 13. System implementation using myRIO

design was improved to include direct communication between the Shimmer sensing device and the myRIO board using the pmodBT2 Universal Asynchronous Receiver Transmitter (UART) Bluetooth module. The bluetooth communication assured preserving time for data processing to satisfy the real-time performance as illustrated in Fig. 13.

5   Results and Analysis

5.1   Software Implementation

The testing phase for fall detection involves eleven subjects who participated in performing the same scenarios done with first training group. This group's results are used to evaluate the system performance. With the aim of achieving high accuracy, it is necessary to choose a suitable value of K and data training percentage, this process of trying different values of K and different percentages is referred to as cross-validation. Cross-validation can be done as follows: Fix the training percentage and vary the value of k (k = 3, 4 and 5) then fix the value of K and vary the training percentages (20, 30, and 50%)

The accuracy results for each case are shown in Table 1. After testing the system with different scenario cases, it is shown that some values of K and training percentages provide more accurate results than others. For instance, in the case when K = 5, and training percentage = 30%, the system provides the least overall accuracy of 77.14%. Whereas in the case when K = 3, and training percentage = 30%, the system provides the most overall accuracy of 90.00%. However, it is evident that the accuracy for almost all cases is below 90%. As a result of that, the system needs to be improved to increase the accuracy, this can be done in several steps: First, the database should be increased to involve at least twelve people, second, wavelet domain analysis can be used before calculating the resultant acceleration.

Table 1. Accuracy results of testing the algorithm with different values of k and different training percentages

Training percentage		K = 3		K = 4		K = 5	
		Fall	ADL	Fall	ADL	Fall	ADL
50%	Accuracy (%)	82.14	88.10	82.14	88.10	78.57	90.48
	Overall (%)	85.7		85.71		85.71	
30%	Accuracy (%)	82.14	95.23	75	92.86	50	95.23
	Overall (%)	90		85.71		77.14	
20%	Accuracy (%)	67.86	88.10	75	90.48	82.14	90.48
	Overall (%)	80		84.29		87.14	

5.2 Hardware Implementation

Vivado HLS tool supports several optimization techniques, such as loop unrolling, array portioning, and pipelining. The "Unroll Loop" approach provides dedicated hardware resources for each iteration of a loop, hence allows iterations of a given loop to be executed in parallel. The second approach, which is "Array Partitioning" allows each entity of an array to have its own data ports rather than considering it as one array entity and limiting data ports. The last approach is "Pipeline" which is applied in the system to allow pipelining of instructions and sub-functions, hence optimize the system [5]. The C code is compiled, then synthesized and a report is generated, this report contains information about timing, latency, and resource usage. This report is used to optimize the design using the previously mentioned techniques. A comparison report is generated as shown in Tables 2 and 3. Solution 1 represents the system with no optimization while solution 2 and 3 represents the design using Pipeline with different parameters. Since the pipelining approach consumed more resources than the available one, only a small part of the design was pipelined (one iteration of the loop used to compute the Euclidian distance).

Table 2. Implementation reports for solutions 1, 2 and 3 in terms of resource usage

	Used recourses			
	BRAM_18K	DSP48E	FF	LUT
Solution1	6	27	5581	9635
Solution2	6	27	5577	9608
Solution3	2	28	24015	40955

Table 3. Implementation reports for solutions 1, 2 and 3 in terms of time consumption

	Clock(ns)	Latency (clock cycles)		Interval (clock cycles)	
		Min	Max	Min	Max
Solution1	8.62	1334	16958	1335	16959
Solution2	9.53	403	16027	404	16028
Solution3	8.74	176	176	10	10

When the design is finalized, it can be used as an IP in Vivado, the design includes four IPs: the KNN block, the PS block, the AXI Interconnect block to connect the KNN IP block to the PS and finally the System Reset block.

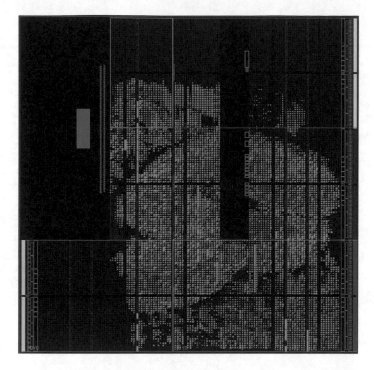

Fig. 14. Chip layout

Figure 14 shows the chip layout for the implemented best solution 3. Vivado does further optimizations when implementing the entire design on Zedboard since it uses less flip flops (FF) and lookup tables (LUTs). Table 4 shows the resource usage for the whole system that include the Processor System Reset, AXI Interconnect and the HLS Predict IP cores when the frequency is set to 667 MHz for the PS and 100 MHz for the PL. The power consumption is shown in Table 5.

The execution time is measured for the KNN algorithm on various platforms including a PC, the PS and the PL of the Zynq. A summary of the execution time can be seen in Table 6. It shows that execution on the programmable logic is the fastest. On the other hand, the execution time on the PS in reasonably similar to the one on PC.

5.3   Alerting System

A user friendly interface has been developed for the alerting system, it can be seen in Fig. 15.

Table 4. Implementation report for solution 3

	BRAM_18K	DSP48E	FF	LUT
HLS Predict	1	28	22423	25030
AXI Interconnect	0	0	475	363
System Reset	0	0	26	15
Total	1	28	22924	25408
Available	140	220	106400	53200

Table 5. Power consumption report for solution 3

Power consumption (W)						
Dynamic 1.663 W (92%)						Static (8%)
PS (82%)	Clocks (4%)	Signals (13%)	Logic (9%)	BRAM (<1%)	DSP (1%)	0.173
1.529	0.091	0.284	0.185	0.003	0.020	

Table 6. Execution time

Implementation platform	Execution time
PC	6 ms
Zynq PS	3 ms
Zynq PL	131 μs

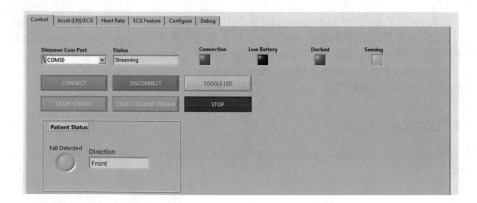

Fig. 15. User interface

6 Conclusion

The work described in this chapter aims at presenting a connected health monitoring and alerting system that meets real time performances. A fall detection system that uses the Shimmer accelerometer data is presented and implemented on the Zynq SoC. A classification accuracy of 90% is reached by using k-Nearest Neighbors with k = 3 while meeting real-time performances using only 48% of LUTs and 22% of FFs available on chip. In addition, an alerting system is implemented, it is capable of sending reports to ambulances and hospital using Wi-Fi. The report includes vital information such as ECG signals which are collected using the same Shimmer sensor and encrypted using the Advanced Encryption Standard for increased privacy.

Acknowledgements. This paper was made possible by National Priorities Research Program (NPRP) grant No. 5-080-2-028 from the Qatar National Research Fund (a member of Qatar Foundation). The statements made herein are solely the responsibility of the authors.

References

1. Beard, J.R., Officer, A., de Carvalho, I.A., Sadana, R., Pot, A.M., Michel, J.P., Lloyd-Sherlock, P., Epping-Jordan, J.E., Peeters, G.G., Mahanani, W.R., et al.: The world report on ageing and health: a policy framework for healthy ageing. The Lancet **387**(10033), 2145–2154 (2016)
2. Bromiley, P., Courtney, P., Thacker, N.: Design of a visual system for detecting natural events by the use of an independent visual estimate: a human fall detector. Empirical Eval. Methods Comput. Vis. **50** (2002)
3. Burns, A., Greene, B.R., McGrath, M.J., O'Shea, T.J., Kuris, B., Ayer, S.M., Stroiescu, F., Cionca, V.: Shimmer-a wireless sensor platform for noninvasive biomedical research. IEEE Sens. J. **10**(9), 1527–1534 (2010)
4. Counsell, S.R.: 2015 updated ags beers criteria offer guide for safer medication use among older adults. J. Gerontological Nurs. **41**(11), 60 (2015)
5. Crockett, L.H., Elliot, R.A., Enderwitz, M.A., Stewart, R.W.: The Zynq book: embedded Processing with the Arm Cortex-A9 on the Xilinx Zynq-7000 All Programmable Soc. Strathclyde Academic Media (2014)
6. Kepski, M., Kwolek, B.: Fall detection on embedded platform using kinect and wireless accelerometer. In: International Conference on Computers for Handicapped Persons, pp. 407–414. Springer (2012)
7. Mubashir, M., Shao, L., Seed, L.: A survey on fall detection: principles and approaches. Neurocomputing **100**, 144–152 (2013)
8. Nait-Charif, H., McKenna, S.J.: Activity summarisation and fall detection in a supportive home environment. In: Proceedings of the 17th International Conference on Pattern Recognition ICPR 2004, vol. 4, pp. 323–326. IEEE (2004)
9. Nguyen, H.T.K., Belleudy, C., Van Tuan, P.: Fall detection application on an arm and fpga heterogeneous computing platform. Int. J. Adv. Res. Electr. Electr. Instrum. Eng. **3**(8), 11349–11357 (2014)
10. Ozcan, K., Velipasalar, S.: Wearable camera-and accelerometer-based fall detection on portable devices. IEEE Embed. Syst. Lett. **8**(1), 6–9 (2016)

11. Özdemir, A.T., Barshan, B.: Detecting falls with wearable sensors using machine learning techniques. Sensors **14**(6), 10691–10708 (2014)
12. Patel, K.V., Brennan, K.L., Davis, M.L., Jupiter, D.C., Brennan, M.L.: High-energy femur fractures increase morbidity but not mortality in elderly patients. Clin. Orthop. Relat. Res. **472**(3), 1030–1035 (2014)
13. Stevens, J.A., Hasbrouck, L., Durant, T.M., Dellinger, A.M., Batabyal, P.K., Crosby, A.E., Valluru, B.R.: Kresnow, M.j., Guerrero, J.L.: Surveillance for injuries and violence among older adults. MMWR CDC Surveill. Summ. **48**(8), 27–50 (1999)
14. Toreyin, B.U., Soyer, E.B., Onaran, I., Cetin, A.E.: Falling person detection using multisensor signal processing. EURASIP J. Adv. Signal Process. **2008**, 29 (2008)
15. Wu, F., Zhao, H., Zhao, Y., Zhong, H.: Development of a wearable-sensor-based fall detection system. Int. J. Telemedicine Appl. **2015**, 2 (2015)
16. Xilinx, Inc.: Zynq-7000 all programmable soc: technical reference manual. ug585, vol. 8.1. http://www.xilinx.com/support/documentation/userguides/ug585-Zynq-7000-TRM.pdf (2014). Accessed 30 Sep 2016

Erratum to: Intelligent Systems and Applications

Yaxin Bi[1(✉)], Supriya Kapoor[2], and Rahul Bhatia[2]

[1] School of Computing, Ulster University at Jordanstown, Newtownabbey,
County Antrim, UK
[2] The Science and Information (SAI) Organization, Bradford, UK

Erratum to:
Y. Bi et al. (eds.), *Intelligent Systems and Applications*,
Studies in Computational Intelligence 751,
https://doi.org/10.1007/978-3-319-69266-1

In the original version of the book, the following corrections have been incorporated:
In Chapter "ARTool—Augmented Reality Human-Machine Interface for Machining Setup and Maintenance", incorrectly published author's name "Amedeo Ragni" has been corrected to read as "Matteo Ragni" and in reference list, citation 30, "Setti, A., Bosetti, P., Ragni, M.: ARTool—Augmented Reality Platform for Machining Setup and Maintenance, vol. 1, pp. 273–281. Springer (2016)" has been changed as "Setti, A., Bosetti, P., Ragni, M.: ARTool—Augmented reality platform for machining setup and maintenance. In: Bi, Y., Kapoor, S., Bhatia, R. (eds.) Proceedings of SAI Intelligent Systems Conference (IntelliSys) 2016. IntelliSys 2016. Lecture Notes in Networks and Systems, vol. 15, pp. 457–475, Springer, Cham (2018)". In Chapter "Integration of Fuzzy C-Means and Artificial Neural Network for Short-Term Localized Rainfall Forecasting in Tropical Climate", mail id of "Noor Zuraidin Mohd-Safar" has been changed from "noorzuraidin.mohdsafar@port.ac.uk" to "UP680382@myport.ac.uk ".
The erratum book has been updated with the changes.

The update online version of these chapters can be found at
https://doi.org/10.1007/978-3-319-69266-1_7
https://doi.org/10.1007/978-3-319-69266-1_16

Author Index